AF803380

Environmental Toxicology of Pesticides

Proceedings of a
United States–Japan Seminar
Held in Oiso, Japan
October, 1971

Environmental Toxicology of Pesticides

edited by

Fumio Matsumura
Department of Entomology
University of Wisconsin, Madison, Wisconsin

G. Mallory Boush
Department of Entomology
University of Wisconsin, Madison, Wisconsin

Tomomasa Misato
Institute of Physical and Chemical Research
Wako-shi, Saitama, Japan

Academic Press New York and London 1972

COPYRIGHT © 1972, BY ACADEMIC PRESS, INC.
ALL RIGHTS RESERVED
NO PART OF THIS BOOK MAY BE REPRODUCED IN ANY FORM,
BY PHOTOSTAT, MICROFILM, RETRIEVAL SYSTEM, OR ANY
OTHER MEANS, WITHOUT WRITTEN PERMISSION FROM
THE PUBLISHERS.

ACADEMIC PRESS, INC.
111 Fifth Avenue, New York, New York 10003

United Kingdom Edition published by
ACADEMIC PRESS, INC. (LONDON) LTD.
24/28 Oval Road, London NW1

LIBRARY OF CONGRESS CATALOG CARD NUMBER: 72-82044

PRINTED IN THE UNITED STATES OF AMERICA

CONTENTS

PART III. CHLORINATED HYDROCARBON INSECTICIDES IN THE ENVIRONMENT

PART IV. FUNGICIDES, HERBICIDES, ORGANOPHOSPHATES, AND CARBAMATES

PART V. MICROBIAL INFLUENCE ON PESTICIDE DEGRADATION

PART VI. PHOTODECOMPOSITION OF PESTICIDES

PART VII. TOXIC EFFECT OF PESTICIDE RESIDUES ON WILDLIFE

PART VIII. DESIGN OF NEW PESTICIDES

CONTRIBUTORS

M. Alexander, Department of Agronomy, Cornell University, Ithaca, New York 14850

John C. Batterton, Department of Entomology, University of Wisconsin, Madison, Wisconsin 53706

G. Mallory Boush, Department of Entomology, University of Wisconsin, Madison, Wisconsin 53706

G. T. Brooks, Agricultural Research Council, Unit of Invertebrate Chemistry and Physiology, The University of Sussex, Brighton, England

Donald G. Crosby, Department of Environmental Toxicology, University of California, Davis, California 95616

Hiroshi Fukami, Pesticide Research Institute, College of Agriculture, Kyoto University, Kyoto, Japan

Kazuo Fukunaga, National Institute of Agricultural Sciences, Nishigahara, Kita-ku, Tokyo, Japan

K. Furukawa, Fermentation Research Institute, Inage, Chiba-shi, 280, Japan

Leonard J. Goldwater, Duke University Medical Center, Durham, North Carolina 27706

Steven G. Herman, Department of Zoology, University of California, Davis, California 95616

Yasuo Homma, Institute of Physical and Chemical Research, Wako-shi, Saitama, Japan

Henry Hurtig, Environmental Quality Research Branch, Canada Department of Agriculture, Ottawa, Canada

Yorinao Inoue, Laboratory of Plant Physiology, Institute of Physical and Chemical Research, Wako-shi, Saitama, Japan

Mitsuo Ishida, Sankyo Co., Ltd., Agricultural Chemicals Research Laboratories, Yasucho, Yasugun, Shigaken, Japan

Hidetsugu Ishikura, Science and Technology Agency, Kasumigaseki, Tokyo, Japan

Jun Kanazawa, National Institute of Agricultural Sciences, Nishigahara, Kita-ku, Tokyo, Japan

Eugene E. Kenaga, Ag-Organics Department, The Dow Chemical Company, Midland, Michigan

Katsuko Konishi, Laboratory of Plant Physiology, Institute of Physical and Chemical Research, Wako-shi, Saitama, Japan

Shozo Kuwatsuka, Department of Agricultural Chemistry, Nagoya University, Nagoya, Aichi, Japan

Louis Lykken, Division of Entomology, University of California, Berkeley, California 94720

Fumio Matsumura, Department of Entomology, University of Wisconsin, Madison, Wisconsin 53706

Shooichi Matsunaka, National Institute of Agricultural Sciences, Konosu, Saitama, Japan

Calvin M. Menzie, Office of Environmental Quality, Bureau of Sport Fisheries and Wildlife, U. S. Department of the Interior, Washington, D. C.

Tomomasa Misato, Institute of Physical and Chemical Research, Wako-shi, Saitama, Japan

Junshi Miyamoto, Research Department, Pesticides Division, Sumitomo Chemical Co., Ltd., Osaka, Japan

Kenneth W. Moilanen, Department of Environmental Toxicology, University of California, Davis, California 95616

Toshinobu Murai, National Institute of Agricultural Sciences, Nishigahara, Kita-ku, Tokyo, Japan

Masayuki Nakagawa, Department of Environmental Toxicology, University of California, Davis, California 95616

Tsutomu Nakatsugawa, Department of Forest Entomology, State University College of Forestry at Syracuse University, Syracuse, New York 13210

Peter A. Nelson, Department of Forest Entomology, State University College of Forestry at Syracuse University, Syracuse, New York 13210

Teruhisa Noguchi, Nisso Institute for Life Science, Oiso, Kanagawa, Japan

Teruo Ogawa, Laboratory of Plant Physiology, Institute of Physical and Chemical Research, Wako-shi, Saitama, Japan

Tsunehiko Ogawa, College of Agriculture, Ehime University, Matsuyama, Ehime, Japan

Joseph D. Rosen, Department of Food Science, Rutgers, The State University, New Brunswick, New Jersey 08903

Robert L. Rudd, Department of Zoology, University of California, Davis, California 95616

Kazuo Shibata, Laboratory of Plant Physiology, Institute of Physical and Chemical Research, Wako-shi, Saitama, Japan

Hiroya Tanabe, National Institute of Hygienic Sciences, Tokyo, Japan

Ryo Tatsukawa, College of Agriculture, Ehime University, Matsuyama, Ehime, Japan

Chojiro Tomizawa, National Institute of Agricultural Sciences, Nishigahara, Kita-ku, Tokyo, Japan

K. Tonomura, Fermentation Research Institute, Inage, Chiba-shi, 280, Japan

Yutaka Tsukano, National Institute of Agricultural Sciences, Nishigahara, Kita-ku, Tokyo, Japan

Yasuhiko Uesugi, National Institute of Agricultural Sciences, Nishigahara, Kita-ku, Tokyo, Japan

Tyunosin Ukita, Faculty of Pharmaceutical Sciences, University of Tokyo, Tokyo, Japan

Tadaaki Wakimoto, College of Agriculture, Ehime University, Matsuyama, Ehime, Japan

Anthony S. Wong, Department of Environmental Toxicology, University of California, Davis, California 95616

Nakatsugu Yaginuma, Laboratory of Plant Physiology, Institute of Physical and Chemical Research, Wako-shi, Saitama, Japan
Japan

M. Yamada, Fermentation Research Institute, Inage, Chiba-shi, 280, Japan

Isamu Yamaguchi, Institute of Physical and Chemical Research, Wako-shi, Saitama, Japan

PREFACE

The need for an academic discipline to cover the area of pesticide toxicology in relation to the environment is apparent. What is not so apparent, however, is the scope of such a discipline, as well as the areas of study or approaches involved. No doubt gradual evolution will settle this matter as it so often has in the past, for many fields have developed somewhat spontaneously through the needs and interests of society. However, we intend to bypass this enigma at the present by assuming that all scientists who are interested in problems of environmental contamination by pesticidal chemicals are "environmental toxicologists."

This treatise developed as a result of a United States-Japan seminar on "Environmental Toxicology of Pesticides" held in Oiso, Japan, in October, 1971. The purpose of the seminar was to discuss and exchange ideas and technology on the problems associated with pesticidal contamination in these two countries.

The problems of environmental contamination are the result mainly of modern industrialization, urbanization, and highly developed agricultural practices. It is not surprising, therefore, that such problems would be especially acute in the United States and Japan. Upon examination of the past use of pesticides in these two countries, several intriguing differences are noted, particularly with regard to the way problems have arisen and were subsequently treated. The scientific communities of both nations have also taken different paths, due partly to the differences in public demands as well as historical and geographical factors, and partly to historical differences in approach to problem-solving.

The major concern in the United States has been with pesticide residues in food for human consumption, as well as mammalian toxicities of pesticidal compounds. Only recently has serious attention been focused on other problem areas such as effects on wildlife, water quality, and environmental alteration of pesticides. Particular emphasis has been placed on the chlorinated hydrocarbons, with DDT singled out as the principal target of public and scientific concern.

In Japan, the major concern has been on development and study of chemicals for rice production, including insecticides, herbicides, and fungicides for rice-blast control. Organomercurials, used in controlling rice blast, have been the major factor in increasing rice yields. In addition, they have created severe environmental problems. As a result, studies on the effects of organomercurials on higher animals as well as their behavior in the environment have been in progress in Japan for many years. In addition, the widespread use of BHC for control of the rice stem borer created problems somewhat similar to those caused by DDT in the United States. Interestingly, Japan has been quick to suspend the use of these chemicals, and her scientists have been highly successful in developing alternative means, such as the use of biological control, antibiotics, and other chemicals with less undesirable properties for pest control.

The intended purpose of this book is to report the present state of knowledge in the major pesticidal subject areas and to describe the efforts and approaches underway in solving or understanding these problems. This is particularly helpful to those of us with English as our primary language as many of the studies of our Japanese colleagues have been previously reported in Japanese. It is apparent that this book is the compilation of individual work, and thus its strength, or weakness, depends upon the individual contribution.

Pollution is a global problem, and, as such, exchange of ideas and data among scientists from different countries and disciplines should certainly help all of us. With the interests shown by people today on the subject matter of environmental pollution, we hope that publication of this book does not need further justification.

The seminar was financed by the National Science Foundation and the Japan Society for Promotion of Science. We extend our thanks to both for their support and cooperation in making the seminar possible.

Fumio Matsumura
G. Mallory Boush
Tomomasa Misato

Environmental Toxicology of Pesticides

PART I

INTRODUCTION: PATTERNS OF PESTICIDE USAGE AND OCCURRENCE OF RESIDUES

IMPACT OF PESTICIDE USE ON THE JAPANESE ENVIRONMENT

Hidetsugu Ishikura

Former Science Counsellor
Science and Technology Agency
Kasumigaseki, Tokyo

Introduction

The consumption of pesticides by Japanese agriculture has increased tremendously during the past two decades. The Report on the World Food Problem prepared for the U.S. President by his Office stated that Japan had consumed 10.8 kilograms of pesticides, in terms of active ingredients per hectare, in 1963. Goto (1970) estimated that a comparable figure for 1967 would be 12.0 kilograms.

The consumption of pesticides at such a high level in Japanese agriculture seems to have been brought about by some characteristics of Japanese agriculture. The territory of Japan is only 370,000 square kilometers, which is about 10% smaller than the state of California. In addition to this small territory, the four major islands of Honshu, Shikoku, Kyushu, and Hokkaido are mostly mountainous and only 16% of the land area is arable. The actual cultivated area amounts to only 5.8 million hectares, and as such the circumstance dictates a heavy use of agricultural lands.

In addition, these agricultural fields of 5.8 million hectares are divided into 5.2 million farms. Many farms are small in size, usually less than one-tenth of a hectare. The total areas of cultivation for several major crops in Japan are summarized in TABLE 1.

To increase the yield of crops per unit area, farmers have been relying heavily on chemical fertilizers for some time. The consumption of chemical fertilizers increased strikingly after World War II along with the restoration and expansion of the manufacturing capacity. As of 1967 the total domestic consumption of fertilizer was estimated to have reached 889,000 tons for nitrogenous material (on nitrogen basis), 671,000 tons for phosphates (on P_2O_5 basis), and 651,000 tons for potash (on K_2O basis). As is shown in TABLE 2, these figures mean a 50, 73, and 68% increase over the 10-year span for nitrogen, P_2O_5, and K_2O consumption, respectively.

The increased use of fertilizers along with adoption of denser spacing methods for cultivating more plants per unit area brought vigorous growth for many crops which made these crops more vulnerable to pest infestation.

The Japanese Archipelago stretches south to north between 28 and 45 north latitudes and is located in the Pacific Ocean off the Asian Continent. Because of this geographical location, climate varies considerably by the district and by season. Summer temperatures are high enough, particularly in the southern half of the country, for growing crops of tropical origin such as rice, oranges, and tea. Rice is grown even in the northern part of Hokkaido under intensive cultural managements. High summer temperatures, along with high humidity, are favorable for fungal and bacterial growth on many crops. The levels of annual precipitation range from around 1,000 milimeters in the eastern part of Hokkaido and in the Seto Inland Sea area to more than 3,000 milimeters in several Pacific coastal areas of Kyushu, Shikoku, and Honshu. Most of

TABLE 1

Planted Area[a] of Major Crops Exceeding 10,000 Hectares

Crop	Area	Crop	Area
Irrigated rice	3,173	Pumpkin	17
Dry field rice	101	Egg plant	27
Wheat	322	Tomato	20
Barley		Strawberry	13
(husked and naked)	316	Cabbage	45
Oat	41	Chinese cabbage	49
Corn	18	Leaf crucifers	23
Potato	169	Spinach	24
Sweet potato	154	Stone leek	28
Soy bean	103	Head onion	33
Kidney bean	68	Radish	87
Peanut	59	Carrot	25
Azuki bean	100	Burdock	18
Buckwheat	23	Dasheen	37
		Garden pea	15
Oranges	188	Kidney bean	
Grapes	23	(vegetable use)	12
Apple	61	Sweet corn	24
Peach	21		
Japanese apricot	16	Oil rape	30
Japanese pear	18	Tea	50
Persimmon	37	Tobacco	75
Chestnut	38	Sugar beet	55
		Sugar cane	13
Cucumber	32	Konnyaku	17
Watermelon	39	Mulberry	162

[a]Data taken in 1968 and expressed in 1,000 hectares.

the precipitation occurs in June, July, and September on the Pacific Coast, frequently as heavy showers. In the north, a cold summer, which is not infrequent in Tohoku and Hokkaido, can cause an epidemic of insect pest infestation and plant diseases.

TABLE 2

Consumption[a] of Chemical Fertilizers in Japanese Agriculture

Year	Nitrogen as N	Phosphate as P_2O_5	Potash as K_2O
1957	592 (100)	389 (100)	386 (100)
1962	699 (118)	493 (129)	508 (132)
1967	889 (150)	671 (173)	651 (168)

[a]Figures indicate amounts of nutrients in 1,000 tons (figures in parenthesis show rates of increase).

More than 1,200 diseases and 2,500 insect pests are known to exist for major economic crops in Japan (TABLE 3). In order to sustain high yields, farmers in Japan must rely heavily upon the use of pesticides.

Trends in Pesticide Use in Japan

The introduction of a wide variety of modern synthetic pesticides in Japan started soon after World War II, as in other countries. The agricultural production of Japan dwindled greatly during World War II due to the shortage of farm labor and farm materials. There was a big shortage of food when the war was over and increased food production was the most important and urgent national need. Food production had also been of high priority during the War, and pest control was encouraged by the Government in order to prevent loss and to secure the greatest possible food supply. A pest forecasting program was

TABLE 3

Important Plant Disease and Pest Insect Species affecting Major Crops in Japan

Crop	Number of diseases	Number of insects	Crop	Number of diseases	Number of insects
Rice	27	122	Pumpkin	8	
Wheat	28	80	Egg plant	11	39
Barley	31		Tomato	25	
Oat	8		Strawberry	2	23
Corn	7	36	Cabbage	12	
Potato	19	30	Chinese		63[c]
Sweet potato	14	36	cabbage	12	
Soy bean	23		Spinach	5	13
Kidney bean	15	145[a]	Stone leek	9	
Peanut	7		Head onion	9	24
Azuki bean	8		Radish	12	-[d]
			Carrot	8	17
Oranges	23	84	Burdock	5	22
Grapes	19	49	Dasheen	6	8
Apple	27	104			
Japanese			Oil rape	11	-[d]
apricot	9	24	Tea	18	49
Japanese pear	20	83	Tobacco	19	37
Peach	20	39	Sugar beet	9	34
Persimmon	12	54	Sugar cane	7	68
Chestnut	12	50	Konnyaku	5	3
			Mulberry	20	73
Cucumber	14				
Watermelon	8	41[b]			

[a]Including other beans and pea.

[b]Including other cucurbits.

[c]Including other crucifers.

[d]Included in "c".

[e]Anonymous (1965).

initiated as early as 1940 in order to predict the occurrence of pests and to conduct control measures by proper timing. Statistical and ecological knowledge concerning the major pest species, particularly of food crops, were obtained by this program, and the knowledge obtained facilitated the establishment of chemical controls with modern pesticides. The government encouraged the chemical control of pests in order to increase food production.

DDT, released to the market in 1947, was the first of the synthetic pesticides. This compound was soon found effective in controlling the paddy stem borer, *Tryporyze oryzae* Walker, which was especially destructive in the western part of the country. Stimulated by this success with DDT, a great number of synthetic pesticides were introduced and tested for their possible effectiveness in controlling local pest species. Like most Japanese industries, pesticide manufacturing and formulating firms are small physically but are numerous. Strong competition existed to increase market sales by finding more potent pesticides. This condition helped to stimulate the development of pesticides.

Production of all pesticides for domestic consumption, in terms of the monetary value at the producer's sales price, soared from two billion yen in 1950 to 24.7 billion in 1960 and 82.8 billion in 1970 (TABLE 4). The average price of pesticides did not rise, but gradually fell during the past two decades. The number of chemical compounds employed as pesticides increased from about 30 in 1950 to 163 in 1960, and to 411 in 1970. The number of registered brands increased from 1733 in 1955 to 3,004 in 1960, and 5,698 in 1970.

Since the contamination of food and the natural environment with pesticide residues drew public attention and criticism, pesticide consumption was lowered. In five years, from 1960 to 1965, production of pesticides almost doubled in terms of the sale price at the producer's plant. However, in the following five years, from 1965 to 1970, production increased only by 1.7-fold. It should be mentioned that sales in 1970 decreased by 2% as compared

TABLE 4

Production of Pesticides in Japan from 1950 to 1970

Year	Number of compounds used as pesticide	Number of formulated products registered	Production in billion Yen
1950	ca 30	-	2.0
1955	102	1,733	12.8
1960	163	3,004	24.7
1965	314	4,355	50.2
1970	411	5,698	82.8

with sales in 1969.

During the past two decades, there has been a considerable change in the pattern of consumption of pesticides, as well as the compounds and formulations employed. Studies on the various adverse effects caused by these chemicals are necessary in understanding the extent and mechanisms of the impact of pesticide use to the environment.

Insecticides - Following the introduction of DDT in 1947, BHC and other chlorinated insecticides were introduced, and by 1955 marketed (TABLE 5). BHC was consumed in large quantities for the control of stem borers and other rice insects as well as insect pests of other crops. Dust formulations produced from technical BHC were used extensively for control of the rice stem borers as well as leaf- and planthoppers. Aldrin and heptachlor were widely employed for soil insect control in upland fields, and endrin was used for the control of vegetable insects. Endrin was prohibited for use in rice fields because of its toxicity to fish. Insect pests having less than one generation per year--such as the rice grasshoppers, *Oxya spp.*, and the black rice bug, *Scotinophara lurida* Burmeister, among rice insects; and wireworms and mole crickets, among soil insects--were remarkably

TABLE 5

Yearwise Introduction of Major Insecticides into Japanese Pesticide Market

Year	Pesticide	Year	Pesticide
1947	DDT	1960	Dichlorvos
1948		1961	Dimethoate, Fenthion
1949	BHC	1962	Fenitrothion, Mecarbam, Thiometon, VC
1950		1963	Cidial, Vermodothion
1951	TEPR, EPN	1964	Arprocarb, Ethylthiometon, Ethion, Telodrin, CPMC
1952	Parathion, Parathion-methyl	1965	IPSP
1953	Malathion	1966	
1954	Chlordane	1967	MPMC, MTMC, Cartap, Chlorphenamizin
1955	Demeton-S-methyl, aldrin, dieldrin, endrin, diazinon	1968	Salithion, Surecide, Vinyphate
1956		1969	BPMC, XMC
1957		1970	Mesomil
1958	Heptachlor, Trichlorfon		
1959	Carbaryl		

suppressed by the increased use of chlorinated insecticides.

Parathion was introduced in 1951 and soon proved to be much more effective against the stem borers than the chlorinated hydrocarbon insecticides. Parathion also proved to be more effective against shoot and fruit moths, aphids, leaf rollers, and scale insects, among fruit insects. Control of rice insects with parathion was encouraged by the government and consumption increased greatly in a very short time. In 1955, sales of parathion formulations (e.g., emulsion and dust) accounted for 39% of the total sales of insecticides.

The extensive use of this highly hazardous insecticide resulted in more than 7,000 cases of human intoxication, including 1,600 intended cases of suicide in the first five years after introduction. This situation stimulated the search for safer organophosphorus compounds to replace this hazardous pesticide. Malathion, diazinon, trichlorfon (dipterex), cidial, fenthion, and several other compounds of lesser mammalian toxicity were introduced from abroad and fenitrothion was developed domestically in the latter half of the 1950's. These safer organophosphorus insecticides gradually replaced parathion which was withdrawn from the market in 1970 at government request.

The incidental discovery of the systemic action of BHC on rice plants led to the development of broadcasting methods of BHC granular formulation over the surface water of rice fields as a new labor-saving method of controlling the rice stem borer (Koshihara and Okamoto, 1957). Beset by a labor shortage, farmers quickly adopted this new practice. BHC granules containing technical BHC with 6% gamma-isomer appeared on the pesticide market in 1962 and their consumption increased rapidly (TABLE 6).

In addition to BHC granules, a number of mixed granular products containing BHC were also commercialized. Accordingly, the consumption of BHC, particularly in rice fields, increased remarkably after the introduction of original BHC granule products. It is estimated from the consumption of various formulated products and domestic production and export of technical BHC that about 40,000 tons of technical BHC was consumed each year in 1967 and 1968.

TABLE 6

Consumption of BHC Dust and Granular Products and the Abnormal Build-up of the Rice Green Leafhopper Population

Year	Consumption of BHC products in 1,000 ton		Rice green leaf-hopper occurrence in 1,000 ha.	
	Dust	Granule	Infested	Treated
1956	41	-	484	102
1957	41	-	503	105
1958	47	-	724	213
1959	54	-	701	254
1960	54	-	919	339
1961	60	-	697	265
1962	49	1	385	316
1963	32	15	875	450
1964	37	23	1,172	471
1965	32	18	1,157	589
1966	46	18	1,303	775
1967	53	22	1,273	867
1968	44	25	1,145	983
1969	29	24	1,348	1,105
1970	16	14		

As one of the reactions to the widespread use of BHC for stem borer control, abnormal build-up of the rice green leafhopper, Nephotetix cincticepts Uhler, became apparent (TABLE 6). This leafhopper was originally tolerant to BHC but predators preying on this leafhopper, spiders in particular, are susceptible. Furthermore, this leafhopper developed resistance to a number of organophosphorus compounds employed for its control and for the control of the rice stem borer. New compounds need to be developed.

Carbamates are a promising group of compounds. Although carbaryl was introduced in 1959, extensive search for carbamate compounds was made several years afterwards and arprocarb (suncide), CPMC (Hopcide), MTMC (Tsumacide), MIPC (Mipcin), BPMC (Bassa), EMPC (Meobal), and several other carbamates were either imported or domestically developed. Goto (1970) stated that carbamates made up 16% of the consumption of all insecticides.

Public concern regarding pesticidal residues was created primarily by sensational reporting by news services since around 1965. The Ministry of Health and Welfare, in consultation with the Ministry of Agriculture and Forestry, established tolerances of pesticides on and in foods in 1967. The first announcement made in 1967 covered the tolerance limits of BHC, DDT, parathion, arsenicals, and lead for apple, grape, cucumber, and tomato. Although the establishment of tolerances was very limited at first, it stimulated research on the residues of pesticides on and in other agricultural commodities and environmental components. This research was facilitated by the development and domestic production of microanalytical apparatus. Research disclosed widespread contamination at high levels in foods, and even in the human body, with BHC. Among four isomers analyzed, the beta-isomer seemed to be very stable in nature, and high residue levels were detected in cows milk, human lactation, and in soil. In order to answer public criticism, the Ministry of Agriculture and Forestry put a restriction on the use of DDT and BHC in rice insect control in 1970, and the use of this insecticide was almost banned early in 1971. Only limited use in forest insect control remains permissible at present, but it will probably soon be banned entirely.

Use of aldrin and endrin for soil and vegetable insect control were also restricted for the 1970 season, and accordingly all major chlorinated insecticides will disappear in the very near future. Sales of BHC dust fell from 34,000 tons (formulation base) in 1969 to 17,000 tons in 1970, and that of BHC granules from 22,000 tons (formulation base) to 14,000 tons. Chlorinated insecticides are being replaced by organophosphorus compounds and other compounds more recently developed such as cartap (Padan) and chlorphenamidin.

Acaricides - Introduction of DDT, BHC, and parathion into fruit insect control by 1955 resulted in a drastic change in the pest species. Parathion almost eradicated the oriental fruit moth, *Grapholitha molesta* Busck, the peach fruit moth, *Carposina nipponensis* Walsingham, and the arrowhead scale, *Unaspis yanonensis* Kuwana, from apple orchards. However, due to simultaneous destruction of predatory species of leaf mites, phytophagous mites became serious pests. The improved nutritional condition of the host plant brought about by the increased application of fertilizers seemed to encourage the build-up of mite populations. A similar increase in leaf mite populations was observed in citrus orchards after fumigation with cyanogas for scale control.

In order to control the increased mite population in orchards after the introduction of parathion, it was necessary to add an acaricide to the spray calendar. Chlorfenson was released to the market in 1954, and by 1960 chlorbenzilate, tetradifon (Tedion), dicofol (Kelthane), phenkapton and several other acaricides had been marketed. However, because of rapid development of mite resistance to acaricides, only a few of these compounds survived more than a few years (TABLE 7). Rotational use of different acaricides was recommended in many prefectures, but most of the acaricides to which resistance had developed were neglected and scarcely reused.

Nematocides - Among halogenated aliphatic nematicides, DD was released to the market as early as 1950. However, it failed in actual use because of the poor

TABLE 7

Yearwise Introduction of Major Acaricides into Japanese Pesticide Market

Year	Acaricide	Year	Acaricide
1951	(TEPP)	1961	Dioxathion, Thioquinox
1952	DN	1962	Mikazin, Milbex
1953	(Malathion)	1963	Tetrasul
1954	Chlorfenson	1964	Chloropropylate
1955	Demeton-S-methyl chlorbenzilate	1965	
1956		1966	Smite
1957	Tetradifon, Dicofol	1967	
1958	Phenkapton	1968	Omite
1959		1969	Chlorphenamizin
1960	Aramite		

recognition of the damage caused by plant parasitic soil nematodes. In 1959, the Ministry of Agriculture and Forestry launched the Upland Crop Encouragement Program in order to improve the earnings of upland farmers to a rate comparable to that of rice farmers. Nematode control was selected as one of the major goals. A Survey Program on Nematode Infestation was initiated in order to identify upland fields which required nematode control by fumigation. Financial supports were given for the purchase of nematocides and mechanical applicators. The support lasted for five years (1959-64). A total of 223,000 hectares of upland fields were thus treated for nematode control (TABLE 8).

TABLE 8

Upland Fields[a] Treated for Soil Nematodes

Year	Areas treated, in 1,000 ha.		
	Total	Vegetable	Industry crops
1959	8.8	4.4	3.6
1960	16.9	10.4	4.3
1961	22.2	12.7	6.3
1962	31.2	17.6	7.8
1963	39.0	21.3	10.0
1964	20.3	12.6	3.1
1965	20.0	13.8	2.3
1966	24.3	16.5	1.7
1967	23.9	16.0	2.0
1968	23.0	15.1	2.0
Total	229.6	140.4	43.1

[a]Besides vegetables and industrial crops, wheat, beans, potatoes, orchards, and ornamental crops were also treated.

In addition to DD both EDB and DBCP were employed for soil fumigation. It was observed that fumigation of the soil with these fumigants affected the soil microorganismic community. Some alterations in recommendations for fertilizing were required to offset these adverse effects of fumigants.

Fungicides - Bordeaux mixture and lime sulphur were the two leading fungicides for controlling fruit and vegetable diseases before World War II. Organomercury compounds were only used for seed treatment. In 1952, dusting with a mixture of phenyl mercury acetate and slaked lime proved to be highly effective in controlling rice blast, the most serious and widespread disease of rice. In the years following, dusting of rice with an organomercury fungicide became widespread throughout the country. In 1960 organomercury fungicides constituted 52% of all fungicide sales. Soon after the introduction of organomercury compounds for rice blast control, translocation of mercury to rice grain was demonstrated by radioautographic methods (Kanazawa and Sato, 1965; Moriya et al., 1965; Tomizawa, 1965). The toxic effect of mercury in the grain was recognized. Blasticidin, an antibiotic developed exclusively for agricultural use, was commercialized in 1961 as a mixture formulation with an organomercury compound. Along with progress in the study of the Minamata disease, foliar application of organomercury compounds for control of rice blast was criticized by medical and toxicological groups. This criticism against the use of organomercury compounds stimulated the search for substitutes. Following the finding of pentachlorobenzyl alcohol as a substitute for organomercury compounds for rice blast, a number of pentachlorobenzene compounds were commercialized. Of the antibiotics, Kasugamycin was added, and since 1967, Kitazin (S-benzyl diisopropyl phosphorothiolate), Inezin (S-benzyl ethyl phenylphosphonothiolate), and Hinosan (0-ethyl diphenyl phosphorodithiolate) were developed and commercialized as organophosphorus fungicides.

Since the chemical control of the rice stem borer with parathion was successful, early-season growing of rice was preferred in the western part of the country in order to escape the damage caused by typhoons. However, early-season growing induced a high incidence of sheath blight of rice. TUZ dust, a mixture of TMTD, Ziram, and urbazid (methylarsenic dimethyldithiocarbamate) proved to be effective and was commercialized in 1957. Several other arsenic compounds such as poly (methyl-bis-9-thiocyanoto)arsine and calcium and iron methanarsonates were also developed and commercialized.

For fruit and vegetable disease control, maneb, zineb, and other dithiocarbamates were commercialized in the early half of the 1950's (TABLE 9). Other compounds such as dichlone, captafol (Difoltan), Sclex (dichlozorine,

TABLE 9

Yearwise Introduction of Major Fungicides into Japanese Pesticide Market

Year	Fungicide	Year	Fungicide
1952	PMA-lime dust Zineb	1962	Amobam
		1963	
1953	Captan, Sulphur dust		
		1964	Morestan, Cellocidin, Dexon, Sanipa, Binapacryl
1954	Ferbam, Ziram, Wettable sulphur		
1955	Streptomycin	1965	Kitazin, Sankel, captafol, bisdithane
1956	Dinocap		
		1966	Blastin, Kasugamycin, Deconil
1957	TUZ, Dichlone-thiuram, Maneb, Triazin, PCNB		
		1967	Hinosan, Inezin, Oryzon, Rabcon, Phenazin, Polyoxin
1958			
		1968	
1959	Organoarsenic compounds cycloheximid	1969	Konen, Topzin, Sclex, Anthracol
1960	Blasticidin S, Organotin compounds	1970	Novobiocin, Ezonomycin

3-(3, 5-dichlorophenyl)-5, 5-dimethyloxazo lizine-dion-2. 4), and Topzin (1, 2-bis(3-ethoxycarbonyl-2-thioureid) benzene and many other compounds are another recent development.

Herbicides - Although 2, 4-D was recommended for weed control by U.S. agronomists working for the Occupation Force soon after the War, herbicides were not actually used until early 1960 when a shortage of farm labor became apparent. Manual or mechanical weeding was conducted when ample labor was available in rural areas. In addition to 2, 4-D and allied chemicals, simazine and allied compounds were in the pesticide market in the late 1950's. However, their sales remained small. The history of introduction of major herbicides into Japan has been summarized and illustrated in TABLE 10.

The first widely used herbicide was PCP. This compound was tested in order to eradicate the freshwater snail which served as the intermediate host of *Schistosoma japonicum* Katsurada, and incidentally was found to have excellent herbicidal effect on weeds in rice fields, including the barnyard grass, *Echinochloa crusgalli* Beauv.--the most noxious grass in rice fields. PCP was marketed in 1959 as a herbicide for transplanted rice. However, immediately after marketing, fish and shell fish kills, suspected to have been caused by the use of this herbicide, were reported from various parts of the country. The kills became so serious in the 1962 season that the Ministry of Agriculture and Forestry had to amend the Agricultural Chemical Registration Act to add an article restricting the use of this herbicide.

The restriction on the use of PCP encouraged the search for herbicides less toxic to fish and shell fish yet effective to weeds in rice fields. A number of new herbicides such as nitrophen (NIP), propanil (Stam), and MCPCA (Mapica, 4-chloro-o-tolyl) oxylaceto-o-chloranilide) were found through screening. The labor shortage in rice-growing areas created a good market for herbicides in the late 1960's. New herbicides were introduced and consumption increased. Recently, symetrine and

TABLE 10

Yearwise Introduction of Major Herbicides into Japanese Pesticide Market

Year	Herbicide
1950	2,4-D sodium and amine salts
1951	
1952	
1953	Sodium chlorate
1954	2,4-D ester, MCP sodium, and potassium salts
1955	
1956	2,4-D ses
1957	CMC
1958	MCP ester, DNOC, PCP (dry field use) simazine
1959	
1960	MCP-TCBA mixture
1961	PCP-MCP mixture, MCPB-ethyl, diuron, dalapon-sodium, propanil (dry field use)
1962	Propazine, amitrol, dinoseb
1963	PCP-MCPB mixture, MCPCA, nitrophen, diquat, dichlorobenil
1964	Prometryne, chlorothiamid, dinoseb-acetate, cyanate-diuron mixture
1965	PCP-DCBN mixture, PCP-MCPP mixture, linuron, atrazine, bromasil, CNP, paraquat, ammonium sulphamate, 2,4,5-T, 2,4-D-2,4,5-T mixture, 2,4-D-Amitrol mixture
1966	Pentanochlor, diphenamid, trifluralin, lenacil, swep, MCP-CNP mixture, naptalam
1967	Propanil-carbaryl mixture, propanil-cyclohexanone mixture, CNP-dichlobenil mixture, pyrazon-chlorbufam mixture
1968	Swep-MCP mixture, amino-chloronaphtoquinone, sodium chlorate-sodium fluorate mixture, sodium ethyl xanthogenate

TABLE 10 - Continued

Year	Herbicide
1969	PCP-MCPE mixture, prometryne-MCPB mixture, ametryne, amitrol-diuron mixture, sodium chlorate-2, 4, 5-T mixture
1970	PCP-DCMU mixture, PCP-linuron mixture, PCP-phenothiol mixture, symetrine, benthiocarb-symetrine mixture, siduron

other triazines, and their mixture with other compounds, are being used extensively.

In labor-saving in reforestration programs, sodium chlorate, and mixtures with related compounds, were used extensively without any noticeable harm. This compound is effective in killing the rhizomes of Sasa, and the success with this herbicide triggered the use of herbicides in reforestration programs. Amitrol and 2, 4, 5-T were introduced as vine and brush killers. However, reports on the possibility of amitrol as a carcinogen, and on 2, 4, 5-T to contain the teratologic agent dioxin were known among laborers. Thus, the labor union rejected their use. Encouraged by the appeal made by conservationists who claimed that over-all application of herbicides in reforested areas deprived wildlife of food, the use of both herbicides was terminated.

General Trend of Pesticide Consumption

Goto (1970) analyzed the trend of consumption of major groups of pesticide as shown in TABLE 11. This table illustrates the trend that the share of the market by insecticides decreased remarkably during the past two decades from 71.5 to 45.1%, and that during the same time period those for fungicides and herbicides increased from 21.6 to 30.3% and 13.2 to 21.2%, respectively. Of

TABLE 11

Trend of Consumption[a] of major groups of Pesticides in Japan during the past two Decades

Pesticide group	1950	1955	1960	1965	1969
Chlorinated insecticides	29.0	26.2	19.2	16.5	10.8
O-P insecticides	-	31.2	27.3	21.5	19.4
Arsenate insecticide	14.3	3.8	2.3	1.2	0.3
Carbamate insecticides	-	-	-	-	7.3
Other insecticides	14.8	10.3	10.0	9.7	7.3
Insecticides total	58.1	71.5	58.8	48.9	45.1
Copper fungicides	6.2	4.9	2.9	3.3	1.6
Inorganic S fungicide	11.9	1.8	2.2	1.2	0.7
Organic mercury fungicides	3.5	10.4	16.0	12.3	0.5
Organic sulphur fungicides	-	2.5	4.5	4.3	6.2
Antibiotics	-	0.1	0.2	2.4	8.0
Organic arsenic fungicides	-	-	3.0	2.4	3.4
O-P. fungicides	-	-	-	-	4.7
Other fungicides	-	0.9	1.2	2.1	5.2
Fungicides total	21.6	20.6	30.0	28.0	30.3
Herbicides total	13.2	4.4	7.8	19.5	21.2
Other pesticides	7.1	3.5	3.4	3.6	3.4
Total pesticide production in billion Yen	2.1	12.8	24.7	50.2	88.6

[a]Figures indicate the percentage of production in value to the total production value.

the insecticides, arsenates and chlorinated insecticides decreased substantially, though the use of organophosphates and carbamates increased. Of the fungicides the use of copper, inorganic sulphur, and organomercuric compounds declined, and that of organosulphur, organochlorine fungicides, and antibiotics increased. More recently organophosphorus fungicides have been found to be effective in controlling rice blast.

Changes in the Formulation and Method of Application

As mentioned previously, almost half the agricultural area is devoted to the cultivation of rice under irrigated conditions. Under such a condition, spraying pesticides with hand-operated sprayers was not practical, therefore dust formulations were introduced. Spraying was limited to orchards and vegetable fields, where growers preferred that method for cost saving and for avoiding incompatibility problems arising from the use of different pesticides. As dusting of rice fields became more extensive by collaborative pest control operation, or by the introduction of aerial application by helicopter, local hazards to silkworm culture, fish pond, and apiary were frequently observed because of the drifting of dust products.

A granular formulation was developed in order to apply BHC and PCP to rice fields by a method to save on labor costs and also prevent phytotoxicity, in the case of PCP. Although granular products could reduce the drift of pesticides, it seemed to have contaminated the environment more seriously, since the product actually contained higher amounts of active ingredient than did dust products. TABLE 12 indicates the production of pesticides according to the different formulations during the past two decades.

TABLE 12

Formulation-wise Production[a] of Pesticides in Japan during the Past Two Decades

Formulation	1950	1955	1960	1965	1970
Dust	7,264	77,090	185,423	243,874	364,118
Granule	-	-	1,955	94,851	175,742
Emulsion and liquid	874	3,597	7,503	15,832	42,389
Wettable	4,980	11,268	18,233	27,677	37,324
Others	14,105	19,261	30,454	46,931	33,272

[a]See Anonymous (1971).

Some Examples of the Impact of the Use of Pesticides on the Japanese Environment

As the chemical control of pests on various crops became extensive, many direct and indirect effects of the use of pesticides became gradually apparent. In the early stages of the development of chemical control, such effects, although adverse, were neglected or underestimated, especially in view of the more important need for food production. Lack of microanalytical methods also prevented meaningful studies. Increased use of pesticides accentuated the effects and improved analytical methods made studies easier and more accurate. However, it should be mentioned that studies on the impact of pesticides to the Japanese environment are still fragmentary, and results obtained to date are far from conclusive. The following are brief descriptions of some of the major findings observed in recent years.

Upset of the Pest-Natural Enemy Equilibrium

As in most countries, the first incidence of the impact of pesticide use on the environment was the disturbance in the equilibrium existing between insect pests and their parasites or predators. Soon after the introduction of DDT into apple insect control, particularly for the fruit moths and leaf rollers, abnormal increases of the wooly aphid, Eriosoma lanigerum Hausmann, occurred. Its imported parasite, Aphelinus mali Haldeman, was badly affected. An abnormal increase of the apple leaf miner (Kinmon-Hosoga), Lithocolletis ringoniella Matsumura gradually became apparent because of the destruction of the Hymenopterous parasite, Copidosoma sp., in orchards where DDT was employed.

The introduction of DDT and parathion for apple insect control induced build-up of red spider populations to the extent that acaricides were required. As mentioned previously, a number of acaricides were introduced into the spray calendar for apples. Kiritani demonstrated that an increase in the number of DDT and parathion sprays contributed to the increase of acaricide applications.

In rice insect control, extensive use of BHC for the control of the rice stem borer was first observed by Oho to be detrimental to the Veliid water slider (Keshi-Katabiro Amenbo) Microvelia douglasi Scott which preys on the leaf- and plant-hoppers of rice. Increase of the rice green leafhopper, Nephotettix apicalis Motschulsky, became apparent since 1955 in Saga Prefecture in northern Kyushu. Kobayashi (1961) studied the abnormal increase of the leaf- and planthopper populations in rice fields where either parathion or BHC was sprayed or dusted. He observed that although populations of the leaf- and planthoppers decreased immediately after application of either parathion or BHC, their populations were restored to original levels 3 weeks after BHC and 5 weeks after the application of parathion, respectively. He further attributed the rapid build-up of the leaf- and planthopper populations, after the application of BHC, to the destruction of the spider population feeding on both the adults and larvae of the leaf- and planthoppers. BHC was found particularly toxic to arachnid species found in rice fields. Modification of the application of BHC from dusting to the broadcasting of granules was expected to reduce this adverse effect on spiders because the broadcasting of granules would not contaminate the outer surface of the rice plant on which spiders occur. However, the spider population seems to be affected by the intake of BHC through their feeding on contaminated prey.

The abnormal increase of leafhoppers, in particular the rice green leafhopper, required the introduction of a carbamate insecticide for their control. Malathion and other organophosphorus insecticides were previously known and recommended for the control of the rice green leafhopper. However, repeated application of parathion and other organophosphorus insecticides for the control of the rice stem borer induced the resistance of the green leafhopper to most of the organophosphorus insecticides. Carbamates were favored for their high effectiveness against the green leafhopper and other planthoppers, and for their safety to the spiders in the rice fields. Application of carbamates is accordingly expected to suppress more effectively the green leafhopper population without affecting their natural control agents.

Impact on Non-Target Species

Together with the target pest, a number of non-target insect species are reduced. Extensive use of insecticides in rice fields have been criticized because they also reduce aquatic beneficial insects such as spiders (TABLE 13), dragonflies, and fireflies. However, no quantitative records of changes in density are available. It is also very difficult to analyze the role of pesticidal use apart from other environmental changes, and the changes in agricultural practices.

Coping with the increased use of pesticides in rice growing in 1955, it was repeatedly stated that local people ceased to use mosquito nets during the summer season. Also, during this period the number of cases of malaria, which sporadically occurred in the western part of the country, drastically decreased although no direct control measures were employed.

Comparative studies on the composition of fauna in the treated and untreated fields or orchards have been conducted by a number of ecologists. Hukushima made many observations on the changes in the faunistic composition in apple orchards in association with pesticide applications. However, most of the observations covered a short term, and the long-term effects of pesticide applications is not yet clear.

TABLE 13

LD-50 Concentration of Insecticides to Rice Field Spider, Oedothorax insecticeps

Insecticides	LD-50 in ppm
BHC	21
Fenitrothion	3,200
Fenthion	500
Diazinon	2,450
Carbaryl	840
MIPC	3,200
CPMC	4,800

Hazards to Aquatic Organisms

Soon after the cooperative control for the rice stem borer was conducted in the Chikugo basin in northern Kysushu in the summer of 1952 by using parathion sprays, the neighboring coastal fishery people met with a very poor harvest of Euphausid krill the following fall. The poor harvest was suspected to be due to the pollution of the offshore waters with parathion originating from rice-growing areas.

Because of this suspected harm, the Fishery Agency began to collect reports of fish kills the following year.

Cases were reported from 22 prefectures out of 46 in 1954, and from 27 prefectures in 1955. A total of 55.7% of the cases were caused by parathion and 21.9% by BHC. It would have been conceivable to attribute most of the cases to either parathion or BHC since both insecticides had been used in large quantities.

Hazards to fish and shell fish caused by pesticidal application became very serious when the sodium salt of PCP was released for weed control in rice fields. Since PCP was known to be very toxic to shell fish and fish, registration of its use in rice fields required the deliberation of the pesticide control officer and an agronomist. The agronomists were very eager to use this herbicide because it is highly effective against barnyard grass in rice fields. Since it was known that PCP was degraded under solar irradiation (Munakata et al., 1966), its use was permitted with the restriction that it not be spilled into irrigating water from the field for 10 days. However, since application is made in the rainy season, heavy showers frequently caused it to flood from treated fields.

The consumption of PCP for rice weed control increased rapidly. In 1959 only 15,000 hectares of rice were treated with PCP, but in 1963, 930,000 hectares (e.g., nearly one-third of all rice fields) were treated. Also, in 1959 large numbers of fish kills were reported from 34 prefectures. The damage amounted to 89,000 tons--2.6 billion yen in value (Agricultural Promotion Bureau, 1962).

Fisheries people were greatly concerned about these serious kills of fish and shell fish caused by PCP and requested the Ministry of Agriculture and Forestry to absolutely ban the use of this herbicide. Its use was restricted to areas in which it would not be likely to cause kills to fish and shell fish.

The restriction on the use of PCP in rice fields encouraged the search for other herbicides safer to fish and shell fish. Rice fields treated with PCP and mixed formulations containing PCP for weed control began to decrease in 1965.

Since the restrictions on the use of PCP in rice fields

were enforced, all registered pesticides as well as pesticides applied for registration were investigated in respect to their toxicity to fish. Hashimoto (1963, 1964, 1967) developed a standard method for evaluating the toxicity of fish to pesticides, employing common carp as test fish. He further classified the conventional pesticides into three categories as follows: (a) Fish kills will not occur if used in accordance with the directions described on the label. (b) Fish kills will not occur if used in accordance with the directions described on the label. However, considerable care should be taken if used on a large scale, and (c) Fish kills will occur unless great care is taken not to contaminate bodies of water.

However, fishery research workers criticize the evaluation of pesticide toxicity using only a single fish species. It is well known that sweet fish and trout are much more susceptible to most pesticides than the common carp. Water quality standards for the fishery industry established by the Japan Fishery Resources Protection Society are based on the toxicity of pesticides both to common carp and sweet fish or trout.

Hazards to Wildlife

When the rodenticide 1080 became available for rat control in 1952, baiting with this compound was urged in many prefectures. Because of the high toxicity of this compound to canine animals, dead wild foxes were frequently found in treated areas. They were suspected to have eaten rats killed by the poisoned bait.

Spraying rice crops with parathion for the control of the rice stem borer became widespread in 1952-55 and the swallow population was reported to have decreased in treated areas. Dead swallows were reportedly observed in the area as well. However, no detailed study was made as to whether the swallow population did actually decrease.

In 1968, ultra-low volume applications of concentrate formulations of organophosphorus insecticides were tested. Application was made by helicopter, and in areas treated with fenthion, dead sparrows were found. Accordingly, fenthion along with parathion was not included in the pesticides recommended for ultra-low volume application.

With respect to the chronic effects of pesticide residues on wildlife, particularly on birds, studies have been conducted on the mercury content of dead birds found in rice-growing areas. A number of bird species frequenting rice fields have been studied so far, and contamination with mercury was detected and reported as the cause of death in some cases.

Contamination of Foods and the Human Body

As the consumption of pesticides in various phases of crop production increased, the Ministry of Health and Welfare launched a program to monitor the possible pesticide residues on and in agricultural products in 1963. Based on the data obtained from this program, the Ministry commenced to establish tolerances for pesticide residues in 1967. By the end of 1970, the Ministry established residue tolerances for nine pesticides (DDT, BHC, aldrin, dieldrin, endrin, parathion, arsene, lead, and EPN), and for 14 crops (rice, summer orange, Japanese pear, grape, peach, apple, strawberry, cabbage, cucumber, radish, tomato, spinach, potato, and tea). However, this coverage is just a fraction of the total in relation to the number of pesticides used in large quantities and the kind of agricultural commodities consumed.

The Ministry of Agriculture and Forestry and the Research Council carried out a five-year program on the pesticide residues on major crops in order to develop guidelines for establishing waiting periods from application to harvest. The Ministry has been establishing the waiting period for crops and pesticides. As of July of this year, a new Environment Protection Agency was established and the authority for setting the waiting period was transferred from the Ministry of Agriculture and Forestry to this new Agency.

Milk was also studied in the course of monitoring

pesticide residues on and in food commodities. From these studies, it was demonstrated that milk produced in the western part of the country was heavily contaminated with various isomers of BHC. Among four isomers identified, the beta isomer was found at very high levels. The origin of this contamination was then traced to rice straw (TABLE 14) which had been fed to lactating cattle in increasing quantities in recent years. From straw, too, high level residues of the beta isomer were detected.

TABLE 14

Residue of BHC in Rice Straw and Grain and in Paddy Soil

	Number of samples		
	Straw 25	Grain 10	Soil 18
Residue of			
alpha BHC	1.914	0.152	0.539
beta BHC	8.146	0.079	1.029
gamma BHC	0.989	0.044	0.321
delta BHC	3.635	0.097	0.220
Total	14.684	0.372	2.019
Proportion of			
alpha BHC	13	41	27
beta BHC	56	21	51
gamma BHC	7	12	11
delta BHC	25	26	11
Total	100	100	100

[a]From Kanazawa (1970).

Suspected from the finding of high contamination of milk with BHC isomers, human lactation was also recently analyzed by medical workers of the Japan Rural Medical

Institute. Samples from 32 women living in Nagano, Tokyo, Kanagawa, and Ibaraki Prefectures all contained some residues of the alpha, beta, gamma, and delta isomers of BHC. Residues of the beta isomer were extraordinarily high, as in the case of cows milk.

It is very clear that the general environment and biotic populations in this country are highly contaminated with BHC, which was employed for more than two decades to control insect pests. Strict restrictions on the use of this compound were enforced in 1970, and efforts are now being made to follow the fate of residues remaining in soil and other environmental constituents.

Contamination of Soil

As mentioned previously, large quantities of a ceresan-lime mixture had been dusted on rice plants in order to control rice blast. When this compound and its relatives were employed most extensively, 120 to 150 tons of mercury were consumed annually. In connection with the expected accumulation of mercury in rice field soil and the consequent possible heavy contamination of rice grain produced from such fields, some investigations were conducted on the fate of mercury in soil. It was demonstrated that if the field were kept irrigated, mercury disappeared gradually, but if the soil was kept without irrigating water, it was found to disappear rather quickly. It was not clear how mercury disappeared from the soil, but a fairly large portion of the loss was suspected to be due to evaporation.

In recent years, attention has been directed to the contamination of soil with chlorinated insecticides, in particular with BHC and aldrin. High levels of the beta isomer of BHC was detected from soil of rice fields. The relative composition of four isomers detected in the paddy field soil was quite different from the composition of isomers in technical BHC. In other words, beta BHC shares 51% of the residue in the soil, but in technical BHC, it constitutes only 5-14%. Residues of aldrin in the soil were particularly noticeable as cucumbers grown on aldrin-treated soil contained high residues. It may require two to three years before aldrin-treated soil will become safe for growing cucumbers.

Concluding Remarks

Since the introduction of modern synthetic pesticides into Japanese agriculture, consumption of these pesticides expanded tremendously in a relatively short period and contributed to the intensification of agricultural productivity of this country. However, the education of large numbers of farmers to rational and safe use of these pesticides was so difficult that many adverse effects of pesticidal use were experienced. Those pesticides which caused serious hazards were not fully investigated as to the actual causes of hazards but usage quickly switched to newer and possibly safer pesticides. Accordingly it has been very difficult to investigate the adverse effects brought about by the use of a given pesticide.

On the other hand, since many pesticides of different chemical groups are being employed in this country, it would be fruitful to make comparative studies on these pesticides in respect to their possible impact on the general environment.

References

Agricultural Promotion Bureau, Ministry of Agriculture and Forestry: Use of PCP Herbicide in 1962 Rice Crop. Mimeographed, 77 pp. (1962).

Anonymous: Decade Advance in Pesticides. Pesticide Yearbook 1963-71 (1971).

Anonymous: Major Pests of Economic Plants in Japan. Plant Protection Assoc., Tokyo (1965).

Goto, S.: Plant Protection Jl (Shokubutsu Boeki) 24, 501 (in Japanese) (1970).

Hashimoto, Y.: Pesticide and Technique (Noyaku seisangijitsu) 8, 59 (1963).

Hashimoto, Y.: Pesticide and Technique (Noyaku seisangijitsu) 10, 49 (1964).
Hashimoto, Y.: Pesticide and Technique (Noyaku seisangijitsu) 18, 43 (1967).
Kanazawa, J.: Presentation at the Symposium of Japan. Soc. Sanit. Zool. (1970).
Kanazawa, J., and R. Sato: Report on Residue Analysis of Organomercuric Fungicides Sprayed on Rice and Fruits. Japan Plant Protec. Assoc., pp. 11-25 (1965).
Kobayashi, T.: Special Rep. No. 6, Plant Disease and Insect Pest Forecasting Program. Ministry Agr. Forest. (1961).
Koshihara, T., and D. Okamoto: Japan. J. Appl. Entomol. Zool. 1(1), 32 (1957).
Moriya, S., T. Tomizawa, and M. Suwanai: Report on Residue Analysis of Organomercuric Fungicides Sprayed on Rice and Fruits. Japan Plant Protec. Assoc., pp. 5-8 (1965).
Munakata, K. et al.: Agr. Biol. Chem. Japan 32, 232 (1966).
Tomizawa, C.: Report on Residue Analysis of Organomercuric Fungicides Sprayed on Rice and Fruits. Japan Plant Protec. Assoc., pp. 2-4 (1965).

CURRENT PESTICIDE SITUATION IN THE UNITED STATES

F. Matsumura

Department of Entomology, University of Wisconsin
Madison, Wisconsin U.S.A.

Introduction

Pesticides hold a unique position among environmental contaminants, being present in the environment in such small quantities as compared to other contaminants such as industrial wastes and fertilizers. The major factor which accounts for public and scientific concern is their high biological activity. Therefore, it is important to mention here that the scope of "Environmental Toxicology of Pesticides," in general, covers the effects of environmentally-occurring pesticides on biological systems. This has sometimes been referred to as "effects upon nontarget organisms." The term "nontarget" is important, since the hazards of environmental contamination are usually associated with unexpected side effects of pesticides or pesticide-derived compounds. Appearance of such side effects can be caused both by physical translocation and biological magnification of the pesticidal compounds.

General Patterns of Pesticidal Use

Figure 1 indicates the trends of total production of pesticides in the United States. Of all pesticides, insecticides account for roughly one-half of the total tonnage. Herbicides are, on the other hand, the fastest growing group of pesticides. They are expected to reach levels higher than the total output for insecticides in the near future.

According to Mitchell (1966) there are 457 million acres of farmland and 758 million acres of forest land in the United States. Approximately 15% of the farmland and 0.3% of the forest land is thought to be sprayed with pesticides. Altogether, 75% of the pesticides are applied to less than 2% of the land area in the United States.

According to Kearney et al. (1969), on the other hand, approximately 119 million acres were sprayed with herbicides, 97 million acres treated for insect control (insecticides), and about 25 million acres treated with fungicides and other minor pesticides in the United States in 1964. Since it is estimated that there are 340 million acres of cultivated land in the continental United States, this means that large percentages of such fertile land areas are treated by one or more applications of pesticides.

The latest available information (USDA, 1971) concerning the amounts of pesticides produced in the United States indicates that one billion pounds are produced yearly in the United States (TABLE 1). Production and sales of synthetic organic pesticides in 1969 actually declined for the first time since 1957. DDT continues its decline and now is only 62% of the peak high of 1959. Similarly, the production of 2,4-D and 2,4,5-T declined drastically in 1969. Cyclodienes also showed a modest decline in total production. Similar figures for exported pesticides are also shown in TABLE 2. It should be noted here that relatively large percentages of insecticides are exported, in comparison with other pesticides. Herbicides, for instance, are mainly produced for domestic consumption. This is particularly true with 2,4-D and 2,4,5-T. Export figures indicate only a few percent of the total production is exported.

The amounts of pesticides actually used in the United States in any particular year are difficult to estimate as certain amounts of produced pesticides are not immediately used but are stocked for the next season. Also, in

TABLE 1

Total Organic Pesticides Production Figures in the United States: in 1,000 Pounds

	1966	1967	1968	1969
Total Fungicides	178,910	177,886	190,773	182,091
Total Herbicides	272,021	348,274	402,781	371,764
2,4-D and 2,4,5-T	90,581	110,939	136,658	68,624
Total Insecticides	562,230	503,764	581,619	579,522
DDT	141,349	103,411	139,401	123,103
cyclodienes	130,470	120,183	115,974	107,311
methyl and ethyl parathion	55,306	44,705	58,163	50,572
Grand Total	1,013,161	1,029,924	1,175,173	1,133,377

TABLE 2

Exports of Pesticides[a] to Other Countries: in 1,000 Pounds

	1966	1967	1968	1969
Total Fungicides	37,629	36,248	21,390	19,691
Total Herbicides	72,865	69,482	65,746	59,199
2,4-D and 2,4,5-T[b]	3,391	7,287	1,234	2,400
Total Insecticides	322,523	250,540	137,573	108,333
DDT[b]	109,148	82,078	20,695	14,646
Other chlorinated hydrocarbons[b,c]	76,217	44,897	29,641	19,956
Organophosphorus[b]	63,819	64,895	50,562	40,293

[a]Including inorganic pesticides.

[b]Including technical materials. Actual amounts of exports should be less.

[c]Including technical BHC and p-dichlorobenzene.

certain cases the amount of imported pesticides should be taken into consideration.

The figures summarized in TABLE 3 indicate levels of pesticides estimated to have been used in the United States in 1969. In most cases the amounts used in the United States are slightly less than those produced. However, 67% of the DDT produced was for export in 1969.

TABLE 3

Actual Amounts of Pesticides Used[a] vs. those in Stock in 1969: in 1,000 Pounds

	Domestic disappearance	Exported 1969	Total production 1969	Stocks Dec. 31 1969
Cyclodienes	89,721	23,312	107,311	39,899
DDT	30,256[b]	82,077	123,103	18,292
2,4-D	49,526	6,558	47,077	17,407
2,4,5-T	3,218	729	4,999	3,245
Lead arsenate	7,721	2,964	9,394	4,037

[a]Disappearance calculated by subtracting the amount of pesticides in stocks in 1969 from the total figures of the similar stock figures for 1968 plus production figures for 1969.

[b]Total disappearance, including imports of 532,000 pounds in 1968 and 440,000 pounds in 1969.

The general pattern of domestic usage versus production of DDT has been summarized both by Woodwell et al. (1971) (Fig. 2) and the DDT Advisory Committee (1971).

Assuming that most of the organic pesticides produced have been used in the United States, one can estimate that altogether about 10 billion pounds of pesticides have been used in the United States since 1945. If all the pesticides were spread evenly throughout the above farm and forest lands, the average concentration of pesticides would be about 8 lb/acre. Westlake and Gunther (1966) have given another set of calculations. They contend that if the cumulative total amount of DDT produced in the world (approximately 3.5 billion pounds) were applied to the total arable land in the world (2.3 billion acres), it should amount to about 1.5 lb/acre. While such calculations give rough levels of the average amount of pesticides introduced, there is no question about the fact that some areas are intensively treated by pesticides.

Of approximately 400 million pounds of insecticides used in the United States (1964) roughly 170 million pounds are used for agricultural purposes. Particularly important is the fact that 90 million pounds are used on cotton. As an example of such high usage, 1.8 million pounds of pesticides were applied to 700,000 acres of cotton in the San Joaquin Valley of California in 1963 (i.e., 2.6 lb/acre per year) according to Westlake and Gunther (1966). Cotton is grown in the same area year after year, resulting in the accumulation of pesticides. Pesticide application per acre in southern Arizona averaged 7.59 lb of toxaphene, 1.35-4.00 lb of DDT, 1.2 lb of endrin, 0.58-1.02 lb of endosulfan and undetermined quantities of aldrin, BHC, heptachlor, and chlordane in 1965, covering 345,000 acres of cotton (Johnson and Lew, 1970). This use increased by a factor of five for DDT and three for toxaphene in 1967 in the state of Arizona. Following the restriction and decline of the use of DDT in 1968, the use of Strobane and Telone (8 and 7 times) and parathion (11 times) drastically increased. Another source of concentrated pesticidal usage in the United States is in orchards. While the total acreage of orchards is relatively low, concentrations of pesticides in such soils are known to be very high (see later description on soil residues).

While the methods of introduction can vary, the majority of pesticides applied eventually reach the surface of the earth, mainly soil surfaces. The soil generally acts as a reservoir for pesticides from which they degrade or are gradually spread or translocated to other environments. In TABLE 4 I have taken two sets of data showing the general speed of disappearance of pesticides from soils in the United States. Kearney et al. (1969), in addition,

TABLE 4

General Persistence of Chlorinated Hydrocarbon Insecticides in Soils

	95% Disappearance[a]		75-100% Disappearance[b]
	Years	(Average)	Years
Aldrin	1-6	(3)	3
chlordane	3-5	(4)	5
DDT	4-30	(10)	4
dieldrin	5-25	(8)	3
heptachlor	3-5	(3-1/2)	2
lindane	3-10	(6-1/2)	3
Telodrin	2-4	(4)	-

[a]From Edwards, 1966.

[b]From Kearney et al., 1965.

have summarized data for many pesticides to show the over-all time periods required to lose 75 to 100% of the original pesticides (TABLE 5).

The rate of disappearance is comparable to a simple exponential decay curve (in analogy to the radioisotopic decay curves). A combination of two major steps of decay reactions is often observed: an initial fast decay reaction, sometimes called the initial "dissipation" phase; and the second slow decay reaction called "persistence." In any case the speed of disappearance becomes slower as time passes.

Not all pesticides disappearing from the site are degraded, however. Water and air are two major physical factors which facilitate the movement of pesticides. Other means of transportation are mainly of biological origin: e.g., plants can absorb pesticides directly from the soil, and soil invertebrates and microorganisms can accumulate pesticides which in turn can be taken up by other animals.

TABLE 5

General Persistence of Pesticides in Soils: Expressed in Time Periods Required to Lose 75 to 100% of the Original Pesticide[a]

	Time Period
Organophosphate	
Diazinon	12 weeks
Malathion and parathion	1 week
Herbicides	
Propazine and picloram	18 months
Simazine	12 months
Atrazine and monuron	10 months
2, 4, 5-T	5 months
MCPA	3 months
2, 4-D	1 month
Dalapon and ClPC	8 weeks
1 PC, EPTC	4 weeks

[a]From Kearney et al., 1965.

Characteristics of Environments in Relation to Pesticidal Use

The bulk of the agricultural and urban area in the United States lies in the plains and coastal areas. The rivers serving these areas are generally slow flowing. These river systems cover great areas as compared to the rivers in Japan. The Mississippi River system, for instance, covers at least one-third of the land mass of the United States. Efforts have been made in the United States to monitor the levels of pesticides in water systems (Anonymous, 1971). The distribution of such monitoring stations throughout the U.S. is shown in Fig. 3. Riseborough et al. (1968) calculated that roughly 10,000 kg/year of pesticides are transported by the above river system to the Gulf of Mexico. As some pesticides adhere tightly to soil particles and could be expected to be washed out into the river, the total pesticides present in the river system could be much higher than the figure shown above.

TABLE 6 indicates the levels of chlorinated hydrocarbon insecticides found in various rivers in the United States. The striking accumulation of endrin and DDT in the Mississippi delta is quite noticeable. However, the levels of all other insecticides in the delta were also high. As compared to levels in the delta, those found in other water systems, particularly in California rivers, appear to be low. A British survey also clearly demonstrates the differences in the pattern of residues resulting from the difference in agricultural practices (Lowden et al., 1971).

Another characteristic of the waterways in the United States is the abundance of adjacent lakes, particularly in the northern portions of the country. Water in lakes is usually exchanged very slowly. The Great Lakes in particular are susceptible to pollution because of their extremely slow rates of discharge, and because of the close proximity of many urban (and some agricultural) areas. DDT levels in Lake Michigan are in the order of 1 ng per liter (1 part per trillion), yet this low level seriously affects the hatching of coho salmon eggs. Larger coho salmon from Lake Michigan, though they are relatively resistant to DDT, contain 3-13 ppm to 12-13 ppm depending largely on the size of the fish (Reinert, 1969).

Pesticide levels in water systems do change gradually from year to year. Lichtenberg et al. (1970) summarized the results of synoptic surveys on chlorinated hydrocarbon insecticides in surface waters of the United States for the

TABLE 6

Mean Concentrations of Chlorinated Hydrocarbon Insecticides Found in Water Systems of the United States: Expressed in Parts per Trillion

Location	Source of references	No. of sites tested	Insecticides			
			DDT and analogs	γ-BHC	Dieldrin	Endrin
U.S. major river basins	Breidenbach et al. (1967)	99	8.2	trace	6.9	2.41
	Weaver et al. (1965)	97	9.3	--	7.5	5.5
	Green et al. (1966)	109	8.3	2.2	5.9	3.6
Mississippi delta	USDA (1966)	10	112	28.0	10.0	541
California rivers	Keith and Hunt (1966)	82	0.62	0.01	--	--
Western streams	Brown and Nishioka (1967)	11	10.3	2.8	2.3	1.4
British rivers	Lowden et al. (1971)[a]	7	1.6	18.7	3.3	--

[a]Indirectly cited from Edwards, C. A.: Rev. Environ. Control 1, 7 (1970).

period 1964 through 1968 and showed that the number of occurrences of these compounds (i.e., frequency of detecting residues among water samples) reached a peak in 1966 and declined sharply in 1967 and 1968 for all chlorinated hydrocarbons tested, except for BHC which showed only a slight decline (Fig. 4). Dieldrin remained as the most serious pollutant in the surface waters of the United States.

Another important route of transportation is by air movement. This includes carrier transport with water particles and dust in addition to true evaporation of pesticides through volatilization. The return of pesticides to the earth is, on the other hand, mainly through rain, and to a lesser extent by falling dust.

Cohen and Pinkerton (1966) collected dust-rain which was brought to Ohio by a mammoth dust storm (documented as well by the Air Pollution Division, U.S. Public Health Service) originating in the Texas-Oklahoma-New Mexico area (Fig. 5). Their data showed that high amounts of chlorinated hydrocarbon insecticides and an organophosphate (Ronnel) could be detected in such dusts (TABLE 7). In other studies, the same authors studied pesticide concentrations in dusts throughout the year. DDT was found in concentrations of 3 to 90 ppm during the application season of June and July (1965). Since the average monthly dustfall in Cincinnati, Ohio, is about 15 tons per square mile, this gives a figure of 45 to 270 g of DDT falling per square mile. The amounts of pesticides carried by air-borne dust in the continental United States appears to be in excess of that found in other places. For instance, Riseborough et al. (1968) collected dusts over Barbados and estimated the total pesticide content in the dust to be 41 ppb, which according to their calculations should yield 600 kg/year of pesticides to the area of the Atlantic Ocean between the equator and 30° north.

Perhaps another important aspect of the usage picture in the United States is the heavy consumption of pesticides in urban areas. The American elm, one of the most important shade tree species in northern and midwestern cities, is currently faced with destruction by Dutch elm disease. The control of this plant disease is dependent upon controlling the bark beetles which transmit the disease-causing fungus. Until the last several years, DDT has been the main source of weaponry against the beetles.

TABLE 7

Pesticide Content of Dust Collected in Ohio[a]

Pesticide	Concentration, ppm[b]
DDT (tech)	0.6
chlordane (tech)	0.5
DDE	0.2
Ronnel	0.2
heptachlor epoxide	0.04
2,4,5-T	0.04
dieldrin	0.003
Total organic chlorine	1.34
Total organic sulfur	0.5

[a]From Cohen and Pinkerton, 1966.

[b]Based on air-dried weight of dust.

It generally required 3 to 4 pounds of DDT, per medium sized tree, to keep the beetle populations under control. According to Norris (1961) there are 300,000 elm trees in Chicago, in addition to 100,000 more in the surrounding cities. There are 154,000 elms in Milwaukee and 77,000 trees in small cities in southern Wisconsin. Assuming that DDT had been applied to these trees, 1.2 to 1.6 million pounds of DDT would be given to the Chicago area and 69,000 to 92,000 pounds of DDT sprayed in Milwaukee and other cities in Wisconsin. In a 15-year time span the total amount of DDT given to such areas could reach as much as 20 million and 12 million pounds, respectively. In terms of the intensity of application, such a spraying program would correspond to between 20 to 40 lb/acre per application, which exceeds the general agricultural levels of DDT application. In other cases of pesticide use in urban areas, cyclodiene insecticides (against grubs of Japanese beetles and June bugs, etc.) and herbicides have been heavily used on lawns.

These pesticides can find their way to lakes and streams through storm sewer systems which are present in many urban areas. Scientists from Michigan State University have found that the DDT concentration was highest in river areas originating from city discharges, in contrast to the agricultural areas (Zabik et al., 1971). In the southern states and Hawaii, cyclodiene insecticides have been heavily applied to soils (and in some cases to the structures themselves) adjacent to the house foundations. Another set of figures indicating heavy use of pesticides in urban areas are shown by Tabor (1966) who found that air samples from an eastern city contained 10 to 10,000 times higher levels of DDT and related compounds than those from non-urban to agricultural communities.

Industrial sources are also important enough to be considered. There are two different types of industrial sources for pesticide discharges into the environment: first, pesticides manufacturing and formulating plants; and second, those which utilize pesticides in their manufacturing processes. Nicholson (1967) points out that the latter source, such as mothproofing chemicals, could be a heavy contaminating source in limited areas. In paper-producing areas, phenylmercuric acetate has been used extensively to control algal slimes associated with raw materials. In northern Wisconsin, for instance, mercury residues are found in areas where there are no other industries but paper mills (e.g., the Chippewa River System). Phenylmercuric acetate has been suspected as the major source of contamination in these areas (Report, Department of Natural Resources, State of Wisconsin).

Finally, the most important source of introduction of pesticides in environments of the United States is probably by aerial spraying. The amount of fallout by this type of application can be great. Cohen and Pinkerton (1966) observed that dusts collected during the spray season contained much higher levels of pesticides. While ultra-low volume (ULV) spraying methods developed in the United States make maximum use of the airplane to cover wide areas, this application method increases the chances of wind drift of fine pesticidal particles. In addition to agricultural applications, air spraying has been extensively used in forest areas and on open water for mosquito control programs.

General Levels of Pesticide Contamination

What is the extent of pesticidal contamination in the United States as compared to other countries? The question can only be answered in vague terms. TABLE 8 is a summary of several surveys made on the levels of pesticidal contamination of soils. It can be seen that despite the differences in locality and years, the level of DDT, for instance, is remarkably constant, with the notable exception of orchard areas. Most of the figures fall between 1 and 3 for DDT, 0.02 to 0.08 for heptachlor-heptachlor epoxide, and 0.03 to 0.1 ppm for the aldrin-dieldrin group. Levels of these pesticides in streams and rivers has already been shown (TABLE 6).

These data, while they are important in assessing general levels, do not themselves indicate the mobility or availability of pesticides. In other words, most of these residues found in the soil represent the non-mobile portion of all pesticides. Perhaps soils serve as the largest reservoir of environmentally present pesticides. However, the speed of release is controlled by several factors such as moisture, organic matter, rain, etc. Similarly, most of these pesticides in water would be present in the form of particles bound in sediment at the bottom of water systems.

The levels of availability of environmentally present pesticides to various biological systems are the most

TABLE 8

Over-all Levels of Chlorinated Hydrocarbon Residues in the Soil in the United States: Data Expressed in Average ppm Values

Type (Location)	No. of samples tested	Year	DDT and related compounds	Aldrin and/or dieldrin	Heptachlor-heptachlor epoxide	γ-BHC
Potato fields (N.J.)	8	1954	2.3	--	--	--
(Eastern states)	5	1967	2.8	0.1	0.08	--
Arable fields (Midwest)	24	1958	1.2	--	--	--
(West)	41	1968	0.7	0.06	t	t
Cotton fields (Texas)	3	1966	2.4	--	--	0.26
Soybean fields[a]						
(Midwest, South)	27	1968	1.7	0.1	0.02	--
Corn fields (N.J.)	10	1954	4.0	--	--	--
(Ill.)	35	1965	--	0.5	--	--
Turf and cultivated[b]						
(Mich.)	227	1965	2.9	0.03	0.03	0.04
Orchards (Midwest)	14	1958	29.4	--	--	--
(Ind.)	35	1964	74[c]	--	--	--
Desert and prairie (Texas)	5	1966	1.6	--	--	0.2

[a]Chlordane 0.24 ppm.

[b]Chlordane 1.2 ppm.

[c]Estimated from pound per acre data.

important criteria in any country. Interestingly, the most convenient criterion of such assessment can be found in the levels of pesticides in human beings. TABLE 9 is a brief summary of the levels of some of the pesticides found in human fat. Among the countries surveyed, U.S. residents have been found to contain levels of DDT and DDE slightly above average values of other nations. The data from human subjects are trustworthy because of the ease by which the case history, as well as the distribution, age, etc. can be selected. They do have several shortcomings, however. First, human beings mainly consume agricultural products and few naturally occurring food sources. Second, their food habits vary greatly from country to country. The data, however, can indicate two important aspects of pesticidal contamination: first, it shows how efficient the regulatory mechanisms of one country are in controlling the total intake by its citizens; and second, the general levels of contamination of food commodities which are mainly obtained from pesticide-treated, cultivated lands.

While the levels of pesticidal contamination in various ecosystems can be gradually determined by coordinated efforts of nationwide surveys such as the National Monitoring Program (Anonymous, 1971), it will be extremely important to be able to assess the general tendency of changes in levels, and resulting consequences of such changes in any nation. Judging by the figures for pesticides content in human fat (TABLE 8), in the surface water (Fig. 2), and in fish (Henderson et al., 1971), it appears that levels of pesticides in such systems are declining since a 1966-67 peak period. This does not mean that the levels of pesticides in animals that are at the top of food chains will show an immediate decline. It is well known, for instance, that the levels of pesticidal residues increase as the size of the fish increases (Cope, 1971).

On the other hand, DDE residues in moribund or dead bald eagles from four midwestern states (Iowa, Michigan, Minnesota, and Wisconsin) appear to have declined since 1966: the average residues found in 1966, 1967, and 1968

TABLE 9

Levels of Chlorinated Hydrocarbon Insecticides Stored in Human Fat[a]
(mg/kg, mean values)

Survey by	Year surveyed	Country	DDT and related	DDE	Aldrin-dieldrin	γ-BHC
Dale and Quinby (1965)	1961-2	USA	4.9	3.8	0.15	0.20
Hoffman et al. (1964)	1964	USA	7.6	7.0	0.14	0.48
Hayes et al. (1965)	1964	USA	10.0	6.9	0.29	0.06
Dale et al. (1965)	1964	India	28	11.6	0.03	1.7
	1964	India	12	6.4	0.06	0.9
Egan (1965)	1965	Britain	3.1	2.0	0.21	0.34
Bick (1967)	1967	Australia	1.7	0.93	0.05	--
de Vlieger et al. (1968)	1968	Netherlands	2.0	1.7	0.17	0.1
Hayes et al. (1963)	1963	France	5.2	3.2	--	--
del Vecchio and Leoni (1967)	1967	Italy	8.2	7.5	0.45	0.06
Wasserman et al. (1967)	1967	Israel	6.4	7.8	--	--

[a]All public samples.

being 18.7, 15.8, and 5.0 ppm per wet weight, respectively (data from 36 birds). Analysis of variance on these data, however, showed no significant differences between years (Mulhern et al., 1970).

The animals and plants which face the least number of biological magnifications are the ones which are likely to show immediate decrease of residue levels. The preliminary report of Ware et al. (1971) from Arizona is interesting as the use of DDT was banned there in 1969. It was found that the levels of DDT in crops and milk declined within two years, while the average residue levels of DDT in soils remained constant at about 0.75 ppm. In this case DDT derivatives (mostly DDE) appear to be bound to soil components which are not available at least to those plants. Doubtlessly soils, including bottom sediments of water systems, act as the largest reservoir for such pesticides. The factors controlling the rate of release to biological systems, therefore, are the most crucial items to be studied.

Are the levels of chlorinated hydrocarbon residues actually declining in the environment? Woodwell et al. (1971) suggest that concentration of DDT residues in the atmosphere and in the mixed layer of the oceans lag by only a few years behind the amount of DDT used annually throughout the world. The annual production of DDT in the United States reached a peak in 1963 and continued at a high level until 1969 (8.13×10^{10} g in 1963; 6.0×10^{10} g in 1969). The maximum concentration of DDT in air occurred in 1966 and is likely to occur in the mixed layer of the oceans in 1971. According to their projection, even with a drastic decrease in the levels of the use of DDT in the seventies the levels of DDT residues in atmosphere and mixed layer of oceans will continue to be high: above 1 ppt until the year 2000.

On the other hand, the level of DDT on food commodities is definitely declining (Duggan et al., 1971). This appears to be caused by two factors: first DDT, despite its affinity to biological materials, is not readily picked up by plants (i.e., it is not known to be systemic), even in the presence of relatively high levels of DDT residues in soils (e.g., Ware et al., 1971). Second, food commodities have been carefully controlled by the regulatory agencies (TABLE 10).

TABLE 10

DDT Levels in Market-Basket Diet[a]

	1965	1966	1967	1968	1969	1970
DDT + DDE	0.031	0.041	0.026	0.019	0.016	0.015
DDE	--	0.028	0.017	0.015	0.011	0.010

[a]Expressed in milligrams/day. Food and Drug Administration: Duggan et al., 1971. Indirectly cited from Report of the DDT Advisory Committee, Environmental Protection Agency, September 9, 1971.

O'Neill and Burke (1971) developed an interesting model to predict the levels of DDT residues in human adipose tissue for several decades to come on various assumptions (TABLE 11). The first assumption was that DDT usage will decrease at the present rate (by 7 million pounds per year; column 1, TABLE 11), and the second assumption was the complete cessation of DDT use in 1972 (column 2). These two sets of projections gave almost insignificant differences in the predicted amount of DDT accumulation in man throughout the early part of the 21st century. The third example was based upon the use at current levels which resulted in much higher levels of DDT residues.

It is apparent from both models (Woodwell et al., 1971; O'Neill and Burke, 1971) that the continued use of DDT at a high level would result in the massive build-up of DDT residues in man and his environment for several decades. Such predictions have undoubtedly influenced the DDT Advisory Committee to recommend the curtailment of the use of DDT in the United States.

TABLE 11

DDT + Concentrations in Human Adipose Tissue Predicted by the Model for Various Assumptions about DDT Usage[a]

Year	Continued reduction of DDT usage at present rates[b]	Zero future usage of DDT[c]	DDT usage maintained at 5 x 10^6 lb/year	DDT usage maintained at 1966 levels
1970	5.14	5.14	5.14	5.30
1974	4.18	4.13	4.20	5.08
1978	3.41	3.35	3.53	5.32
1982	2.78	2.73	3.02	5.60
1986	2.27	2.23	2.60	5.84
1990	1.86	1.82	2.25	6.03
1994	1.52	1.49	1.98	6.20
1998	1.24	1.21	1.75	6.32
2002	1.01	0.99	1.56	6.43
2006	0.82	0.81	1.41	6.52
2010	0.67	0.66	1.29	6.59
2014	0.55	0.54	1.18	6.65
2018	0.45	0.44	1.10	6.70
2022	0.37	0.36	1.03	6.73

[a]From O'Neill and Burke, 1971. The data are expressed in ppm of DDT + DDE in human adipose tissue.

[b]Decrease by 5 million pounds per year.

[c]Zero application in 1972.

On the other hand, there are a number of places and ecological systems where the immediate withdrawal of persistent pesticides may result in a quick reduction of their residues (e.g., food commodities). Also, in small enclosed systems the rates of reduction following the ceasing of the pesticidal use would be much faster than in the larger systems.

Herman et al. (1969), for instance, showed that the levels of DDD residues in the western grebes decreased by one-half on an average over a 10-year period. DDT residues in salmon in New Brunswick and Maine decreased yearly, reflecting the reduction of the use of DDT (Keenleyside, 1959, 1967; Ide, 1967; Anderson and Fenderson, 1970). Ware et al. (1971) demonstrated that levels of DDT derivatives in alfalfa and beef fat immediately declined during the 2-year moratorium period on agricultural use of DDT in Arizona.

There are a number of pressing problems in the United States that require immediate attention. One is the current gypsy moth infestation that started from the northeastern U.S. in 1970. The infestation has now spread through most of the eastern forests. Although there are a number of newspaper accounts attributing the outbreak to the latest restrictions in the use of DDT, the relationship is not quite apparent as the actual use of DDT in these forest areas gradually decreased during the 1960's. While carbaryl (Sevin) can be used as a substitute (somewhat less effective than DDT), public opposition to any kind of spray program makes the control of this pest very difficult. California is facing a serious problem of mosquito control. The mosquito species in central California have developed chlorinated hydrocarbon, organophosphate, and carbamate resistance, and they are extremely difficult to control. *Culex tarsalis* in particular poses a serious problem from the viewpoint of a public health hazard inasmuch as this species has the potential of becoming the vector of Western equine encephalites--a disease which in 1952 caused 800 cases and 50 deaths in California alone. Similarly, Venezuelan equine encephalitis, which is also carried by a mosquito vector, has recently killed thousands of horses in the southern United States.

The use of pesticides will continue for the foreseeable future. Thus, the potential hazards of environmental contamination are always there, particularly in countries such as the United States with large areas of mono-cultural crop lands (e.g., cotton, corn, wheat, etc.) and vast forest areas. In addition, many plants are especially susceptible to introduced pests (e.g., Dutch elm disease, Gypsy moth, European corn borer, etc.). The important point is that we analyze the current situation carefully and assess the factors contributing to potential hazards. With such information we can determine the impact of pesticides and develop proper plans and methods to protect our environment.

References

Anderson, R. B., and O. Fenderson: J. Fish. Board Can. 27, 1 (1970).

Anonymous: National pesticides monitoring program (revised). Pesticides Monit. J. 5, 35 (1971).

Bick, M.: Med. J. Austral. 1, 1127 (1967).

Breidenbach, A. W., C. G. Gunnerson, F. K. Kuwahara, J. J. Lichtenberg, and R. S. Green: Chlorinated hydrocarbon pesticides in major river basins 1957-65. Publ. Health Dep. Washington 82, 139 (1967).

Brown, E., and Y. A. Nishioka: Pesticide Monit. J. 1, 38 (1967).

Cohen, J. M., and C. Pinkerton: Organic pesticides in the environment. Adv. Chem. Ser. 60, 163 (1966).

Cope, O. B.: Ann. Rev. Entomol. 16, 325 (1971).

Dale, W. E., and G. E. Quinby: Science 142, 593 (1965).

Dale, W. E., T. B. Gaines, and W. T. Hayes, Jr.: Bull. World Health Organ. 33, 471 (1965).

Duggan, R. E., G. Q. Lipscomb, E. L. Cox, R. E. Heatwole, and R. C. Kling: Pesticide residue levels in foods in the United States from July 1, 1963 to June 1969. Indirectly cited from Report of the DDT Advisory Committee, Environ. Protection Agency, September 9, 1971 (1971).

Edwards, C. A.: Residue Rev. 13, 83 (1966).
Egan, H.: Brit. Med. J. 2, 66 (1965).
Green, R. S., G. C. Gunnerson, and J. J. Lichtenberg: Agriculture and the Quality of our Environment. Amer. Ass. Adv. Sci. 137 (1966).
Hayes, W. J., W. E. Dale, and R. Lebreton: Nature 199, 1189 (1963).
Hayes, W. J., Jr., W. E. Dale, and V. W. Birse: Life Sci. 4, 1611 (1965).
Henderson, C., A. Ingus, and W. L. Johnson: Pesticides Monit. J. 5, 1 (1971).
Herman, S. G., R. L. Garrett, and R. L. Rudd: Chemical Fallout. M. W. Miller and G. G. Bergs, Eds. Thomas, Springfield, Ill., p. 24 (1969).
Hoffman, W. S., W. I. Fishbein, and M. B. Andelman: Arch. Environ. Health 9, 387 (1964).
Ide, F. P.: J. Fish. Res. Board Can. 24, 769 (1967).
Johnson, D. W., and S. Lew: Pesticide Monit. J. 4, 57 (1970).
Kearney, P. C., J. R. Plimmer, and C. S. Helling: Encycl. Chem. Technol. 18, 515 (1969).
Keenleyside, M. H. A.: Can. Fish Cult. 24, 17 (1959).
Keenleyside, M. H. A.: J. Fish. Res. Board Can. 24, 807 (1967).
Keith, J. O., and E. G. Hunt: Levels of insecticide residues in fish and wildlife in California. Trans. 31st North Amer. Wildl. Res. Conf., 150 (1966).
Lichtenberg, J. J., J. W. Eichelberger, R. C. Dressman, and J. E. Longbottom: Pesticides Monit. J. 4, 71 (1970).
Lowden, G. F., C. L. Saunders, and R. W. Edwards: Organochlorine insecticides in water (Pt. II). Proc. Soc. Water Treat. Exam. (1971).
Mitchell, L. E.: Organic pesticides in the environment. Adv. Chem. Ser. 60, 1 (1966).
Mulhern, B. M., W. L. Reichel, L. N. Locke, T. G. Lamont, A. Belisle, E. Cromartie, G. E. Bagley, and R. M. Prouty: Pesticides Monit. J. 4, 141 (1970).

Nicholson, H. P.: Science 158, 871 (1967).

Norris, D. M.: Use of pesticides in regard to Dutch elm disease. Report of the subcommittee of relation of chemicals to forestry and wildlife for the State of Wisconsin, January 1, 1961, State of Wisconsin (1961).

O'Neill, R. V., and O. W. Burke: A simple systems model for DDT and DDE movement in the human food-chain. Indirectly cited from Report of the DDT Advisory Committee, Environ. Protection Agency, September 9, 1971 (1971).

Reinert, R. E.: Limnos 2, 3 (1969).

Riseborough, R. W., R. J. Huggett, J. J. Griffin, and E. D. Goldberg: Science 159, 1233 (1968).

Tabor, E. C.: Trans. New York Acad. Sci. Ser. 2, 28, 569 (1966).

USDA: Pesticides and their effects on soils and waters. Soil Sci. Soc. Amer. 150 (1966).

USDA: The Pesticide Review, 1970. U.S. Department of Agriculture, Agricultural Stabilization and Conservation Service, Washington, D. C. (1971).

del Vecchio, V., and V. Leoni: Nuovi Ann. Ig. Microbiol. 28, 107 (1967).

Vlieger, M. de., J. Robinson, M. K. Baldwin, A. N. Crabtree, and M. C. van Dijk: Arch. Environ. Health 17, 759 (1968).

Ware, G. W., B. J. Estesen, and W. P. Cahill: Pesticide Monit. J. 5, 276 (1971).

Wasserman, M., D. Wasserman, L. Zellermayer, and M. Gon: Pesticide Monit. J. 1, 15 (1967).

Weaver, L., C. G. Gunnerson, A. W. Breidenbach, and J. J. Lichtenberg: Chlorinated hydrocarbon pesticides in major U.S. river basins. Publ. Health Dep. Washington 80, 481 (1965).

Westlake, W. E., and F. A. Gunther: Organic pesticides in the environment. Adv. Chem. Ser. 60, 110 (1966).

Woodwell, G. M., P. P. Craig, and H. A. Johnson: Science 174, 1101 (1971).

Zabik, M. J., B. E. Pape, and J. W. Bedford: Pesticide Monit. J. 5, 301 (1971).

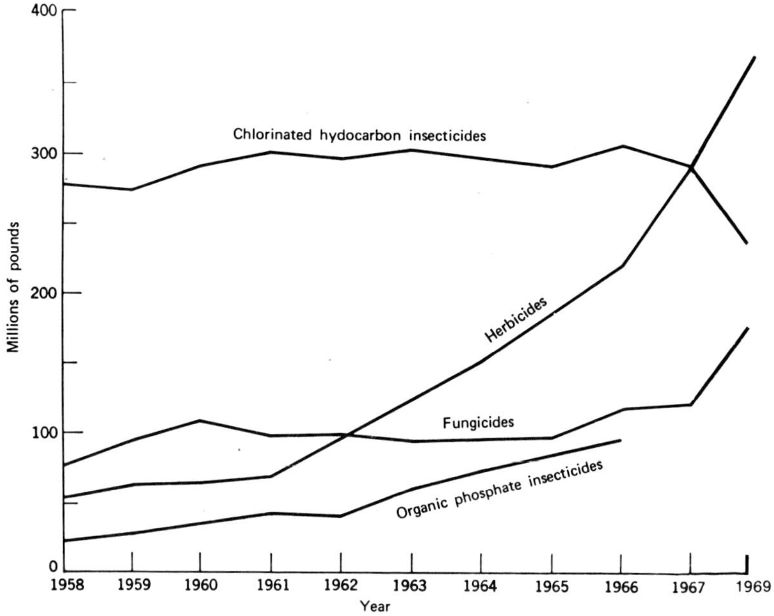

Fig. 1. Total production of pesticides in the United States; expressed in the amounts produced per year. (Modified from Kearney et al., 1969.)

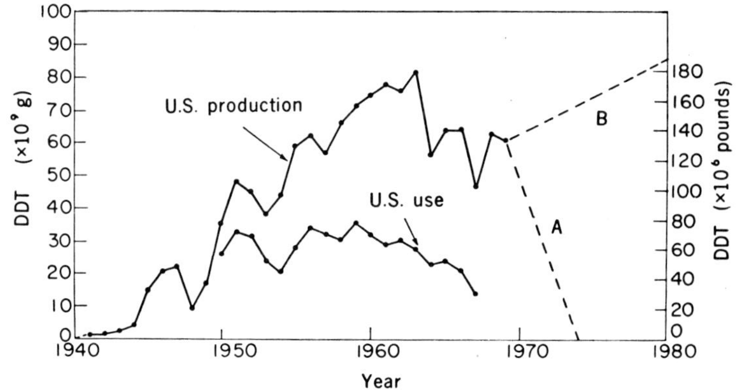

Fig. 2. Amounts of DDT production and usage in the United States. (From Woodwell et al., 1971.)

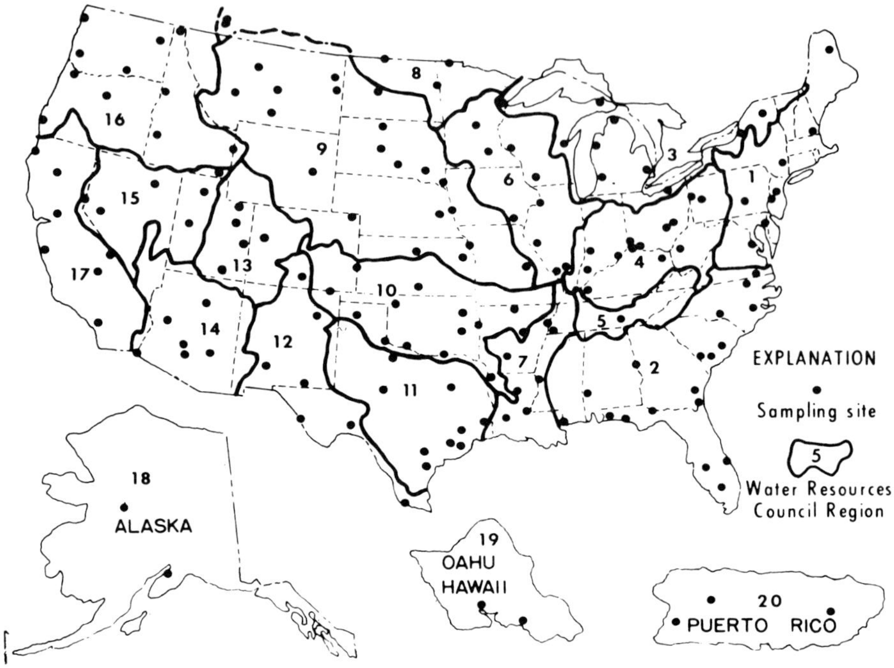

Fig. 3. Geographical distribution of pesticide monitoring stations for water samples, according to the National Pesticide Monitoring Program (1971).

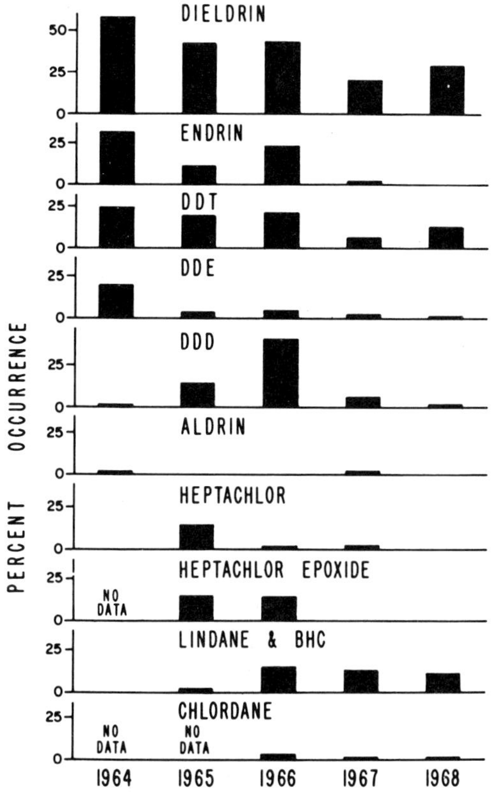

Fig. 4. Per cent occurrence of chlorinated hydrocarbon insecticides in the surface water samples of the United States. (From Lichtenberg et al., 1970.)

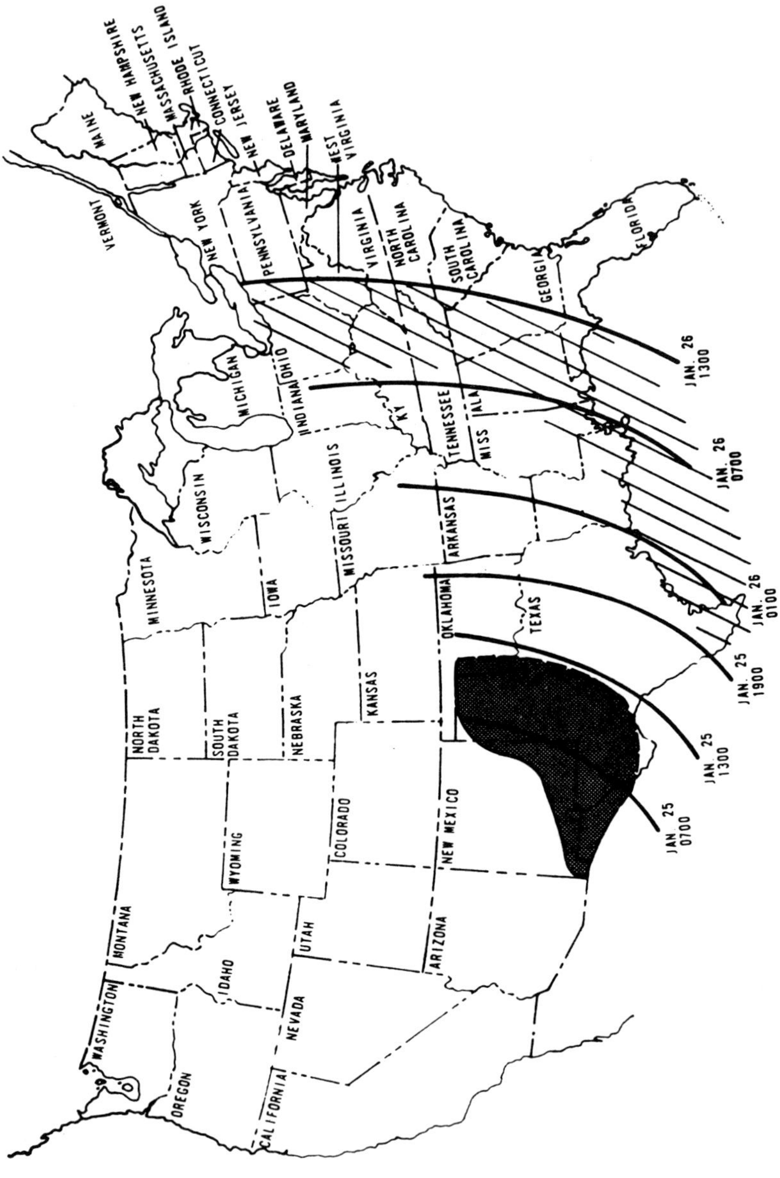

Fig. 5. Origin and movement of dust mass which carried pesticides across the U.S. continent. (From Cohen and Pinkerton, 1966.)

PESTICIDES IN BRITAIN

G. T. Brooks

Agricultural Research Council
Unit of Invertebrate Chemistry and Physiology
The University of Sussex
Brighton, U.K.

In this account, the main emphasis is inevitably placed on data for the organochlorine compounds, since these are most persistent and, to date, have caused most concern. The information available for these compounds is still the most comprehensive, although the pace of environmental research on other pesticides has increased greatly in recent years.

Regulation of Pesticide Usage

Public awareness of the possible ill effects of pesticides in non-target organisms appears to date from the early 1960's. In Britain, this development was probably related on the one hand to the occurrence of wild bird poisoning episodes involving organochlorine-dressed seeds, and on the other to the advent of electron-capture gas-liquid chromatography and the consequent newly available evidence for the presence of these compounds in animals and in the environment generally.

However, the situation in Britain has been under observation since the 1950's (Fig. 1) when the three Working Parties under Sir Solly Zuckerman on 'Toxic Chemicals in Agriculture' (1951), 'Residues in Food' (1953), and 'Risks to Wildlife' (1955), together with the later Report of the Research Study Group on 'Toxic Chemicals used in Agriculture and Food Storage' (1961), provided the basis for action to avoid unwanted side effects and to indicate areas for further investigation.

From 1956 onwards the Royal Society for the Protection of Birds and the British Trust for Ornithology (voluntary bodies) and the Nature Conservancy (official body) provided the main initiative in studying side effects of pesticides on wildlife. The Nature Conservancy set up its research division to deal with Toxic Chemicals and Wildlife at Monks Wood in 1960 and in that year the Scientific Sub-committee of the Ministry of Agriculture, Fisheries, and Food (MAFF) Advisory Committee set up an advisory Wildlife Panel. In 1962 the Research Committee on Toxic Chemicals (Agricultural Research Council, 1970) succeeded the review bodies of the 1950's and 1961 (Fig. 1) and continues to keep under review researches on the detection, evaluation, and prevention of unwanted side effects of toxic chemicals. Its terms of reference include the 'review of research on pesticides, especially that under the auspices of the Research Councils, and the formulation of proposals for future research.' A further function is 'to review the research in progress or required to ensure the safe use of any toxic chemical (including any toxic chemical on sale to farmers for veterinary use and veterinary medicines prescribed for use by veterinary surgeons, but excluding toxic effects in the animals being treated) specifically referred to it by the Advisory Committee on Pesticides and Other Toxic Chemicals' (Wilson Committee). The latter Committee makes recommendations regarding the safe use of pesticides. It published a 'Review of the Persistent Organochlorine Pesticides' in February 1964, with a Supplement later that year, a 'Review of Present Safety Arrangements' in January 1967, and a 'Further Review of Certain Persistent Organochlorine Pesticides Used in Great Britain' in December 1969 (Department of Education and Science, 1969). Its findings have been instrumental in reducing the usage of the persistent organochlorines wherever this is possible at present.

As a result of these activities, voluntary limitations

on the uses of pesticides have been successfully arranged informally since 1954 and formally since 1957 under the Pesticides Safety Precautions Scheme (PSPS) and its predecessor, the Notification of Pesticides Scheme. In this voluntary scheme, agreed between the associations representing the pesticide manufacturers and the Agriculture and Health Departments in Great Britain, manufacturers have undertaken to notify the MAFF before marketing new chemicals or recommending new uses of an existing chemical. Manufacturers provide comprehensive data regarding the physical, chemical, and biological properties of chemicals, their persistence, breakdown products, and mode of action. Details are furnished relating to toxicity to mammals and (where available) to Man, and also information regarding possible effects on wildlife. Included in the submission are methods of analysis, medical data, and any information concerning use in other countries.

The Wilson Committee and its Scientific Subcommittee examines these data, and, if satisfied that the user of the chemical, the consumer of treated crops and wildlife can be adequately protected, makes recommendations for the safe use of the chemical. If accepted by the Government Departments concerned, the recommendations are publicised, a condition of this clearance being that manufacturers include them on the label of the product. The Committee may review the safe use of any chemical at any time after its clearance in the light of new evidence.

A Veterinary Products Safety Precautions Scheme operates on similar lines to the PSPS in relation to such products on direct sale to farmers, and advice on such products is provided by a Veterinary Sub-committee.

The Agricultural Chemicals Approval Scheme is a voluntary scheme under which proprietary brands of crop protection chemicals can be submitted for official 'approval' of their biological efficiency. This Scheme enables users to select from an approved list recommended by advisers, efficient and appropriate crop protection chemicals for particular uses, and discourages the use of unsatisfactory products. The Scheme covers only crop protection chemicals and does not include rodenticides, or pesticides used for food storage, veterinary or domestic purposes. It is operated by the Agricultural Chemicals Approval Organisation on behalf of the Agricultural Departments of the United Kingdom and 'approval' is only given if the product has been first cleared for safety under the PSPS and if the organisation is satisfied that it fulfills the claims made for it on the label, which then carries an 'approved' symbol. The status of products so approved is kept constantly under review.

Finally, the Agriculture (Poisonous Substances) Act, 1952, is designed to ensure that agricultural workers are protected by the use of appropriate clothing when they use the more toxic pesticides listed in the Regulations under the Act, and these Regulations are under constant review by the Wilson Committee.

Pesticide Usage

Use of organochlorines and alternatives on crops

Figure 2 gives some idea of the acreages treated with organochlorine insecticides in relation to the total areas of crops grown in England, Wales, and Scotland in 1966. Farm crops (cereals, roots, potatoes, legumes, and brassicas for stockfeed) were grown on 10 million acres and accounted for the major use of aldrin, dieldrin, and BHC. DDT was less extensively used on these crops, its use increasing through vegetable crops (legumes, brassicas, roots, onions, etc., for human consumption) to fruit and hops, in England and Wales. In Scotland, BHC was most widely used on farm crops and there was also a wider use of DDT on these crops than in England and Wales. Despite its large area (about 17 million acres) grassland appears to be treated only rarely with organochlorines, DDT being used when necessary. Some approximate tonnages of organochlorines and other insecticides used in Great Britain in the early 1960's are given in

TABLE 1. Comparison of these figures with the areas treated (Fig. 2) shows that the tonnages used of cyclodienes plus BHC roughly equalled those of DDT and relatives in England and Wales, but use of the former group was much more widespread. BHC was also used much more widely than DDT in Scotland, but the amounts of the DDT group used were very much greater.

TABLE 1

Estimated Tonnages (of active ingredients) of Organochlorine[a] and Other Insecticides used Annually in the 1960's

Insecticide	England and Wales (1962-64)	Scotland (1964-66)	Great Britain (1966-68)
Aldrin	135	3	)
Dieldrin	28	3	) 80[b]
Heptachlor	5		)
BHC	101	11	
DDT group	262	57	220
Endrin	1		
Endosulfan	6		
Organophosphates			200 (1960-64)
			240 (1966-69)
Carbamates			10
Others			50

[a]In the U.S.A. about 1963, the total annual use of organochlorines was over 60,000 tons. In 1966-67, world production of DDT was about 85,000 tons, of which the U.S.A. used 27,000 tons.

[b]Includes endrin.

Data from Strickland (1966), Agricultural Research Council (1970), J.M.A. Sly (personal communication).

It has been estimated that between 1964 and 1969 there was a fall of about 25% in the total tonnages of aldrin, dieldrin, DDT, and DDD used in agriculture and horticulture annually in Great Britain. A further reduction of 20% was anticipated to result from the Wilson Committee's 1969 recommendations, so that we should presently be using about half the quantities of these compounds that were used annually in the early 1960's. At the same time there has been a progressive increase in the use of organophosphorus insecticides. The Ministry of Agriculture's Pesticide Survey Group at the Plant Pathology Laboratory, Harpenden, has in hand a number of reports on current pesticide usage in the U.K., and these will be published in late 1971 or early 1972.

Cyclodienes and BHC - The use of aldrin and dieldrin seed dressings on winter-sown wheat has been estimated to save production equivalent to about 84,000 acres of winter wheat valued at some 3.3 million British pounds when the wheat bulb fly (*Leptohylemyia coarcta*) is unusually abundant. Gamma-BHC seed dressings have proved less effective than aldrin or dieldrin dressings for control of this insect, especially on peat soils. There has been widespread use of aldrin and dieldrin, especially as seed dressings, sprays, dusts, or root dips, for the control of other pests attacking cereals (leather jackets, fruit fly), potatoes and sugar beet (wireworm), brassicas (cabbage root fly, *Eroischia brassicae*), legumes (bean seed fly, *Delia platura*), onions (onion fly, *Delia antiqua*), and narcissus (narcissus fly, *Merodon equestris*), and many of these uses are being retained pending the full evaluation of a number of organophosphorus compounds as alternatives. Chlorfenvinphos and ethion appear to provide effective seed dressings against wheat bulb fly, but there is still some concern about possible hazards to personnel engaged in seed dressing operations, and the use of the organochlorine dressings is likely to be retained for some time. The use of heptachlor as a seed dressing has been subject to the same restrictions as aldrin and dieldrin since 1964, but this compound has been little used in Great Britain for some years.

Several organophosphorus insecticides (fenitrothion, chlorfenvinphos, parathion) have been under examination as alternatives to DDT or aldrin for control of leather jackets attacking barley. Parathion spray is comparable in cost to aldrin, but unacceptable on safety grounds; parathion granules are acceptable and are under evaluation for this purpose. Since DDT is phytotoxic to a number of strains of barley and BHC gives inconsistent results, the use of aldrin will continue as the only entirely suitable alternative at present.

In the late 1950's to early 1960's, fertilisers containing aldrin were used to protect potatoes against wireworm damage. Although these compounds are still available as sprays and dusts, voluntary withdrawal of the fertiliser mixtures made control using aldrin more complicated, and the use of both aldrin and dieldrin on potatoes declined during the 1960's.

Past trials indicate that fenitrothion, diazinon and chlorfenvinphos seed dressings are as effective as dieldrin and more effective than gamma-BHC for the control of bean seed fly, which is a serious pest of green bean crops in the U.K. Formulations of these compounds suitable for use with beans are not yet commercially available, however, and the use of dieldrin seed dressings continues, although cyclodiene resistance is now occurring in this insect. Low rates of dieldrin application are used as a dry seed dressing for sugar beet in areas of high density wireworm populations, and there is currently no satisfactory alternative. Dieldrin seed dressings are retained at present for the control of onion fly, but cyclodiene resistance has been found in this insect and several organophosphorus compounds show promise as alternatives.

Endrin and endosulfan are used on soft fruits (black currants, strawberries, blackberries), narcissus and ornamentals grown under glass, for the control of various mites, and, since there are no satisfactory alternatives at present (except dicofol in some cases), such usage is likely to continue.

DDT (and relatives) - This broad spectrum insecticide has a wide range of uses in Great Britain, as elsewhere, and for some uses, especially aphid control, replacement by other compounds is not difficult. Despite a great deal of work, however, there are many pests for which no satisfactory alternative to DDT has been discovered. Limited experiments have sometimes indicated promising compounds, but have not been followed by commercial trials; in other cases, suitable formulations of promising insecticides have not become commercially available. The Wilson Committee has recommended the withdrawal of a few of the many uses of DDT, while recognising that its retention for the remaining purposes makes usage control difficult, a situation which will present problems if the Pesticide Safety Precautions Scheme becomes mandatory.

DDT has been widely used to protect cereals, and the only suitable alternative for DDT-susceptible crops such as barley is aldrin, when BHC cannot be used because of possible contamination of following crops. A DDT bait non-phytotoxic to barley could replace aldrin, but there are difficulties of application and the method is not popular with farmers. Organophosphorus alternatives are likely to replace DDT for aphid control on potatoes, but the use of DDT is retained for any necessary use against Colorado beetle, since possible alternatives such as carbaryl, azinphos-methyl and phosphamidon are untried for this purpose in Great Britain. DDT has been used to control various pests of brassicas in addition to the commonly used gamma-BHC seed dressings. The use of organophosphorus compounds as alternatives can be effective, but requires correct timing. Pea and bean weevil control with DDT is likely to continue, but for a number of other pests attacking peas, such as moths, azinphos-methyl and carbaryl are satisfactory, though more expensive. Malathion, azinphos-methyl, mevinphos, and carbaryl are as effective as DDT for some other uses, and systemic organophosphorus sprays provide better control of aphids than DDT, so that these usages should now cease.

The organophosphorus compounds used as alternatives to dieldrin for the control of carrot-fly are sometimes supplemented by the use of DDT to control the second generation of this pest during the autumn and winter. The indications are that such supplementation does not give a high degree of control; diazinon might be a suitable substitute but the cost would be much higher. For other root crops such as sugar beet and mangolds, the use of DDT is declining and is being replaced by an increased use of organophosphorus insecticides. Aphids are better controlled by the latter compounds, but there is no satisfactory alternative for leaf-eating beetles and cutworms. DDT is still extensively used to control mangold fly in Scotland, and nearly half the total acreage of mangolds grown was sprayed for this purpose in 1966.

DDT has been extensively used to control a range of pests attacking apple trees at various times during the season, the situation being complicated by the need to integrate such control with the use of fungicides. As in other areas, systemic organophosphorus compounds are replacing former uses of both DDT and gamma-BHC against aphids. Azinphos-methyl, phosalone, and carbaryl are effective alternatives for DDT against Winter moth and tortricid caterpillars, and also give some control of aphids. The organophosphorus compounds will also control fruit tree red spider mite, and in some orchards would replace the post-blossom uses of DDT or DDD. Like the organochlorines, however, all the satisfactory alternatives are toxic to arthropod predators, carbaryl being more toxic than DDT. It seems probable that organophosphates will gradually replace DDT for many control uses against pests of other top fruit, although with some low value crops such as plums, the increased cost of treatment may be uneconomical.

DDT is the best control agent for many pests of soft fruit at present, and in some cases the choice of reliable alternatives is difficult, although malathion, phosmet, azinphos-methyl, trichlorphon, phosalone, and fenitrothion promise to be useful against strawberry pests. Malathion, derris, and azinphos-methyl are reliable alternatives for the control of certain raspberry pests against which DDT has hitherto been used very extensively in Scotland.

DDT has been widely used against pests of glasshouse crops and gamma-BHC, malathion, and diazinon are suitable alternatives for some purposes. Leaf miner is better controlled by gamma-BHC or parathion smokes, but the latter compound is not favoured on account of the toxic hazard. Ornamentals and outdoor herbaceous plants usually have a high cash value and, although infestation may be infrequent, the damage done can be costly; gamma-BHC, malathion, and trichlorphon are alternatives for several uses of DDT in this context. There appear to be no reliable alternatives to DDT for caterpillar control. Other uses of DDT include its formulation with gamma-BHC as a spray against willow foliage pests, against some of which malathion may be a useful replacement, and in forestry, where for economic reasons its persistence has been highly useful.

DDD is an alternative for some of the uses of DDT on fruit crops, but less than one-tenth of the total acreage of apples grown in 1967 was treated with this compound. There are good non-organochlorine alternatives for pre-blossom control of caterpillars and aphids on apples, but it is uncertain whether any of these will be adequate for control of various weevils and capsids. Effective alternatives for DDD are available for codling moth and fruit tree tortrix in June and July. In the U.K., DDT and DDD uses on fruit are to be treated similarly, and it is recommended that their post-blossom use on apples, pears, loganberries, and raspberries should cease, while the uses on strawberries and pre-blossom on apples and pears should continue.

Use of organochlorines in food storage and for veterinary, home, and other purposes

There is no evidence that heptachlor, endosulfan, or camphechlor (toxaphene) are used in food storage practice in Great Britain, but endrin is used occasionally as a rodenticide on official premises. Although some insecticidal lacquers still contain aldrin with dieldrin, the use of aldrin in food storage ceased voluntarily in 1964. Various formulations of dieldrin are used by professional operators against cockroaches and ants, in addition to its incorporation into resins, coatings, lacquers, and paints for storage purposes. Some usage against cockroaches may be replaced by arprocarb, diazinon, and fenitrothion. The small use of chlordane to control ants and cockroaches is also likely to continue.

DDT is no longer much used during the commercial storage of food and it has largely been replaced by less persistent contact insecticides and fumigants. It is, however, formulated for a range of uses against insect pests of food at some stage in its storage or processing prior to consumption, and is included in many preparations for housefly control in restaurants, canteens, and kitchens, shops, food factories, slaughterhouses, and animal houses. Because of the fear that houseflies might become generally resistant, there has been a trend towards the use of dichlorvos, trichlorphon, fenitrothion, diazinon, and pyrethrins instead. The use of DDT in paints and lacquers, in smoke generators in any situation where processed food might be contaminated, and in industrial kitchens, canteens, and restaurants is no longer permitted.

Thermal vaporisers using gamma-BHC, DDT, or mixtures of the two, have been widely used against flies, stored food pests and a variety of others, in food premises, intensive poultry houses, cattle rearing houses, offices, shops, mills, warehouses, hospitals, and in the home. The Wilson Committee recommended in 1969 that the vaporisation of gamma-BHC in sleeping quarters, living rooms and similar situations should be discontinued and that vaporisers should not be continuously operated in the presence of people, animals, or foodstuffs. Food, unless protected by impermeable packaging, should not be exposed to DDT vapour at any time, nor should DDT be vaporised in the presence of persons or animals, or where food is present, or might be later. There has recently been a trend towards intermittent spraying with pyrethrins aerosol formulations, or the use of dichlorvos resin strips, or trichlorphon, diazinon, or gamma-BHC sprays, so that the use of vaporisers may decline.

Chlordane, endosulfan, endrin, camphechlor, and DDD are not used in veterinary practice and the use of aldrin or dieldrin in sheep dips, widespread before 1964, has been discontinued since 1966. There has been a corresponding increase in the use of organophosphorus compounds in sheep dips, although the dipping technique appears to be more critical and the cost greater.

A detailed study of the use of pesticides for hygiene and public health purposes is currently in progress, and a survey in 1964 indicated that gamma-BHC, dieldrin, chlordane, and DDT are widely used. Since it is possible that the use of organochlorines to control flies and other insects on refuse tips may have produced resistant strains, such usage should probably be more restricted. The use of dieldrin to protect sawn timber declined between 1962 and 1966, and there was a considerable increase in the use of gamma-BHC. Large capacity dipping tanks are used for this purpose and the disposal of their spent contents and sludge could present an environmental hazard, although no incidents have been reported.

The 1960 figures indicated that moth-proofing was carried out on a large proportion of wool and wool-blended carpets, hand-knitted textiles, and knitting wool and machine-knitted textiles. There was also some moth-proofing of blankets and other woollen goods, dieldrin being favoured for this purpose. When correctly applied, dieldrin moth-proofing resists normal cleaning treatments and the Wilson Committee consider this use to be acceptable and unlikely to result in significant uptake by Man.

There is no doubt that the mothproofing process has contributed significantly to the contamination of the rivers in Scotland, the West Riding of Yorkshire and Worcestershire which receive the effluents from these processes. There is no effective substitute for dieldrin for this purpose and means to reduce the dieldrin content of the effluents are continually under review. Sewage sludge from towns engaged in mothproofing contains dieldrin (up to 60 ppm has been found) and its occasional use on the land may constitute a source of environmental contamination, as may also the occasional use of wool waste from carpet factories as fertiliser.

A wide range of pesticide chemicals is available in small packs for home garden and interior uses, and there is a general lack of information on the amounts of pesticides used outside agriculture and food storage, especially in regard to the persistent organochlorines. Home garden use of aldrin, dieldrin, and heptachlor was withdrawn in 1964. Endrin and endosulfan have never been available for this purpose; the use of chlordane and camphenchlor for earthworm control and for a few other purposes has not been found to affect wildlife, and continues. Since there are a number of organophosphorus alternatives to DDT (malathion, trichlorphon, dimethoate, menazon) for garden use, its withdrawal for this purpose was recommended in 1969.

There is some anxiety concerning the variety of pesticide formulations, frequently without adequate labels, available to the public for use in the home. In view of the less persistent alternatives now available, organochlorines, except chlordane, are now withdrawn from home use.

Herbicide usage

Herbicide usage in Britain has increased spectacularly since 1964, when some 241 herbicide products had been improved by the M.A.F.F. By 1969, there were 386 approved products, including 24 new materials, and in this period the sales of herbicides (including exports) of the larger British manufacturers increased from 7.8 to 22 million pounds, the latter being more than the total value of all pesticides produced in Britain in 1964. Many of the herbicides introduced since 1965 are mainly soil-acting; their activity, selectivity, and persistence are dependent on soil type and climatic factors, and their safe use requires more detailed evaluation than is normally required for foliar herbicides.

In the last few years most cereal crops have received annual spray treatment and there has been a sharp rise in the use of herbicides with arable crops such as sugar beet and potatoes, with horticultural crops such as fruit and vegetables, and in forestry. Three-quarters of all land under cultivation in Britain, excluding grassland, is now regularly treated with herbicides. There has been a marked increase in the 'convenience' use of herbicides in gardens, sports fields, golf courses, and public parks. The value of herbicides for the control of aquatic vegetation in waterways and lakes is being increasingly recognised and dalapon, 2,4-D-amine, maleic hydrazide, and diquat have been cleared for safe use for aquatic weed control under the Pesticide Safety Precautions Scheme. Herbicide usage for non-agricultural purposes such as weed control on industrial sites and railways is also of some importance.

There is an increased use of cropping techniques in which herbicides have partly or completely replaced soil cultivation and this has reduced the need for crop rotation. Many annual weeds have become less important, but perennial weeds such as couch grass (*Agropyron repens*), bindweed (*Convolvulus arvenis*), and wild oats (*Avena spp.*) are increasing, sometimes seriously. The combined use of chemical and cultivation methods often provides the most efficient and economic weed control in such circumstances.

Distribution of Pesticides in the Abiotic Environment (Soils, Water, Air)

The major source of environmental contamination must clearly arise from pesticide application to plants and soils. In 1966, a survey of English soils showed the presence of dieldrin at rates up to about 1 lb/acre in potato fields (0.5 ppm in the top 20 cm) and up to four times this amount of DDT in other arable soils. Traces of these insecticides and also of gamma-BHC were widespread at levels of a few parts in 10^9 and common at 5 to 30 parts in 10^8 (TABLE 2). Analyses in 1968 revealed dieldrin residues up to 1.5 ppm in some soils (Edwards, 1970). The persistence of dieldrin, DDT, and gamma-BHC was studied at the National Vegetable Research Station in Warwickshire from 1959 to 1965. When applied to the surface, without being worked into the soil, the half-life of gamma-BHC was about 5 weeks, of dieldrin 9 weeks, and of DDT 18 weeks. When the insecticides were worked in, the corresponding half-lives were 15-20 weeks (BHC), 4-7 years (dieldrin), and 5-8 years (DDT).

From a practical point of view, this persistence means that soil treatment with dieldrin at 2 lb/acre in 1956 was still effective in controlling carrot fly in 1964, which was clearly of considerable economic value. This persistence also means that such compounds cannot be leaching out rapidly or contributing much to the contamination of ground waters, streams, and reservoirs. On the other hand, there seems to be no known way of reducing their levels in soils rapidly. Camphechlor (toxaphene), chlordane, endrin, endosulfan, and heptachlor are not likely to contribute much to soil residues from their present usage and there are no figures available on this point.

Since 1964 the quantities of aldrin and dieldrin used in the soil in Britain have been reduced substantially, both by design and because of the appearance of resistant populations of cabbage root fly, carrot fly, bean seed fly, and onion fly. The control of soil pests now relies heavily on the use of organophosphorus insecticides applied to soils.

TABLE 2

Average Concentrations of Gamma-BHC, the DDT group, and the Dieldrin group in parts per 10^{12} by Weight of Air, Water, or Soil, respectively, in Britain

Sample	Insecticides		
	Gamma-BHC	DDT group	Aldrin and dieldrin
London air	5, 11	10, 13	18, 21
Rain water (7 stations): year-round averages and ranges	60 (10-230)	79(12-210) [46(6-130)][a]	7(0-40)
Surface water (449 samples)	34	21	28
Arable soils	$1\text{-}50 \times 10^3$	$50\text{-}2000 \times 10^3$	$10\text{-}1500 \times 10^3$ (dieldrin)
Orchard soils		$3.6\text{-}26.6 \times 10^6$	

[a] p,p'-DDT concentration

Data from Agricultural Research Council (1970), Abbott et al. (1966), Department of Education and Science (1969).

By 1970 there were nine cleared (PSPS) compounds and others are under field trial by industry, by the National Agricultural Advisory Service, and by research institutes. Some, chlorfenvinphos, for example, yield hydrolysis products of low potential toxicity as residues. On the other hand, little information is available about the nature and persistence of the residues of phorate and disulfoton, both of which are being increasingly used in soils. Both can undergo oxidative toxication and it seems significant that although little of these compounds remains in soils 6-8 weeks after application, such application protects carrots from attack by carrot fly 5-6 months afterwards. There has been no general survey for residues of organophosphorus insecticides, herbicides, nematicides, fungicides, or other pesticides.

Apart from any contribution from human uses, most of the pesticide contamination in air must be that arising by evaporation from soils and plants. In 1964, organochlorine insecticides were found in rain water in central England, and in 1965 an analysis of 8000 litres of London air gave the results shown in TABLE 2 (Abbott et al., 1966). Although 10 ppt (parts in 10^{12}) is only 1 g of insecticide in 10^5 tonnes of air, this is equivalent, over the whole of the U.K., to about 1 tonne* in the air up to a height of 300 metres, assuming that the London figures applied generally. With a total organochlorine content in the air of 40 ppt, the daily amount inhaled by Man would be 0.6-1.8 μg, which is 2-5% of the amount ingested with the diet.

The average values for the total organochlorine insecticide content in samples of rain water in Britain, taken from seven stations ranging from the Shetland Islands to Cornwall, fall mostly between 3 and 300 ppt, with an overall average of 150 ppt. Average concentrations with their ranges from 28 quarterly analyses obtained from the seven stations appear in TABLE 2. Insecticide washed

---

*The metric tonne is 1.6% larger than the British ton.

down to the soil from this concentration of gamma-BHC, for example, would deposit less than one-thousandth of the dose corresponding to a moderate annual agricultural treatment, assuming an annual rainfall of 900 mm (36 inches). Thus, these insecticides were generally distributed in rain all the year round and there is no simple relation between concentration and the season of the year. The figures appear roughly consistent with the hypothesis that these compounds are generally present in the earth's atmospheric circulation, being both contributed to by Britain and washed out into Britain by the rainfall.

Since organochlorines are strongly adsorbed on soil particles, it is not surprising that despite their extensive use in agriculture, drainage water is not a serious source of contamination of British surface waters. The general background of organochlorine concentration in inland waters is of the same order as in rain water and in some cases considerably lower. Although heavy rainfall after spraying may produce measurable concentrations in the surface run-off, underground water that has percolated through soil does not contain measurable quantities of organochlorines. The Water Research Association analysed water from boreholes and springs under Kent orchards in which organochlorines had been used for years. No insecticide was detected in any sample (limits of detection 1 ppt gamma-BHC, 3 ppt dieldrin, 10 ppt DDT), which is consistent with results in other situations where rain water percolates through soil.

High figures for dieldrin in surface waters, as compared with rain water (TABLE 2) result from the inclusion of analyses from the rivers Aire, Holme, Calder, Stour, and Severn, all of which receive the effluents of mothproofing processes (concentrations 100-3000 ppt). The distribution of organochlorine concentrations is markedly asymmetrical, with a majority of low figures and a few high ones. Thus, Scottish streams usually contain less than 1 ppt of DDT with a few reaching 10 ppt, while dieldrin concentrations would normally be less than 10 ppt. Before the use of dieldrin in sheep dips ceased, local concentrations from 100 to >10,000 ppt occasionally occurred. However, the concentrations of dieldrin in normal surface waters generally approach those for rain water, while those for gamma-BHC and DDT are normally lower (TABLE 2). Estimates of human intake of organochlorines from drinking water in Britain indicate that the uptakes would be similar to those from air, so that these two sources together contribute very much less than the diet.

Organochlorine pesticides have not been detected in the open sea, but their presence is indicated by the occurrence in marine organisms living and feeding far from land. The concentration of DDT in sea water must be generally lower than the detection limit of 1 ppt, which would be the average concentration in the open sea resulting from admixture of the total world production of DDT.

A few surface samples taken in British coastal waters 10-24 km offshore from the River Crouch in water of depth 10-20 m contained 5-7 ppt of BHC (alpha- plus gamma-isomers) and 3-12 ppt of dieldrin. DDT was not detected offshore, but levels of this compound attained 2-7 ppt inside the Crouch estuary.

What are the probable contributions made by sewage effluents to the concentrations of organochlorines in rivers and in the sea? If 50 million people in Britain each consume daily 1.7 kg of food (a generous estimate based on the average food consumption of a 70 kg man) containing 34 μg of combined main organochlorines, the total of organochlorine ingested is about 600 kg/annum. Concentrations in human fat have been steady for several years, and most of this material must be excreted either unchanged or as metabolites and will enter the sewers. If sewage (annual weight 3.3×10^9 tonnes) treatment removed none of this and it was all unchanged pesticide, the concentration in sewage would be 150 ppt. The average concentration in rivers (annual flow estimated at 6.6×10^{10} tonnes), if they received all of this, would be below 10 ppt. Substantial amounts of organochlorines, especially DDT, are undoubtedly removed by sewage treatments, and actual concentrations in sewage are estimated to be 50-100 ppt. However, a rough estimate indicates that about 10 tonnes per annum are actually carried by rivers at an average concentration of 180 ppt, indicating that domestic sewage contributes only some 5% of the total organochlorines entering rivers and the sea. However, the outflow from sewage treatment plants receiving, along with raw sewage, effluent from wool mills and carpet factories using dieldrin may contain from 100-10,000 ppt, and there are unassessed contributions from pesticide manufacturing factories and possibly from sheep dip discharges containing gamma-BHC at 160 ppm and DDT at up to 6000 ppm.

The polychlorobiphenyls (PCBs) should be mentioned here, since although they are not organochlorine pesticides, they behave in a similar manner and are apparently generally distributed in the environment on account of their use in many industrial processes. A number of investigations of their distributions and effects in Britain are in progress but little is currently known about their occurrence in waters or their effects on aquatic organisms.

Concentrations of hundreds of ppm of these compounds were found in the livers of many of the dead marine birds washed ashore in September-October, 1969, but the significance of these concentrations in relation to the deaths is not yet clear.

As in the case of soils, there is currently little information available about the distribution of other pesticides in British waters. This is due to generally lower persistence, and measurements have usually only been made when fish kills occur. A survey of six English rivers, starting in June 1968, has shown the presence of carbophenothion in the Chelmer in Essex, rising to 8 ppb (parts per 10^9) on one occasion. Diazinon has been detected in five rivers, malathion in two rivers, and chlorfenvinphos, phorate and demeton (or a derivative) in one river each, all at less than 70 ppt. Currently used organophosphorus compounds are relatively non-persistent, but more persistent compounds may be required if they are to substitute completely for some organochlorine uses; this may well cause problems unless persistence can be combined with correspondingly low vertebrate toxicity. Chlorfenvinphos is widely used and Beynon et al. (1971) found that the water of a pond sprayed to contain an average concentration of 6 ppm lost 4 ppm within 5 hr and a further 1.9 ppm after 1 month. They concluded that this compound is taken up by particulate matter and mud and should not be a serious hazard to free swimming aquatic animals.

Samples of ground waters taken from sprayed areas failed to show detectable concentrations (over 10 ppb) of MCPA, MCPB, dichloroprop, DNBP, 2,4-D, and mecoprop. 2,4-D was readily leached out of sandy soil but not out of calcareous soil; about 2 ppm were found in drainage water 4 days after spraying heather, but none was found 3 weeks later. Dinoseb has caused fish kills in Scotland, but apart from accidents resulting in gross pollution, herbicides in general, and those approved for aquatic weed control (diquat, dalapon, 2,4-D amine, and maleic hydrazide) appear unlikely to cause direct harm when properly used; the main danger to aquatic organisms arises from the deoxygenation of waters through decay of the dead weeds. Among fungicides, the dithiocarbamates, mancozeb and maneb, have caused fish kills in Scotland. Heavy metal fungicides such as the organomercurials are under examination at the moment; there is little information regarding their occurrence in British waters at present, mainly because they had caused little concern until events in Japan and Sweden raised the question of their toxic hazard in Britain.

Distribution and Effects in the Biotic Environment

Effects on animal populations that result from widespread, acute, poisoning incidents are readily recognised and once the reason for such events is determined, repetitions can usually be prevented. Of more concern, therefore, are chronic effects that may lead eventually to mortality or to subtle physiological changes. Such effects only become apparent if detailed information is available about animal populations and the way that they respond to physical, as well as chemical, changes in the environment. There has recently been a considerable increase in the amount of appropriate research in progress in Britain. As an example, the effects on wildlife of changes intended to improve farming efficiency, such as the removal of hedges from fields, the removal of farm ponds, and the reduction of the number of small woods, are currently being investigated by the Nature Conservancy.

Effects on soil organisms and plants

A consensus of the available literature on this subject indicates that organochlorines do not affect microbial populations if applied to soil at the recommended rates, but that effects such as reduced nitrification, reduced ammonification, increased carbon dioxide evolution, and reduced root nodulation of legumes may be observed if the recommended rates are greatly exceeded. Most workers agree that the levels of DDT found in British soils at present do not influence significantly the total number of soil microorganisms or the beneficial processes referred to above. There are, of course, known effects *in vitro*; aldrin inhibited the growth of only 2 out of 17 fungi isolated from Wisconsin prairie soils by more than 50%, whereas lindane gave more than 50% inhibition with seven species. Parathion, phorate, and carbaryl inhibited in the majority of cases (Cowley and Lichtenstein, 1970). Clearly, any effects in the field will be complex, being dependent on the rate and frequency of application, the species present, and the soil type.

Edwards has studied the effect of many pesticide treatments on populations of soil invertebrates in relation to soil fertility (Edwards et al., 1967; Edwards, 1968, 1969). He found that many different insecticides are lethal to four particularly susceptible groups of soil animals; predatory mites, pauropods, springtails (Collembola) of the isotomid family, and fly larvae. However, most of the other families of springtails, all symphylids, earthworms and enchytraeid worms have considerable immunity to many pesticides. The duration of ecological imbalance following application of a soil pesticide depends both on the persistence of the chemical and its absolute toxicity. Thus, after 30 days, when the last trace of a soil application of dichloropropane-dichloropropene (DD) had disappeared, the springtail population had disappeared and did not reappear until 90 days later, the population remaining below normal for nearly 2 years. On the other hand, the persistent but less toxic herbicide simazine eliminated only 70% of the population in 30 days, and the population returned to near normal by the time the last residues disappeared 5 months later (Fig. 3). Nevertheless, persistent pesticides in general will have most effect, since the animals are exposed to them indefinitely.

One of the most consistent effects observed was that many species of springtails and some species of non-predatory mites increased when DDT or most organophosphorus insecticides were used--even to the extent of increasing the total biomass. In a 5-year period following a single application of DDT, the populations of predatory mites gradually fell to about 5% of the original level after 4 years and, regardless of seasonal fluctuations, the population of springtails continually increased until it attained 400% of the original level shortly before the time of lowest mite population. The increase in small soil animals produced by such an effect may compensate for the loss of larger invertebrates involved in destroying decaying plant matter; the breakdown of leaf material in soil from which earthworms were excluded was 73% in untreated soil, 90% in DDT-treated soil, and only 43% in aldrin-treated soil after 9 months, figures that correspond closely to the numbers of springtails present in each case.

Aldrin does not, however, affect parasitic mites, although it does reduce the numbers of other mites. In a comparison of DDT and aldrin for 1 year following application at practical rates, both insecticides greatly reduced the biomass of pests; the biomass of predators was reduced by about 50% in each case, but DDT killed more of the mites that prey on springtails and the consequent increase in numbers of the latter species compensated for some of the reduction in total biomass by DDT, giving an overall reduction of 10%. There was a 70% reduction of total arthropod biomass with aldrin, however (Edwards, 1969). During several years of observation, the greatest changes resulting from insecticide action were in the fall, when populations reached their peak numbers.

Following a single treatment with aldrin resulting in a 50% reduction in biomass, the populations increased progressively and reached the control level again after 6 years. Repeated annual applications of aldrin at 2 lb/acre maintained the biomass at about 40% of control and a 60 lb/acre application of aldrin only had twice the effect on populations as a 4 lb. application. In the long-term comparative experiment, aldrin had a much greater effect than DDT on dipterous larvae in the first 2 years, but 4 years after treatment the effects of DDT exceeded those of aldrin. DDT continued to reduce the populations in subsequent years whereas the dipterous larvae increased annually in aldrin-treated plots. When aldrin was added annually to plots at 2 lb/acre, there continued to be no effect on parasitic mite populations, but whereas other species of mites were reduced in numbers to about 20% of control in the first season, their numbers increased again in successive seasons. These effects suggest the development of resistant populations in aldrin-treated soils and seem particularly well illustrated in the case of the cabbage root fly. Not only has this pest developed cyclodiene resistance, but its carabid beetle predators, being more susceptible to dieldrin (Wright et al., 1960), have been eliminated and with them an additional source of control for the pest.

Earthworms appear to make a much larger contribution to soil fertility than the smaller animals (Edwards, 1968) and they are fortunately tolerant to most insecticides except chlordane, heptachlor, phorate, and carbaryl (Edwards, 1969). Since aldrin (and by implication dieldrin) does not reduce earthworm populations, and its effects in reducing soil animals frequently seem little greater than the effects of ploughing pasture land, there is no evidence of any serious direct or indirect effect on soil fertility. Earthworms, particularly those inhabiting orchards where fruit trees are sprayed with DDT, have been found to contain more than ten times the concentrations of organochlorines that are found in the surrounding soil, and, as the lower link in a food chain, they may be an important source of contamination for higher animals. There is so far no evidence for such accumulation of organophosphorus insecticides used as organochlorine replacements (Edwards, 1969).

Like simazine, DNOC (2,4-dinitro-6-methylphenol) has a direct action on soil animals, but usually not beyond the time taken for its disappearance from soil. This may be generally true of herbicides, but they may also have indirect effects on soil populations by destroying the weeds that are the normal habitat and food source of some species. With the object of determining the long-term effects on one soil type of repeated applications of four herbicides, each representing an important chemical group, five long-term field experiments were set up at the Weed Research Organisation on a coarse, sandy loam soil divided into plots planted either with the same crop each year or uncropped (Fryer and Kirkland, 1970). The compounds were MCPA (representing phenoxy alkanoic acids), simazine (substituted s-triazines), linuron (substituted ureas), and tri-allate (carbamates). Crops grown were spring barley (treated with MCPA or tri-allate), maize (treated with simazine), and carrots (treated with linuron). Neither MCPA nor tri-allate influenced the yield of barley during the 6-year period from 1963. There was a reduction in maize foliage and stems but no effect on grain yield with simazine in 1968. Linuron showed some phytotoxicity to carrots and reduced the yield in two of the years, but there was otherwise no effect. MCPA disappeared within

3 weeks at the start of the experiment, but this interval had declined to 4 days in 1968, after ten applications. Following each application of 1.5 lb/acre of tri-allate, residues persisted for 5-6 months; there was no build-up of soil residues during the 6-year period and no residues of tri-allate were detected in barley straw or grain. Simazine applied to maize declined characteristically in the soil; 80% disappeared in 6 weeks, while the remainder declined slowly and some was detectable 50 weeks after application. There was no accumulation of residues from year to year and no indication that repeated treatment influenced breakdown rate. Half the applied dose of linuron disappeared in 8 weeks and there were no soil residues after 6 months; there was no build-up of residues after repeated application and no indication of enhanced breakdown rate in successive years. Linuron was not detected in carrot roots in 1965; in 1966, 0.5 ppm was found following an abnormally late application. Analysis of soil samples taken at regular intervals during 1964-67 showed that herbicide treatments had not resulted in any major changes in inorganic nutrients. The lack of effects on crop yields or on soil nutrients is supported by microbiological work on the soil in indicating no changes in the behaviour of the soil that need cause concern. Even abnormally high applications of these compounds did not cause progressive build-up of soil residues, although such use of tri-allate, simazine, or linuron appeared likely to result in a phytotoxic residue lasting from year to year.

Although obtained on only one soil type and under one set of climatic conditions, these results, together with the general absence of observable residues in the crops or noticeable build-up of pests or diseases, provide evidence that under these conditions the four herbicides can be used continuously at recommended rates without deleterious effects on either soil or crops.

There have recently been a number of reports of damage to following crops due to herbicide residues in soil, especially following dry weather in 1970 (British Crop Protection Council, 1970). It seems likely that some occurrences resulted from incorrect procedures following crop failure, but there is much to be learned about the inherent susceptibility of crops to herbicides. Large amounts of herbicides are now being used for weed control in non-agricultural situations and such changes could result in unwanted and irreversible changes in vegetation patterns and movement of the chemicals in undesirable ways. It may well be that some restraint is needed, because in some ways the situation is very similar to that existing in the early days when organochlorines appear to have been used rather indiscriminately as 'cure all' pest control agents. If our present knowledge of organochlorines could have been applied then, a number of current difficulties might not have arisen.

Effects on aquatic organisms

In 1967, 26 of the 29 River Authorities in England reported fish kills attributable to pesticides; BHC, DDT, and dieldrin being the most common causes. The number of such kills was about 13% of the total cases of poisoning reported. In recent years about 10 to 15 kills (25% of the total observed) due to pesticides have occurred in Scotland, of which half were attributable to organochlorines (mainly dieldrin from sheep dips when it was used for this purpose). There have been no reports of incidents involving endrin or endosulfan. Such events frequently result from accident or ignorance and, hopefully, will tend to become less frequent.

An interesting account of the effects of organochlorine pesticides on the biota of a freshwater lake has been given by Bays (1971). The Chew Valley lake in Somerset is situated in a 'perforated catchment area' in which sheep dips have been used and which feeds into streams entering the lake. Early in 1968, the disappearance of *Daphnia* sp., *Bosmina* sp., and some Chironomid larvae from the lake was followed by successive severe algal blooms of a '*Monodus* sp.' (610,000 per ml), *Asterionella formosa* (50,000 per ml), and *Zygnema* sp. (15,000 per ml).

Detailed investigations in the surrounding catchment area indicated that significant quantities of gamma-BHC, dieldrin, and DDE had entered a feeder stream at a particular point, from an underground source. It was concluded that sheep dip waste discharged in the fields had penetrated to an underground storage system and that a subsequent underground collapse produced a surge of toxicants into the feeder stream and hence into the lake. The resultant reduction in algal predators caused a biotic imbalance lasting for about a year.

These occurrences apart, there is no conclusive evidence of harmful effects of pesticides in the aquatic environment in Britain. On the other hand, the background levels of organochlorines in inland waters would not be expected to produce easily noticeable effects in fish, for example. Since living organisms may be regarded as complexes of lipid and protein exposed to the abiotic environment, it follows that they must come to equilibrium with this environment and that if it contains lipid-soluble materials, these will be taken up until saturation level is reached. The efficiency of any elimination processes available to the organism will therefore be critical in determining the actual level of such 'contaminants' attained in the tissues by the time that intake has become balanced by elimination. The result is the familiar one of accumulation of pollutants to plateau levels in the tissues that depend on the rate of uptake. The toxicological significance of these exposures and of the tissue levels attained has to be assessed by comparison with the tissue levels attained when similar toxic effects or physiological changes are observed in experimentally-treated animals. Though such results are important toxicologically, and may indicate that a pesticide ought not to be used, they do not immediately help with the larger environmental problem of predicting possible effects on population. Clearly, a realistic assessment can only be made when detailed information about population variation can be compared with concurrent pesticide usage.

Experiments conducted by the Ministry of Agriculture

indicate that a DDT concentration of more than 1 ppb, which killed 50% in 100 days or less, is lethal to brown trout (Salmo trutta). The dead fish had DDT concentrations in muscle of 1.1-11.1 ppm, and none of the surviving fish had concentrations greater than 2 ppm, 80% of the levels being less than 1.1 ppm. Thus, the concentration in muscle was about 2000 times greater than the threshold concentration for mortality of 1 ppb in water. Trout muscle from uncontaminated areas contained 0.004-0.13 ppm of DDT and 0.006-2.7 ppm of DDE. The threshold concentration seems to be about the same for dieldrin as for DDT, and dieldrin was found at a concentration of 0.8-17.2 ppm in muscle from experimentally poisoned fish. Muscle from healthy trout in uncontaminated areas contained 0.004-0.42 ppm, while muscle from fish poisoned by sheep dip effluent contained 0.7-7.0 ppm together with 0.2-4.4 ppm of gamma-BHC. Most trout muscle has less than 0.2 ppm of dieldrin in it.

The concentrations found are always greatest in the more fatty tissues and in samples of brown trout, dieldrin concentrations in muscle were two to six times lower than those in the gills and two to eight times lower than those in liver. The distribution in tissues of different fish depends on the distribution of fats and oils in the tissues; Salmonids (trout, salmon) and Clupeoids (herrings) have a relatively large proportion of oil in their muscle and insecticide levels in liver and muscle do not usually differ greatly. In Gadoids (cod), having livers rich in oil, concentrations in liver greatly exceed those in muscle. For this reason, concentration factors derived from trout are difficult to relate to concentrations and effects found in other species. Few Scottish streams have concentrations approaching the threshold levels, and there appears to be a safety factor of about 100 for trout. However, these fish seem to be closer to danger from organochlorines than any human beings are, and in some cases the margin of safety may be quite narrow. Lower down the food chain algae (mainly examined in Scotland) usually contain less than 0.01 ppm of dieldrin, and in stonefly nymphs and

Gammarus sp. the concentration varied from less than 0.01 to 0.6 ppm. Molluscs upstream from an industrial discharge of dieldrin contained 0.02-0.08 ppm, while downstream they contained 2.2 ppm, without apparent damage.

Organochlorine residues have been found in most of the marine organisms that have been examined, but the limited data on invertebrates and fish from British waters show that DDT and dieldrin residues are mostly below 0.1 ppm. The concentrations found in filter-feeding molluscs, such as oysters, mussels, and cockles from British coastal waters, tend to be lower (0.03-0.7 ppm) than those in Crustacea and fish from the same area. This is an interesting observation, since some of the former organisms are known to have a large capacity for concentrating organochlorine compounds in their tissues. There is no conclusive evidence that shrimps or other Crustacea in coastal waters have been affected by organochlorines, but the pink shrimp, *Pandalus montagui*, is particularly susceptible to pollutants and has declined in coastal waters in recent years; it is possible that this is partly related to organochlorine levels in these waters. Species susceptibility varies considerably; thus cockles appear to be unaffected by concentrations of DDT as high as 1 ppm (presumably in colloidal solution), while certain Crustacea (e.g., shrimps) are much more sensitive, since the 48-hr LC_{50} for DDT, gamma-BHC, or parathion was found to be 1 ppb.

The problem of transfer and accumulation of organochlorine pesticides along food chains is a familiar one and the distribution of dieldrin and DDE among the trophic levels in Farne Island ecosystem is shown in Fig. 4 (Robinson et al., 1967). There are usually significant positive correlations between the concentrations of the compounds and the trophic levels, as shown, indicating progressive accumulation and concentration of each successive level. There are exceptions, however, since the concentration of dieldrin in cod (*Gadus morrhua*) (0.009 ppm) is less than that in the sand eel (*Ammodytes lanceolatus*) (0.016 ppm), its main food, and the concentration of dieldrin in planktonic crustacea (0.16 ppm) is greater than in any of the fish examined.

Extensive data on the eggs of seabirds collected from colonies on the east and west coasts of Britain (Robinson et al., 1967; Department of Education and Science, 1969) indicate levels of dieldrin mostly within 0.1-3.0 ppm, and of DDE mostly within 0.1-6.0 ppm. In spite of the high levels found in British seabirds and their eggs, there is no evidence that organochlorine compounds have had deleterious effects on populations.

The situation is somewhat similar in the case of predators of freshwater organisms (Fig. 5); herons (*Ardea cinerea*) and grebes (*Podiceps cristatus*) and their eggs contain much higher levels of organochlorines than vegetable feeders such as the moorhen (*Gallinula chloroplus*). In spite of their large organochlorine content, herons eggs produce viable young and there is no evidence of a decline in the heron population, although suspicions have been voiced that recovery of the heron population from the severe winter of 1962-63 has been retarded by these residues. In laboratory trials, eggs containing 12 ppm of dieldrin from chickens and homing pigeons fed this compound produced viable progeny (Robinson, 1970; Robinson and Crabtree, 1969); the concentration of dieldrin in these eggs is twice that found in heron eggs. Egg-shell weight and thickness were unaffected in the chicken trial, but the shells tended to be about 8-10% thinner in the pigeon trial.

With regard to aquatic organisms, it is difficult to tell from the existing evidence whether the general organochlorine levels are changing or stable, although recent data from seabird (Shag's) eggs suggests that the levels in them have been declining since 1966-68 (Coulson et al., 1971). Certain coastal waters receiving the flow of organochlorine-contaminated rivers may achieve levels that are dangerously high for some species. Thus, the margins of safety may not be high in some fresh waters and certain coastal waters, a position which might be improved by reducing the incidence of locally high contaminations.

Effects on terrestial vertebrates

Birds - Much controversy has surrounded the question of the effects of organochlorine compounds on individual birds and bird populations. Evidence for or against deleterious effects comes from two sources: comparisons of the levels of organochlorine compounds in the tissues of dead birds with levels found in experimentally poisoned birds, and studies of bird population changes in relation to pattern or organochlorine usage.

The main sources of information on residues in birds in the U.K. are the Reports of the Joint Committee for the British Trust for Ornithology and the Royal Society for the Protection of Birds on Toxic Chemicals (Reports 1-5, 1960-65). Residues were determined in connection with investigations on bird deaths by Turtle et al. (1963) and Taylor (1962). Numerous other discussions of residues have been given (Robinson, 1967; and references cited).

Circumstantial evidence initially linked the deaths of grain-eating birds in Britain in the early 1960's with the use of dieldrin-dressed seeds. Substantiation was later provided when it was found that the tissues of these birds contained concentrations of HEOD (dieldrin) or heptachlor epoxide similar to those found in the tissues of experimentally poisoned birds (Turtle et al., 1963). The difficulty of obtaining reliable data for other birds has been extensively discussed by Robinson (1967, 1969, 1970), and the suggested links between organochlorines and the deaths of individual birds or depletion of populations, especially of predators, appear to depend frequently on circumstantial evidence and subjective impressions. Thus, circumstantial evidence for the involvement of organochlorines in the decline of the peregrine falcon in Britain was advanced in 1963 (Ratcliffe, 1963). The onset of decline of the peregrine may have occurred in 1953-54, with a partial recovery in 1955, or during 1955-56. DDT and gamma-BHC have been used in Britain since 1945-46 (the decline of sparrow hawks from about 1952 has been associated with their use), and the cyclodienes from 1955, so that the peregrine decline might have been caused by the cyclodienes (if 1955 was the onset of decline), or a combination of all three organochlorines (if 1953 was the onset). The decline began in southern England, and the occurrence at this time of high cyclodiene residues in pigeons eating dressed grain is consistent with the involvement of organochlorine insecticides. However, there is no evidence that contaminated pigeons were the actual cause, or that the level of contamination was sufficient to affect the peregrines (Robinson, 1969).

As a general rule, flesh-eating birds have higher organochlorine residues in their tissues than non-flesh-eating ones. However, dead grain-eating birds in 1960-61, now believed to have died from poisoning, had much higher total body burdens of organic chlorine than were found in most of the dead predators. When the total concentrations of organochlorine insecticides in bird tissues in 1962-65 are compared with levels accepted to be toxic, it seems unlikely that the deaths of a large proportion of these birds can be attributed to organochlorine poisoning (Robinson, 1967). Thus, results for HEOD (dieldrin), summarised in Fig. 6, indicate that a number of grain-eating birds were poisoned (average dieldrin plus heptachlor epoxide content about 20 ppm in some cases), but that not more than 5-10% of the predators examined after 1961 are likely to have died from dieldrin poisoning unless they are much more susceptible than other birds. From analyses of the tissues of 105 predatory birds in 1963-64, Walker et al. (1967) concluded that it is unlikely that more than 10% of the birds examined die of organochlorine poisoning, even if the effect of different compounds is additive. Walker (1971) notes, however, that this survey post-dates the initial voluntary restrictions of 1961 on spring seed-dressings, and that most specimens were not collected from areas where sharp declines had occurred, so that the residues found are likely to be lower than were carried by birds in the affected areas. He concludes that organochlorines ingested from the prey were a probable cause of decline of the peregrine falcon and sparrow hawk populations.

When considering the toxicological significance of residues in tissues, the effects of combinations of pesticides and species differences have to be considered and there are many other factors that make comparison of results from the field with those from the laboratory difficult (Robinson, 1969).

The results of determinations of organochlorine residues in peregrine eggs were evoked to support the possible association of organochlorines with the decline of this species (Ratcliffe, 1965, 1967a), and reports implicating organochlorines in reduced eggshell thickness for certain birds of prey (Ratcliffe, 1967b) initiated a number of investigations, both in Britain and abroad, of this phenomenon and of possible sub-lethal effects not yet recognised.

Some experimental trials relating residues in eggs to hatchability and progeny survival have already been mentioned. Using the available criteria (both open to the accepted criticism that interspecific variations in sensitivity cannot be allowed for) that (a) residues of organochlorines in eggs of predators were frequently lower than those in other species whose reproductive success seemed only slightly affected and (b) that there is no significant increase in the mortality of chicks of quail, pigeons, or domestic fowl hatched from eggs containing up to 15 ppm of HEOD, Robinson (1969) concluded that the evidence associating organochlorines with declines in breeding success is rather tenuous. Other workers incline to the view that organochlorines are implicated; experiments (Porter and Wiemeyer, 1969) with the American sparrow hawk fed a diet containing 0.3 ppm of dieldrin and 1.4 ppm of DDT, involving a dieldrin intake below that received by grain-eating birds during the British seed-dressing incidents, resulted in thin eggshells and reduced reproductive success. A number of sub-lethal effects such as delayed ovulation and alterations in the thyroid in Bengalese finches (<u>Lonchura striata</u>)

(Jefferies, 1969), alterations in the thyroid with homing pigeons Columba livia (Jefferies and French, 1969), and thin eggshells in Japanese quail (Bitman et al., 1969) have been observed in laboratory trials, often at high dietary levels, with DDT. If they occur in the field, such phenomena may interact with other environmental factors to produce adverse effects, but their significance in relation to population changes is difficult to assess at the moment.

Reviewing all these findings, the Wilson Committee concluded that 'the available evidence suggests that dieldrin was responsible for these reported changes' in predator populations, and following the withdrawal of aldrin and dieldrin for use in seed dressings in 1962 and some other restrictions (1964), 'some population increases have been observed in the peregrine falcon, sparrowhawk and kestrel.' In any case, these events have proved the necessity for detailed studies of population dynamics and the need for careful documentation of pesticide usage.

Polychlorinated biphenyls (PCB) are of interest in the present context because of resemblances between them and the insecticidal organochlorines in regard to environmental distribution. Prestt et al. (1970) have shown that these compounds occur in the eggs and tissues of terrestial, freshwater, and marine birds obtained from all areas of Britain. The highest liver residues are found in freshwater feeding birds (up to about 900 ppm) and bird-feeding raptors (up to 70 ppm), the levels present being generally similar to those of p,p'-DDE. There is no indication from presently available dosage-mortality data or from information concerning PCB usage that these compounds have had acute effects on bird populations. The possibility that they have sub-lethal effects, or have added to the sub-lethal effects of other organochlorines cannot be ignored, however, since they are capable, for example, of inducing steroid hydroxylating enzymes, just as other organochlorines do.

Mammals - Pesticides may be taken up directly when animals eat dressed seeds or treated plants, or other animals. Cases have been reported of the poisoning of foxes and badgers which ate pigeons that had previously consumed aldrin- and dieldrin-dressed cereals. There is no evidence, however, of any effects on the population of these creatures. Endrin is rapidly metabolised in the mammals so far examined and residues are rarely found in tissues. Neither it nor endosulfan have caused problems so far in Britain. Despite the voluntary withdrawal of heptachlor seed dressings, residues of heptachlor epoxide are still found in wildlife tissues and these may be still risks to predators of acute poisoning from this source. There is little information that will permit any correlation between residue levels in mammals and death due to DDT poisoning. Reported total DDT group residues in livers of several mammals, including marine species (Department of Education and Science, 1969) are relatively low (0.01-0.34 ppm), except for various bats (up to 6.0 ppm) which are also known to be rather sensitive to DDT (LD_{50} is around 20 μg for some species).

As with other wildlife species, the situation for mammals is not very satisfactory; extrapolation from the results of laboratory experiments is difficult and examples of unusual sensitivity occur. Although paraquat, for example, appears to be safe to wildlife when properly used, it has been responsible for the deaths of hares in isolated incidents. Hares appear to be about three times as sensitive to paraquat as rabbits, but there is no evidence of effects on populations. The reported teratogenic effects of 2,4,5-T in mammals and the isolation of the highly toxic contaminant tetrachlorodibenzo-p-dioxin are cautionary, but this herbicide has been used in Europe for about 15 years without evidence of ill effects. There is a clear need for more studies in comparative detoxication between mammals, birds, fishes, and insects, which might ultimately provide an explanation for some of the ecological effects.

Only 22 out of 7,500 reported accidents (none fatal) to agricultural workers in Britain in 1968 were due to pesticides, so the hazards to those most exposed to them appear to be small. DDT and dieldrin have been outstandingly safe to man and there have been no reports of accidental poisoning from the use of vaporisers employing gamma-BHC and DDT such as have occurred elsewhere. The major concern, therefore, is with possible sub-lethal effects. In Britain, there are no groups of individuals with an unusual occupational exposure confined to DDT or dieldrin who could be observed for long-term effects. We have, therefore, to rely on the fact that the health records of Dutch dieldrin plant workers, having a daily exposure 50-100 times greater than that of the general population, or of DDT formulators in the U.S.A., do not appear to be different from those of other industrial workers. Further, we know that there was no affect on the health of human males ingesting up to 225 μg of HEOD (dieldrin) daily for 2 years, although at the highest ingestion level, body burdens of toxicant were some ten times higher than those found in the general population of Britain (Hunter et al., 1969).

The major source of organochlorine intake is evidently the diet, and estimates of daily intake in Britain have been made from surveys of residue levels in foodstuffs contained in the average diet, and by calculation from the concentration found in blood and adipose tissue, using the by now well-known pharmacodynamic relationships between intake and tissue levels. The dietary surveys indicate dieldrin intakes of 19.9 μg/man per day (1965), 12.6 μg/man per day (1967) (McGill et al., 1969), 10 μg/man per day (1966-68) (Association of Public Analysts, 1971), and 7 μg/man per day (Department of Education and Science, 1969). Except for the last, these figures are somewhat greater than the Acceptable Daily Intake (A.D.I.) advised by the World Health Organisation (W.H.O.) (TABLE 3), but are generally accepted to be decreasing, and are 160-500 times smaller than the dietary no-toxic-effect intake of dieldrin in the rat, dog, and monkey (50 μg/kg of body weight/day) (Jager, 1970). Estimates of dieldrin intake (2.7-19 μg/man per day) derived from blood and adipose tissue levels are in fair agreement with the other method, and the total DDT group intake (49.7 μg/man per day) calculated from adipose tissue levels (Department of Education and Science, 1969) correlates well with the daily intake (46 μg/man per day) estimated from total diet (TABLE 3). This figure for DDT

TABLE 3

Organochlorine Compounds in the Average Diet in England and Wales (1966-68)

Situation	BHC	Aldrin/Dieldrin	DDT
Acceptable daily intake (A.D.I.) mg/70 kg adult	0.875	0.007	0.700
Average dietary residues based on the 2 years of survey (ppm)	0.010	0.006	0.027
Average daily intake (mg) from 1.7 kg of food for the 2-year period	0.017	0.010	0.046

Data from Association of Public Analysts (1971).

is well below the A.D.I. of 700 μg/man per day advised by the W.H.O. The composition of the average British diet and the contribution to organochlorine residue intake made by each food group is indicated in Fig. 7, which illustrates the relatively large contribution made by meat, fish, and fats, although these are only small proportions of the total diet. It is interesting that the concentrations of HEOD in the kidney fat of sheep reared in Britain declined rapidly between 1964 and 1969, the mean concentration being approximately halved yearly between 1965 and 1969 (Ministry of Technology, 1970).

A survey of adipose tissue levels in Britain conducted between 1965 and 1967 (Abbott et al., 1968) showed no marked differences related to geographical location but levels in men were higher than those in women. The levels of BHC (beta-BHC predominating in all samples) had declined by about 35%, of the levels found in 1963-64. These changes parallel the previously-mentioned trends in the dietary intake and provide hope that the more restricted use of organochlorines in Britain since 1964 is taking effect.

Questions have arisen about the possible effects of these intakes and body burdens of organochlorines on the levels of hepatic and other enzymes. The dietary threshold level for hepatic enzyme induction by cyclodiene epoxides in the rat, a very sensitive species, seems to be about 1 ppm, so that a total organochlorine residue intake at 0.04-0.05 ppm for Man (TABLE 3) seems to be well below the level at which such effects would be observed. In support of this proposition, it has been found that the Dutch dieldrin plant operatives, receiving at least 85 times the daily oral intake of this compound, had normal blood levels of p, p'-DDE and normal 6-beta-hydroxycortisol excretion rates, indicating an absence of enzyme induction.

There is clearly a need for continuing vigilance with regard to residues from other types of compounds, since there have been reports, for example, of relatively high residues of pentachloronitrobenzene in lettuce and of thiocarbamates in other crops (Pagington, 1968). The situation with regard to organophosphorus compounds in foodstuffs appears to be satisfactory (Association of Public Analysts, 1971), in spite of the increase in their usage, since residues (0.06-1.4 ppm) were found in only a very small number of the total samples examined.

Analyses of fruit conducted in this survey indicated that there was no excessive contamination with lead and that 2 out of 34 samples of apples and 1 out of 71 samples of pears contained more than 1 ppm of arsenic. Only a limited number of other foodstuffs was examined, but a high proportion of the samples contained traces of arsenic (0.10-0.80 ppm) and some contained traces of lead (0.14-0.60 ppm). Organomercurial fungicides are used in Britain as seed dressings for root crops such as wheat, barley, oats, beet, and mangolds, for the protection of seed potatoes, and in connection with fruit growing. Out of 140 samples representing 22 different foodstuffs examined (Association of Public Analysts, 1971) none contained mercury at or above 0.1 ppm. One out of 38 samples of eggs examined in 1967-68 contained 1 ppm, and 1 out of 50 samples of hard imported cheese contained 0.42 ppm of mercury. Residues of 0.015 ppm or less were found in food groups from the total diet by Abbott and Tatton (1970), indicating that the total dietary intake of mercury is below 14 μg/day.

In general, the residues of these heavy metals received by humans in Britain from food sources appear to give no cause for concern, although the occasional incidences of relatively high residues emphasise the need for continued surveillance.

Effects on insects

Reference has already been made to some of the effects of pesticides on the ecological balance of soil organisms. If a reduction in the numbers of natural predators of some pest should be accompanied by development of toxicant resistance in the pest, then severe crop destruction may result. This combination of circumstances seems to have arisen in the case of cabbage root fly (*Erioischia brassicae*) in Oxfordshire in 1963 (Coaker *et al.*, 1963). On a particular farm, large scale destruction of brassica crops by this insect was later linked with selection for resistance resulting from the copious use of aldrin. Analysis of soil samples from the infested fields indicated aldrin concentrations as high as 20 lb/acre near plant stems and up to 2 lb/acre between the plants. There had been no previous event of this sort in Britain, the recommended aldrin treatments having given a high degree of control at levels not exceeding 0.4 lb/acre. Since then, cabbage root fly resistance has been observed in several areas where there has been exposure to high aldrin pressure. Such infestations may be remarkably localised, so that a field with a resistant infestation may be surrounded by an area in which the numbers of resistant insects is low (Gostick and Baker, 1968a). Cyclodiene resistance has also been reported in carrot fly (Wright and Coaker, 1968; Gostick and Baker, 1968b), in bean seed fly (Gostick and Baker, 1966), and in onion fly (Gostick et al., 1971), but not so far in wheat bulb fly or the large narcissus fly. Although the cyclodiene resistance is accompanied by the usual cross-resistance to gamma-BHC, organophosphorus alternatives such as diazinon have given satisfactory control.

When searching for compounds to control insects, it has always been easier to pick out those giving acutely lethal effects, and such compounds have obvious economic attractions. Since the lethal effect usually predominates, relatively little attention has been given to the sub-lethal effects of classical toxicants, and separate searches have been made for anti-feeding compounds, chemosterilants, compounds affecting hormone balance, and so on. Since the methods for measuring sub-lethal effects of this sort and for measuring acute toxicity are so different, it may be that many compounds which failed screens for acute toxicity might have other effects equally useful for controlling insect populations. The classical insecticides certainly produce sub-lethal effects in insects, and these are under investigation by the Nature Conservancy in England (Moriarty, 1969, 1971). Such effects must certainly occur in the field; for example, the induction of mixed function oxidases in insects already resistant to soil insecticides such as aldrin is a very likely phenomenon, and may reveal itself in enhanced tolerance to other types of insecticides. The consequences of these effects for wild population levels remains to be determined.

These observations indicate the need for great care in application of pesticides and especially the need for some really effective means to impart to users the knowledge of pesticide-environment interactions acquired during the last 10 years, so that previous mistakes are not repeated. Since it is unlikely that chemical control can be dispensed with in the foreseeable future, greatest hope would seem to lie in the use of combinations of biodegradable chemicals with biological control methods. However, integrated control requires a detailed knowledge of the ecological relationships of the pest to be controlled, as well as of pest-pesticide interactions. Many difficulties attend the application of this technique in the field, but the more stable climate of glass houses favours planned biological control. Research at the Agricultural Research Council's Glasshouse Crops Research Institute illustrates the possibilities and complexities of this technique. Thus, infestations of red spider mite (*Tetranychus urticae*) on cucumbers have been treated by controlled application of this mite together with the predatory mite *Phytoseilus persimilis*. An abundance of the predator has to be ensured during May when the mites begin to swarm upwards into the foliage and in August before they go into hibernation, and re-seeding of the plants with the prey may be necessary throughout the season to maintain the predator.

In the absence of chemicals, minor pests which are normally controlled incidentally may become troublesome, and the control of cotton aphid, *Aphis gossypii*, and glasshouse whitefly by parasites is under investigation. Chemicals are still used to control cotton aphid and mildews, but their choice requires considerable care in order to avoid damage to the predator. Systemic aphicides such as menazon, isolan, and pirimicarb had no effect on the predator when used entirely systemically, whereas demeton-S-methyl, thionazin, and methomyl killed the predator; pirimicarb can be used as a spray provided that this falls on the upper surface only of leaves. *T. urticae* and *P. persimilis* infesting cucumbers were unaffected by dicofol and tetradifon applied to the upper surface of leaves, but were killed by a translaminar effect when demeton-S-methyl was applied in this way. Spider mite eggs contaminated with captan or malathion repelled the predator, whereas dinocap did not have this effect. *Thrips tabaci* infestations on cucumbers can be managed in association with the biological control of *T. urticae*, by the soil application of BHC or diazinon to kill *Thrips* pupae, events in the foliage being thereby unaffected. These observations give some indication of the extent of the investigations required to develop a successful programme of integrated control.

Concluding Comments

It seems clear that even in Britain, where the range of pest problems is perhaps more limited than in some other countries, it will not be possible to dispense with chemical control in the foreseeable future, except in certain well-defined areas. Since the sources of persistent pesticides in the environment cannot all be identified, and their effects cannot be reliably determined at present, the regulation and reduction of current usage, where possible, is clearly prudent. At this point the availability of satisfactory and economic alternatives must be carefully considered, because if priority is given to the removal of those substances the presence of which is known to be harmful to Man, then on such a basis no high priority can be given to the removal of DDT or dieldrin (Department of Education and Science, 1969). Careful surveillance of the pesticide situation in Britain will continue and there is already evidence that the levels of organochlorine compounds in the environment is falling as a result of the measures taken regarding their usage. The fall in total human dietary intake of organochlorines, the fall in concentrations in human adipose tissue and that from British sheep kidney, and the fall in concentrations in Shag's eggs all appear to point in this direction. Fortunately, the general concern about environmental pollution resulting from organochlorine usage has alerted us to possible more immediate dangers, such as those from heavy metals.

As the Wilson Committee points out (Department of Education and Science, 1969), its recommendations regarding the limited retention of organochlorines, and this can be applied to pesticides generally, are a compromise based on a careful consideration of the environmental situation in Britain, especially that resulting from their use in this country. People in temperature countries, with high living standards and no threat from insect-borne diseases, may reasonably be concerned with the undesirable aspects of persistent pesticides. However, some countries face problems in agriculture, public health, and living conditions quite different from our own, and, where food production is vital and mortality from insect-borne disease is high, the hazards to wildlife and the presence of low residues in human fat may rightly be regarded as relatively unimportant.

Acknowledgment

Information in this paper has been freely drawn from the report of the Wilson Committee on organochlorine usage (Department of Education and Science, 1969) and from the report of the Research Committee on Toxic Chemicals (Agricultural Research Council, 1970), as well as from the original literature. I also thank a number of colleagues in Government Service for helpful discussions, C.Walker for the loan of unpublished proofs, and J. Robinson for stimulating discussions and the provision of unpublished data.

References

Abbott, D. C., and J. O'G. Tatton: Pesticide Sci. 1, 99 (1970).

Abbott, D. C., P. B. Harrison, and J. O'G Tatton: Nature 211, 259 (1966).

Abbott, D. C., R. Goulding, and J. O'G Tatton: Brit. Med. J. 3, 146 (1968).
Agricultural Research Council: Third Report of the research committee on toxic chemicals. London (1970).
Association of Public Analysts: Joint survey of pesticide residues in foodstuffs sold in England and Wales, 1st August 1967-31st July 1968 (second year). W. Heffer and Sons, Cambridge (1971).
Bays, L. R.: Environ. Pollut. 1, 205 (1971).
Beynon, K. I., M. J. Edwards, A. R. Thompson, and C. A. Edwards: Pesticide Sci. 2, 5 (1971).
Bitman, J., H. C. Cecil, S. J. Harris, and E. F. Fries: Nature 224, 44 (1969).
British Crop Protection Council: Monograph No. 3, 7th Review of herbicide usage, p. 16 (1970).
Coaker, T. H., D. J. Mowat, and G.A. Wheatley: Nature 200, 664 (1963).
Coulson, J. C., I. R. Deans, G. R. Potts, J. Robinson, and A. N. Crabtree: Private Communication (1971).
Cowley, G. T., and E. P. Lichtenstein: J. Gen. Microbiol. 62, 27 (1970).
Department of Education and Science: Further review of certain persistent organochlorine pesticides used in Great Britain. Report by the advisory committee on pesticides and other toxic chemicals, London, Her Magesty's Stationery Office (1969).
Edwards, C. A.: Proceedings of the Symposium on the Science and Technology of Residual Insecticides in Food Production with Special Reference to Aldrin and Dieldrin, p. 168-174. Shell Oil Co., New York (1968).
Edwards, C. A.: Sci. Amer. 220, 88 (1969).
Edwards, C. A.: Problem of insecticidal residues in agricultural soils. Pesticide Abstr. News Summary 16, 271 (1970).
Edwards, C. A., E. B. Dennis, and D. W. Empson: Ann. Appl. Biol. 60, 11 (1967).
Fryer, J. D., and K. Kirkland: Weed Res. 10, 133 (1970).
Gostick, K. G., and P. M. Baker: Agriculture 73, 70 (1966).

Gostick, K. G., and P. M. Baker: Ann. Appl. Biol. 61, 447 (1968a).

Gostick, K. G., and P. M. Baker: Plant Pathol. 17, 182 (1968b).

Gostick, K. G., D. F. Powell, and C. Slough: Plant Pathol. 20, 63 (1971).

Hunter, C. G., J. Robinson, and M. Roberts: Arch. Environ. Health 18, 12 (1969).

Jager, K. W.: Aldrin, dieldrin, endrin and telodrin: an epidemiological and toxicological study of long-term occupational exposure. Elsevier Publishing Company (1970).

Jefferies, D. J.: Nature 222, 578 (1969).

Jefferies, D. J., and M. C. French: Science 166, 1278 (1969).

McGill, A. E. J., J. Robinson, and M. Stein: Nature 221, 761 (1969).

Ministry of Technology: Report of the Government chemist, 1969. H.M.S.O., London (1970).

Moriarty, F.: Biol. Rev. 44, 321 (1969).

Moriarty, F.: Mededel. Landbouwhogeschool Opsoekingssta. Staat Gent 36, 27 (1971).

Pagington, J. S.: J. Ass. Publ. Analysts 6, 25 (1968).

Porter, R. D., and S. N. Wiemeyer: Science 165, 199 (1969).

Prestt, I., D. J. Jefferies, and N. W. Moore: Environ. Pollut. 1, 3 (1970).

Ratcliffe, D. A.: Bird Study 10, 56 (1963).

Ratcliffe, D. A.: Bird Study 12, 66 (1965).

Ratcliffe, D. A.: Bird Study 14, 238 (1967a).

Ratcliffe, D. A.: Nature 215, 208 (1967b).

Robinson, J.: Chem. Ind., 1974 (1967).

Robinson, J.: Chemical Fallout: current research on persistent pesticides. M. W. Miller, G. G. Berg, C. C. Thomas, Eds. Springfield, Ill. (1969).

Robinson, J.: Bird Study 17, 195 (1970).

Robinson, J., and A. N. Crabtree: Mededel. Landbouwhogeschool Opzoekingssta. Staat Gent 34, 413 (1969).

Robinson, J., A. Richardson, A. N. Crabtree, J. C. Coulson, and G. R. Potts: Nature 214, 1307 (1967).

Strickland, A. H.: J. Appl. Ecol. 3 (Suppl.), 3-13 (1966).

Taylor, A.: Analyst 87, 824 (1962).

Turtle, E. E., A. Taylor, E. N. Wright, R. J. P. Thearle, H. Egan, W. H. Evans, and N. M. Soutar: J. Sci. Food Agr. 14, 567 (1963).

Walker, C. H.: Environmental Pollution by chemicals. Hutchinson Educational Ltd. (1971).

Walker, C. H., G. A. Hamilton, and R. B. Harrison: J. Sci. Food Agr. 18, 123 (1967).

Wright, D. W., and T. H. Coaker: Plant Pathol. 17, 178 (1968).

Wright, D. W., R.D. Hughes, and J. Worrall: Ann. Appl. Biol. 48, 756 (1960).

<u>Working Parties On:</u>

Toxic Chemicals in Agriculture (1951) Residues in Food (1953) Risks to Wildlife (1955) Report of Research Study Group on Toxic Chemicals used in Agriculture (1961)	succeeded by the Agricultural Research Council's Research Committee on Toxic Chemicals (1962)

Royal Society for Protection of Birds (voluntary) British Trust for Ornithology (voluntary) Nature Conservancy (official) (Set up its Toxic Chemicals and Wildlife Research Division at Monks Wood in 1960).	Provided main initiative in studying side effects of Pesticides in Wildlife from 1956.

MAFF Advisory Committee: Scientific Sub-Committee
Wildlife Panel set up in 1960.

Advisory Committee on Pesticides and other Toxic Chemicals (Wilson Committee) (1961).

Fig. 1. Supervision of pesticide usage in the United Kingdom since 1950.

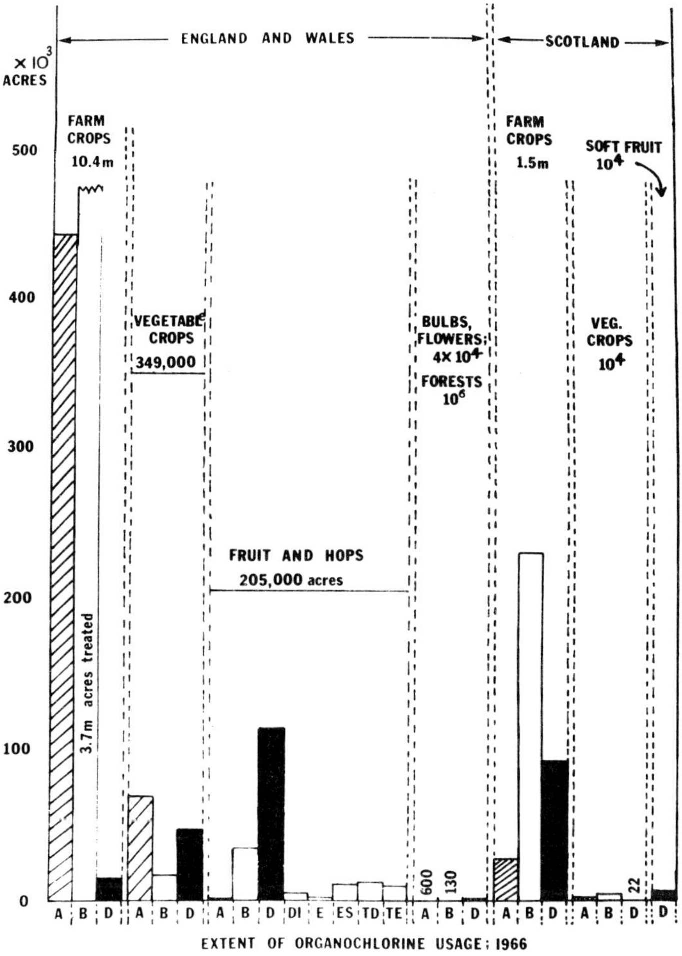

Fig. 2. Crops grown in Britain and acres treated annually with organochlorine insecticides (1966). Figures given with the crop are acres grown (m=million). Where acres treated are off scale, figures are given in the appropriate place. A = Aldrin plus dieldrin; B = BHC; D = DDT; DI = dicofol; E = Endrin; ES = endosulfan; TD = TDE; TE = tetradifon.

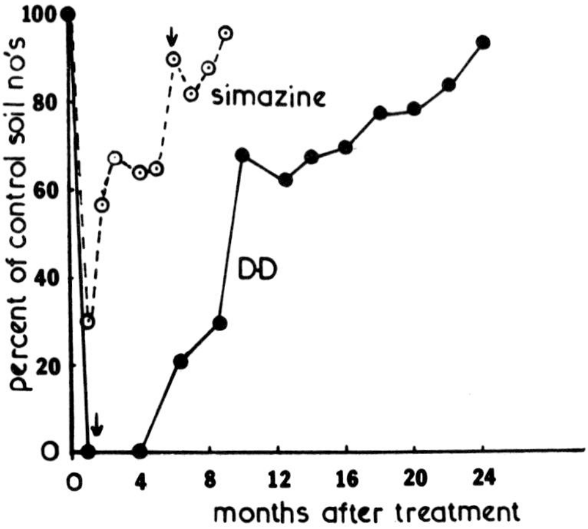

Fig. 3. Effect of the persistent herbicide simazine and the non-persistent soil fumigant D-D on populations of collembola in field plots. Arrows indicate time of disappearance of pesticide residues. (After Edwards, 1969).

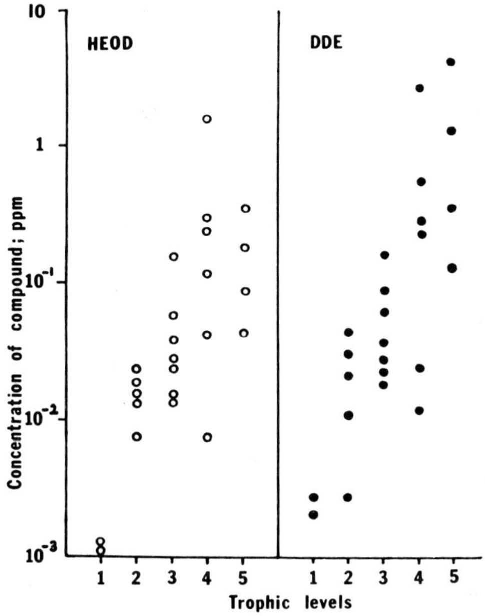

1. Sea weed ; 2. microzooplankton ; 3. Lobster, shore crab, herring, sand eel; 4. Cod, whiting, eider duck, shag, herring gull; 5. Gannet, cormorant, grey seal, dolphin

Fig. 4. HEOD (dieldrin) and DDE concentration in relation to trophic level in the Farne Island ecosystem. After Robinson et al. (1967).

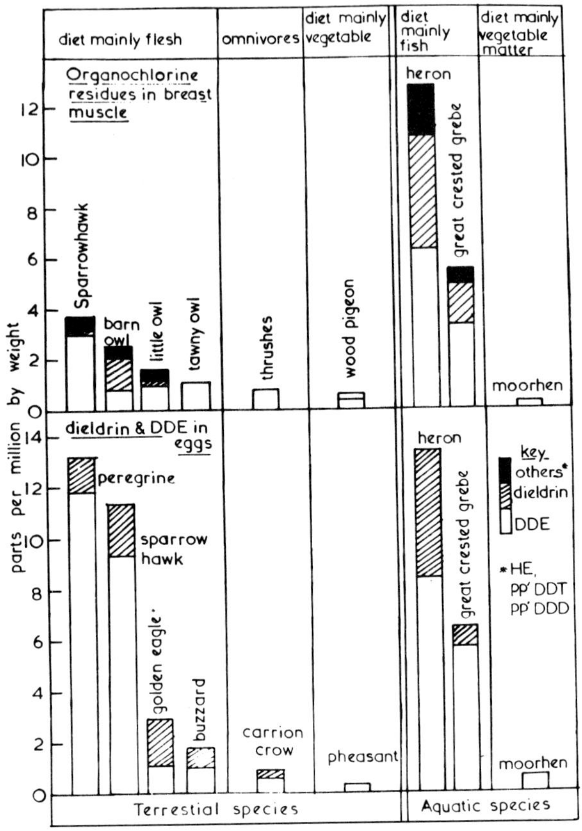

Fig. 5. Organochlorine residues in some British birds and their eggs (1960-1965). HE = heptachlor epoxide. After Walker (1971).

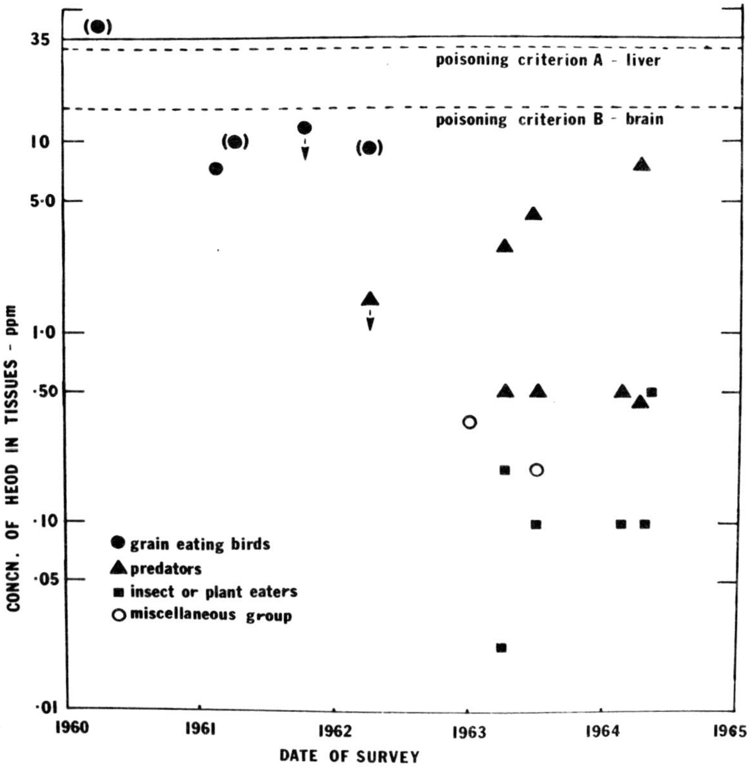

Fig. 6. HEOD (dieldrin) in tissues of dead birds. ( ) indicate values estimated from total organic-chlorine content. indicates that the actual value may be lower because the analytical method did not distinguish between HEOD and p, p'-DDE. Criteria A and B are estimates of toxic concentrations in liver and brain of experimentally poisoned birds. After Robinson (1967).

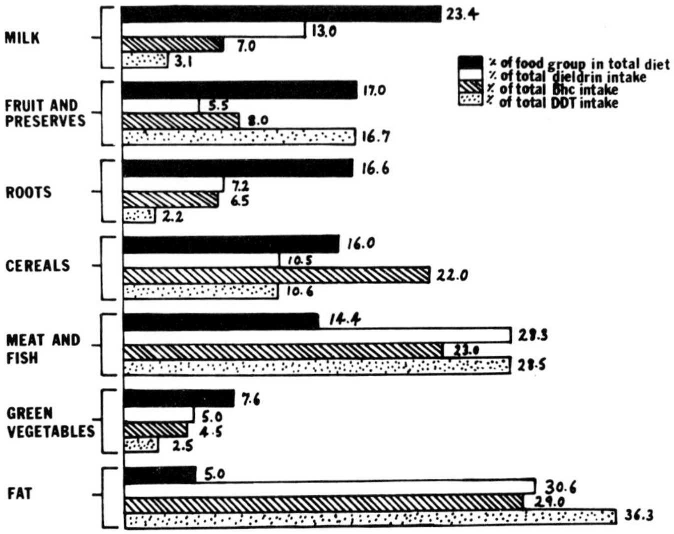

Fig. 7. Organochlorine residues in the whole diet-contribution of the various dietary components to the total intake of HEOD (dieldrin), BHC isomers, and the DDT group.

PART II

MERCURY TRANSFORMATION IN THE ENVIRONMENT

MICROBIAL CONVERSION OF MERCURY COMPOUNDS

K. Tonomura, K. Furukawa, and M. Yamada

Fermentation Research Institute
Inage, Chiba-shi, 280, Japan

Introduction

Environmental pollution by mercury has recently become one of the greatest social problems in Japan. To cope with this problem, it is believed that the most important thing is to first determine what changes occur to mercury compounds in nature. As one of such changes, we have been studying the conversion of mercury compounds by microorganisms. In the present paper two subjects are discussed: first, the reductive decomposition of organic and inorganic mercury compounds by mercury-resistant *Pseudomonas spp.*, and second, the formation of methylmercury from inorganic mercury compounds by an isolated strain of *Clostridium cochlearium*. The former is related to the activity of aerobic bacteria and the latter to the activity of anaerobic bacteria.

Abbreviations used in the text are: PMA - phenylmercuric acetate, EMP - ethylmercuric phosphate, MMC - methylmercuric chloride, and MC - mercuric chloride.

Reductive Decomposition of Mercury Compounds by Mercury-Resistant *Pseudomonas spp.*

A mercury-resistant Pseudomonas and its action on mercury compounds

At the beginning of the study in 1964, we isolated mercury-resistant bacteria from nature. Several strains of such bacteria were isolated from soil. All had red pigments in their cells, and were similar in their characteristics. Taxonomic studies revealed that these organisms belonged to the genus *Pseudomonas*. In this study one strain (K 62) was selected and used. The resistance to mercury compounds in this organism was about a thousand times greater than that in *Escherichia coli* (Tonomura et al., 1968a).

The microbial activity on mercury compounds has been studied in detail. Figure 1 shows the uptake of mercury by the organism, when the organism was incubated with PMA labeled with radioactive ^{203}Hg. In order to determine the intracellular location of bound mercury, mercury-bound cells were mechanically disrupted, and differentially centrifuged. Several fractions were obtained, and mercury was determined by radioactivity and also an analysis was made for protein. Results are shown in TABLE 1, indicating that 76% of the mercury was found in the heaviest fraction. This fraction appeared to contain a greater portion of cell-surface material, and it was thus considered that a greater part of the bound mercury might be located on the cell surface. In order to examine the fate of cell-bound mercury, mercury-bound cells were transferred to fresh culture medium. Figure 2 shows that the bound mercury rapidly decreased, and bacterial growth seemed to begin after mercury had almost disappeared from the culture. The mercury which had disappeared was completely recovered by charcoal in a closed system (Tonomura et al., 1968b).

The fate of mercury was further studied. The culture was run on a much larger scale, and vaporized mercury was obtained in a cold trap. Analysis of the mercury in the trap proved it to be metallic (Furukawa et al., 1969). When the reaction occurred in a closed system, a part of the metallic mercury produced was deposited on the bottom of the flask.

TABLE 1

Distribution[a] of Radioactive Mercury in PMA taken up by *Pseudomonas* Strain K 62

Fraction	Radioactivity		Protein	
	10^2 cpm	%	mg	%
Homogenate	348	100	45.0	100
2,000 x *g*, 15 min pellet 1	264	76	5.2	12
14,000 x *g*, 15 min pellet 2	34	10	6.8	15
37,000 x *g*, 15 min pellet 3	49	14	7.7	17
100,000 x *g*, 60 min pellet 4	15.5	5	1.6	4
Supernatant 4	6	2	23.0	51

[a]The organism was incubated with ^{203}Hg-PMA for 2 hr in the same conditions as in Fig. 1.

The change in the carbon skeleton of organomercury was studied, when decomposed to metallic mercury. The production of methane from MMC, ethane from EMP, and benzene from PMA were demonstrated by gas chromatography (Furukawa et al., 1969). It was also demonstrated that MC was reduced to metallic mercury. Therefore, these reactions are expressed as follows:

$$CH_3 - Hg^+ \longrightarrow Hg^0 + CH_4$$

$$C_2H_5 - Hg^+ \longrightarrow Hg^0 + C_2H_6$$

$$C_6H_5 - Hg^+ \longrightarrow Hg^0 + C_6H_6$$

$$Hg^{++} \longrightarrow Hg^0$$

Enzyme system responsible for the reductive decomposition of organic and inorganic mercury compounds

To elucidate the reaction on an enzymatic basis, enzymes responsible for the reaction were fractionated from cell extract. The cells, grown with PMA, were mechanically disrupted and a cell-free extract was prepared. This was differentially precipitated with ammonium sulfate at a concentration between 0.2 and 0.5 saturation, and dialyzed. The dialyzed preparation was examined for reductive decomposition. TABLE 2 shows the requirements for reductive decomposition of PMA. The reaction mixture contained phosphate buffer, glucose, NAD, thioglycolate, PMA labeled with radioactive mercury, and cell-free extract. The reaction was followed by measuring the radioactivity of mercury remaining in the medium, since the metallic mercury produced quantitatively vaporizes from the medium during aerobic incubation. It was found that glucose, NAD, thioglycolate and extract were all necessary for the reductive decomposition of PMA, and also that NADH could substitute for glucose and NAD. Glucose dehydrogenase was found in the extract, and NADH was produced from NAD as glucose was oxidized by glucose dehydrogenase (Tonomura and Kanzaki, 1969a, 1969b). Cysteine, homocysteine, reduced glutathione, mercaptoethanol, and thioglycolate were all effective as a sulfhydryl compound, but the oxidized form of these compounds was not effective, as shown in TABLE 3 (Tonomura and Kanzaki, 1969b).

TABLE 2

Requirements for the Decomposition of PMA

	PMA decomposed µg/ml/hr
Complete[a]	7.90
-glucose	0.31
-NAD	0.52
-glucose -NAD +NADH	7.75
-thioglycolate	0.26
-extract	0.35

[a]The complete medium contained 0.05 M phosphate buffer (pH 5.8), 0.01 M glucose, 0.06 mM NAD, 1 mM thioglycolate, 0.15 mM ^{203}Hg-PMA, and extract (as protein, 300 µg/ml). The concentration of NADH was 0.6 mM.

TABLE 3

Effect of a Sulfhydryl Compound on the Decomposition of PMA

Additions (10^{-3} M)	Decomposition rate µg/ml/hr
L-Cysteine	3.3
L-Cystine	0.0
DL-Homocysteine	5.2
DL-Homocystine	0.0
Red. Glutathione	5.2
Ox. Glutathione	0.0
Mercaptoethanol	3.6
Thioglycolate	5.5
2,3-Dimercaptopropanol	0.0

Cell-free extract fractionated by ammonium sulfate-precipitation was subjected to column chromatography on sephadex G-150. Figure 3 shows the elution pattern of glucose dehydrogenase, metallic mercury-releasing enzyme (MMR-Enzyme), cytochrome c-1 and c-2 on sephadex G-150 column chromatography (Furukawa and Tonomura, 1971a). The fraction of MMR-Enzyme obtained was further purified by chromatography on DEAE sephadex. Figure 4 shows an elution pattern of MMR-Enzyme on this column. MMR-Enzyme was separated into two fractions, I and II. When MMR-Enzyme I was subjected to re-chromatography on DEAE sephadex, it still released the enzyme II. At the present time nothing is known about the relationship between MMR-Enzyme I and MMR-Enzyme II. MMR-Enzyme I seemed to have a homogeneity on Disc electrophoresis. It was found that the specific activity of MMR-Enzyme I was about 2.7 times greater than that of MMR-Enzyme II. Figure 5 shows the absorption spectrum of MMR-Enzyme I, indicating it is a flavoprotein. The extinction ratio at 277 to 460 nm was 10.3 and the ratio at 365 to 460 nm was 1.00. A flavin compound was extracted from the enzyme, and subjected to thin-layer chromatography. It was found that a flavin component of the enzyme was FAD. This was confirmed by the fact that the component was able to show the activity of a co-enzyme of d-amino acid oxidase (Furukawa and Tonomura, 1971b).

The cytochrome c fraction separated by column chromatography on sephadex G-150 was further purified by chromatography on CM sephadex. Figure 6 shows that cytochrome c-1 was clearly separated from cytochrome c-2. Figure 7 shows that only cytochrome c-1 participated in the decomposition of PMA whereas c-2 did not. Figure 8 shows absorption spectra of cytochrome c-1 and c-2 in a reduced form. The absorption at 547, 519, and 416 nm were observed. The molecular weight of cytochrome c-1 and c-2 was estimated by a gel-filtration method. It showed an approximate mol wt of 26,000 for cytochrome c-1, and 14,000 for cytochrome c-2. In addition, the molecular weight of MMR-Enzyme I was 67,000 and glucose dehydrogenase was 150,000 (Furukawa and Tonomura, 1971a).

As a conclusive experiment, requirements for the reductive decomposition of MMC, EMP, PMA, and MC were examined. TABLE 4 shows that MMC and EMP could be decomposed as well as PMA and MC, although the rate of the decomposition of MMC and EMP was lower than that of PMA and MC. However, the decomposition could not be observed when each one of the components was removed from a complete system. Therefore, it is believed that NAD(P)H, MMR-Enzyme, cytochrome c-1, and the thioglycolate were all necessary for the decomposition, and also that MC was reduced by the same enzyme system as the system involved in the decomposition of organomercury (Furukawa and Tonomura, 1971b).

Figure 9 shows a hypothetical scheme proposed for the decomposition of mercury compounds. NAD(P)H is formed as glucose is oxidized by glucose dehydrogenase. Hydrogen of NAD(P)H is transferred to MMR-Enzyme, and then separated into a proton and an electron. The electron is transferred to the mercury atom through cytochrome c-1. Meanwhile, methylmercury previously reacts with a sulfhydryl compound to form a mercaptide linkage. An electron is supplied to the mercury atom so that the C-Hg linkage is split. This resulted in the formation of methane and metallic mercury. MC is also reduced by the same system as mentioned above. MMR-Enzyme seemed to be a kind of NAD(P)H:cytochrome c oxidoreductase.

It was found that the formation of MMR-Enzyme was inducible. When the organism was grown with PMA or MC, the cell-free extract prepared from such cells could decompose organomercury and reduce mercuric ions. However, when the organism was grown without any mercury compounds, the cell-free extract did not show such an activity. This was supported by the finding that MMR-Enzyme was not present in the cells grown without any mercury compounds, while present in the cells grown

TABLE 4

Requirements for the Decomposition of MMC, EMP, PMA, and MC

	MMC	EMP	PMA	MC
	2×10^{-5}	0.9×10^{-5}	6×10^{-5}	3.7×10^{-5} M
Complete[a] (μg/ml/hr)	1.2	1.9	13.4	16.1
-NADPH[b]	0.2	0.4	0.	0.
-MMR-Enz.	0.	0.	0.	0.
-Cyt. c-1	0.	0.	0.3	0.
-Thioglycolate	0.3	0.	0.2	0.3

[a] Arabinose dehydrogenase was coupled to the decomposition process as a NAD(P)H-generating system. The complete system contained 50 mM phosphate buffer (pH 5.8), 15 μg/ml of MMR-Enzyme for MMC- and EMP-decomposition, or 4.5 μg for PMA- and MC-decomposition, 3 to 4 μg/ml arabinose dehydrogenase, 0.06 mM NADP, 3 mM l-arabinose, 1 μM FAD, 0.01 μM cytochrome c-1, 0.5 mM thioglycolate, 0.5 mM EDTA, and a mercury compound, the concentration of which is indicated in the table. The decomposition of MMC and EMP was determined by measuring methane produced from MMC and ethane from EMP by gas chromatography, respectively. The decomposition of PMA and MC was assayed by measuring radioactivity of mercury remaining in the reaction mixture during aerobic incubation.

[b] An addition of a NADPH-generating system, l-arabinose, arabinose dehydrogenase, and NADP, was omitted.

with a mercury compound. However, cytochrome c-1 was constitutive.

Formation of Methylmercury from Inorganic Mercury Compounds by a Strain of *Clostridium cochlearium*

Jensen and Jernelöv (1969) found that methylmercury is formed when MC is incubated with the bottom sediment of a lake or in an aquarium, and also with a homogenate of rotting fish. Wood et al. (1968) reported an interesting fact that mercury is methylated when MC is incubated with cell-free extract of a methanogenic bacterium and methylcobalamin. Recently, Imura et al. (1971) and also independently Bertilsson et al. (1971) indicated the chemical methylation of mercury by methylcobalamin. The authors confirmed the formation of methylmercury by adding MC to methane fermentation liquid. However, it was not possible to observe the formation of methylmercury in the methane fermentation process. Therefore, a search was made for an organism other than a methanogenic bacterium capable of forming methylmercury. As a result, an anaerobic bacterium was isolated from the soil and identified as a strain of *Clostridium cochlearium* (Yamada and Tonomura, 1971).

The organism was grown in medium containing meat extract, peptone, NaCl, glucose, cysteine, MC, and vitamin B_{12}. The air in the culture flask was replaced by nitrogen before culture. For the assay, methylmercury was extracted from the culture with HCl-acidic benzene after treatment with ammonium hydroxide. The benzene layer was washed with distilled water and then extracted with a cysteine solution. The cysteine layer was washed with benzene and then extracted with HCl-acidic benzene. Methylmercury in the extract was assayed with gas chromatography equipped with an electron capture detector (Nishi and Horimoto, 1968). For the confirmation of methylmercury, the formed methylmercury was decomposed by mercury-resistant *Pseudomonas*, and the evolved methane was assayed by gas chromatography.

Figure 10 shows the effect of vitamin B_{12} on the formation of methylmercury. The addition of vitamin B_{12} was effective. The organism was capable of forming vitamin B_{12} by itself, but vitamin B_{12} added at the beginning of culture actually enhanced the formation. Figure 11 shows the formation of methylmercury during the culture. TABLE 5 shows the formation of methylmercury from various inorganic mercury compounds. It was found that methylmercury was formed from most inorganic mercury compounds but not from HgS.

TABLE 5

Formation of Methylmercury from Various Inorganic Mercury Compounds

Mercury compounds (initial conc. 10 μg/ml)	Methylmercury (μg/ml)
$HgCl_2$	0.121
HgI_2	0.078
HgO	0.068
$Hg(NO_3)_2$	0.068
$Hg(CN)_2$	0.110
$Hg(SCN)_2$	0.052
$Hg(CH_3CO_2)_2$	0.094
HgS	not detected

The relationship between <u>Clostridium cochlearium</u> and other microorganism was examined with reference to the formation of methylmercury. <u>Clostridium cochlearium</u> is usually grown after the air in a culture flask is replaced by nitrogen, and a redox-potential generally shows about +50 mv. If the air is not replaced by nitrogen, the growth cannot be observed. However, when <u>Pseudomonas stutzeri</u> was grown together with <u>C. cochlearium</u>, methylmercury was formed even if the air was not replaced by nitrogen, as shown in TABLE 6. This may be because P. stutzeri used up oxygen in a closed flask so that C. cochlearium

TABLE 6

Formation of Methylmercury by the Combined Culture of Clostridium cochlearium and Pseudomonas stutzeri without replacing Air by Nitrogen

Organism[a]	Gas	Eh, initial, mV, about	Growth	MMC formed, ppm
C. cochlearium	Air	+ 300	-	-
	N_2	+ 50	+	0.06
P. stutzeri	Air	+ 300	+	-
C. cochlearium + P. stutzeri	Air	+ 300	+ +	0.06

[a]The organism(s) was grown in the same culture medium as indicated in the text for 5 days with or without replacing air in a flask by nitrogen.

could be grown. TABLE 7 shows the effect of the combined culture of C. cochlearium and sulfate-reducing bacteria on the formation of methylmercury. In the combined culture a great amount of H_2S was produced from sulfate by sulfate-reducing bacteria, and mercuric ions might be converted to HgS. Methylmercury produced in the combined culture was actually very small compared with that produced in a control. Therefore, it is believed that the environment where H_2S is not produced is more significant for the formation of methylmercury compared with the environment where H_2S is produced.

TABLE 7

Effect of the Combined Culture of *Clostridium cochlearium* and Sulfate-reducing Bacteria on the Formation of Methylmercury

Organism[a]	Gas	Growth	H_2S produced	MMC produced, ppm
C. cochlearium	N_2	+	+	0.121
C. cochlearium + Sulf.-red. Bact.	N_2	+ +	+++	0.003

[a]The organism(s) was grown in the culture medium, shown in the text, supplemented with 0.1% ammonium sulfate for 5 days.

Discussion

The relationship between conversion of a mercury compound by a microorganism and the environment where the organism can be grown is summarized in Fig. 12. The upper direction shows a more aerobic condition, and the lower shows more anaerobic. MMC, EMP, PMA, and MC are decomposed to metallic mercury by *Pseudomonas* strain K 62. Methylmercury is formed by *Clostridium cochlearium*. At lower than -200 mv of a redox-potential, sulfate-reducing bacteria produce a great amount of H_2S from sulfate. Mercuric ions are chemically converted to HgS by the hydrogen sulfide. Since mercury is methylated by cell-free extract of a methanogenic bacterium and methylcobalamin (Wood et al., 1968), a methanogenic bacterium would produce methylmercury. In the process of methane fermentation the authors could not observe the formation of methylmercury. The authors also could not observe the formation of methylmercury in the process of activated sludge. The reductive decomposition of PMA and MC by activated sludge was observed (Yamada et al., 1969; Tonomura, 1971). These observations led us to the viewpoint that mercury undergoes repeated changes in the process of natural circulation.

References

Bertilsson, L., and H. Y. Neujahr: Biochemistry 10, 2805 (1971).

Furukawa, K., T. Suzuki, and K. Tonomura: Agr. Biol. Chem. (Tokyo), 33, 128 (1969).

Furukawa, K., and K. Tonomura: Agr. Biol. Chem. (Tokyo), 35, 604 (1971a).

Furukawa, K., and K. Tonomura: Agr. Biol. Chem. (Tokyo), (1971b), in preparation.

Imura, N., E. Sukegawa, S. Pan, K. Nagao, J. Kim, T. Kwan, and T. Ukita: Science 172, 1248 (1971).

Jensen, S., and A. Jernelöv: Nature 223, 753 (1969).

Nishi, M., and N. Horimoto: Japan Analyst 17, 1247 (1968).

Tonomura, K.: Biochemical and Industrial Aspect of Fermentation. K. Sakaguchi, T. Uemura, and S. Kinoshita, Eds. Kodansha, Tokyo. p. 317 (1971).

Tonomura, K., T. Nakagami, F. Futai, and K. Maeda: J. Ferment. Technol. (Osaka), 46, 506 (1968a).

Tonomura, K., K. Maeda, F. Futai, T. Nakagami, and M. Yamada: Nature 217, 644 (1968b).

Tonomura, K., and F. Kanzaki: Biochim. Biophys. Acta 184, 227 (1969a).

Tonomura, K., and F. Kanzaki: J. Ferment. Technol. (Osaka), 47, 430 (1969b).

Wood, J. M., F. S. Kennedy, and C. G. Rosen: Nature 220, 173 (1968).

Yamada, M., M. Dazai, and K. Tonomura: J. Ferment. Technol. (Osaka), 47, 155 (1969).

Yamada, M., and K. Tonomura: J. Ferment. Technol. (Osaka), (1971), in preparation.

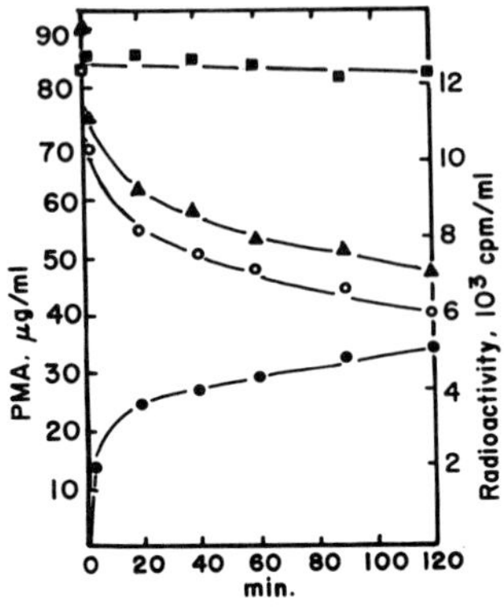

Fig. 1. Uptake of ^{203}Hg-PMA by Pseudomonas K 62. The washed cells (8.8 x 10^8) were incubated in 1 ml of 0.02 M phosphate buffer, pH 7.2, containing 90 ppm PMA. Radioactivity of the whole medium (■), supernatant fluid (o), and cells (●) was assayed. PMA in supernatant fluid (▲) was also assayed colorimetrically.

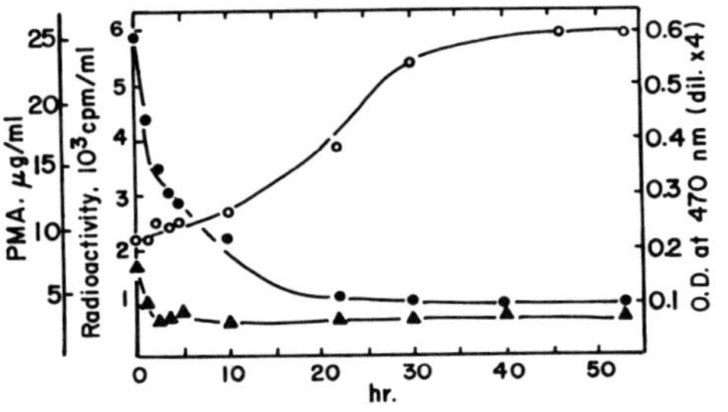

Fig. 2. Vaporization of cell-bound ^{203}Hg-PMA. PMA-bound cells were transferred to fresh culture medium and incubated. Radioactivity in the whole medium (●) and supernatant fluid (▲) was assayed. Growth (o) was measured by reading the optical density at 470 nm.

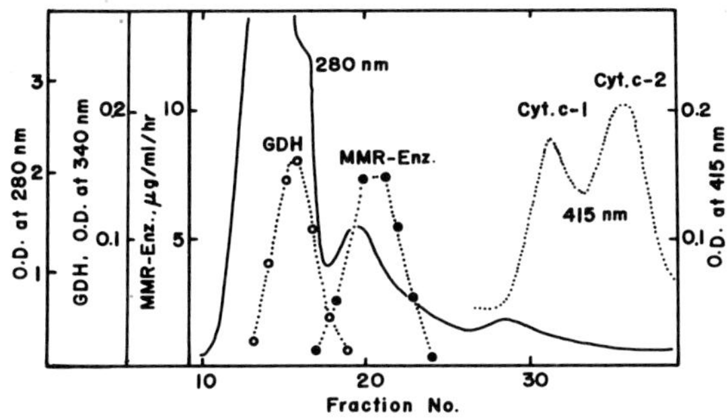

Fig. 3. Elution pattern of glucose dehydrogenase, MMR-Enzyme, cytochrome c-1 and c-2, on a Sephadex G-150 column. The extract fractionated by ammonium sulfate-precipitation was subjected to chromatography on Sephadex G-150. For detail, see Jensen and Jernelöv (1969). (o), glucose dehydrogenase (GDH); (●), MMR-Enzyme. A solid line, absorption at 280 nm. A dotted line, 415 nm.

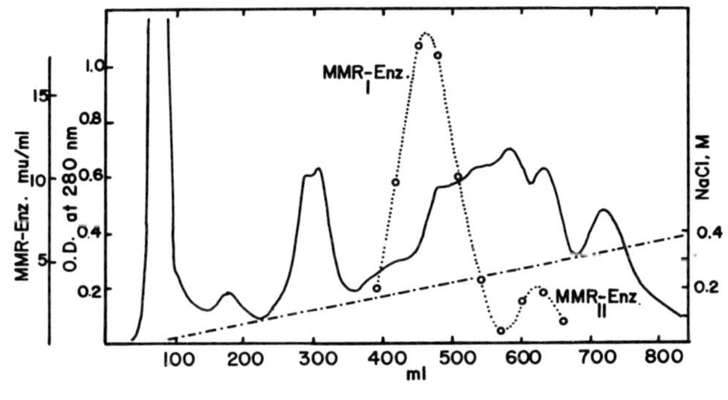

Fig. 4. Chromatogram of MMR-Enzyme on DEAE-Sephadex. The fractions of MMR-Enzyme separated by chromatography on Sephadex G-150 were pooled, concentrated, and applied to a DEAE Sephadex A-50 column (2.5 x 30 cm) which was previously equilibrated with 50 mM phosphate buffer at pH 6.7 containing 0.5 mM EDTA. A linear gradient from 0.0 to 0.5 M NaCl in the same buffer was used to elute MMR-Enzyme.

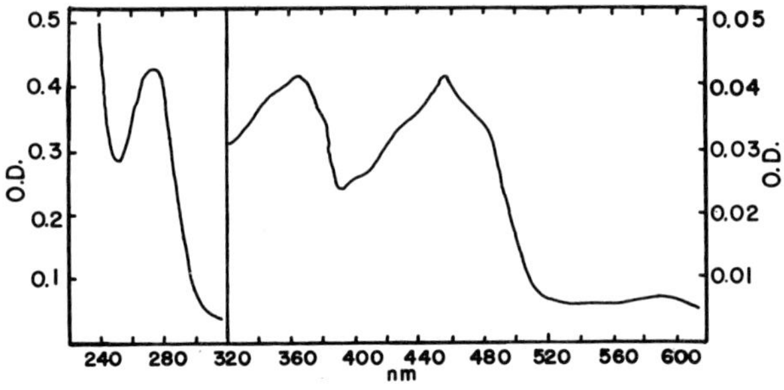

Fig. 5. Absorption spectrum of MMR-Enzyme I.

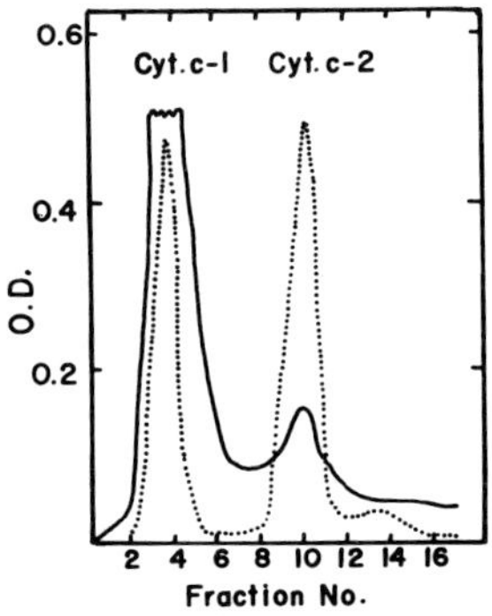

Fig. 6. Fractionation of cytochrome c-1 and c-2 by column chromatography on CM-sephadex. Solid line, absorption at 280 nm. Dotted line, absorption at 415 nm.

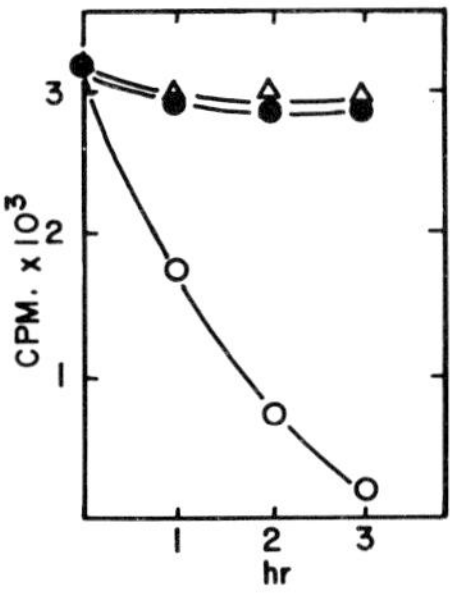

Fig. 7. Effect of cytochrome c-1 and c-2 on the rate of decomposition of PMA. (O), cytochrome c-1; (●), cytochrome c-2; (△), no addition. The reaction mixture contained 50 mM phosphate buffer (pH 5.8), 0.5 mM EDTA, 3 mM glucose, 0.06 mM NAD, 0.5 mM thioglycolate, 0.06 mM ^{203}Hg-PMA, glucose dehydrogenase, and crude MMR-Enzyme. For detail, see Jensen and Jernelöv (1969).

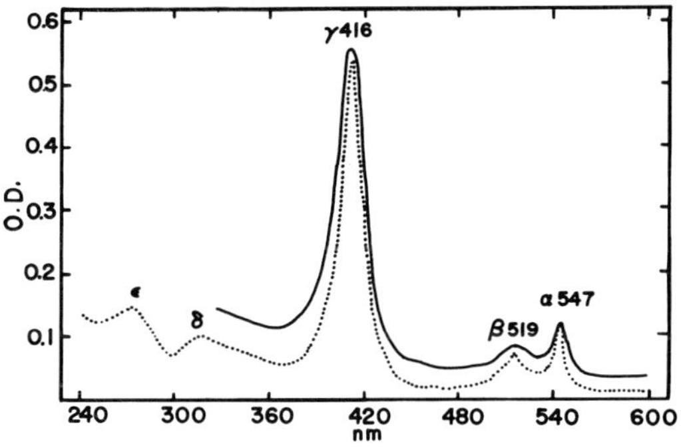

Fig. 8. Absorption spectrum of cytochrome c-1 and c-2 in a reduced form. Solid line, cytochrome c-1. Dotted line, cytochrome c-2.

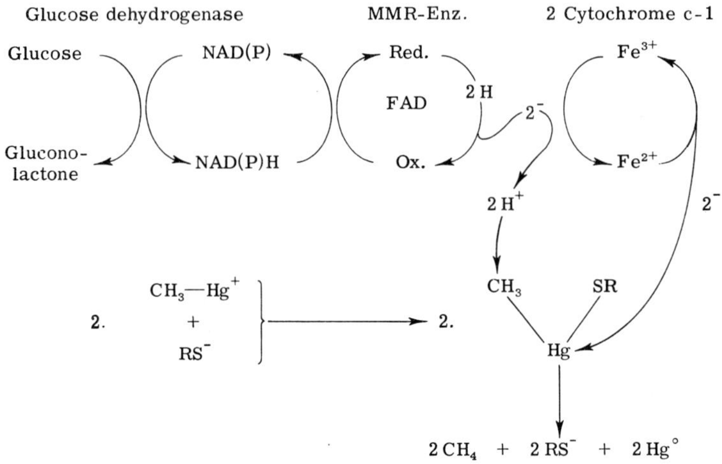

Fig. 9. Hypothetical scheme for decomposition of organomercury. RS^-, a sulfhydryl compound.

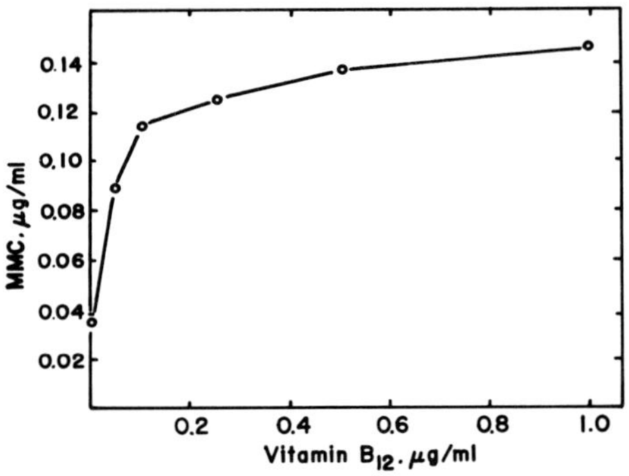

Fig. 10. Effect of vitamin B_{12} on the formation of methylmercury.

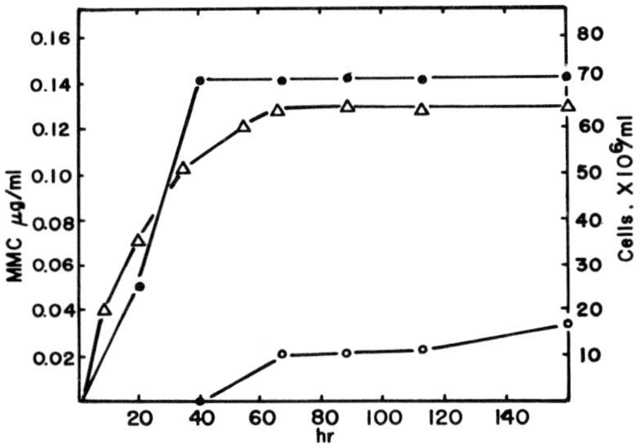

Fig. 11. Formation of methylmercury by Clostridium cochlearium. (●), extracellular MMC; (o), intracellular MMC; (△), total cell count.

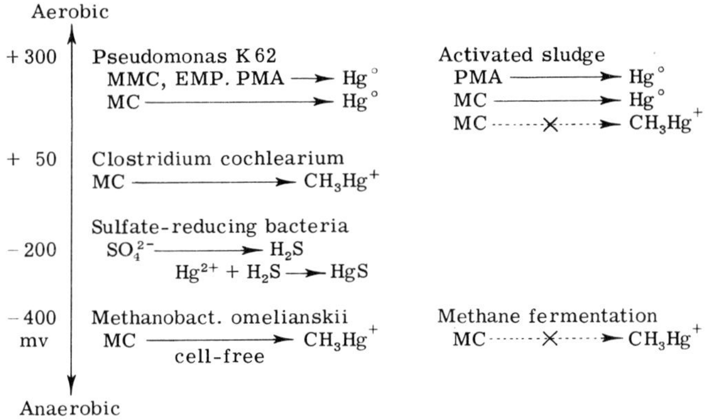

Fig. 12. Microbial conversion of mercury compounds. The reaction shown by a dotted line could not be observed.

RESEARCH ON THE DISTRIBUTION AND ACCUMULATION OF ORGANOMERCURIALS IN ANIMAL BODIES

Tyunosin Ukita

Faculty of Pharmaceutical Sciences
University of Tokyo, Tokyo, Japan

Introduction

Mercury compounds have been widely used as medicines or pesticides for a long time, but it has also been known that they generally are remarkably toxic to living matter. Increasing environmental pollution with mercury caused by unrestricted release of industrial wastes and by use of mercury pesticides has recently been one of the most serious international problems. In Japan, high environmental mercury levels resulting from a wide use of phenylmercuric acetate in rice cultivation has been reported (Ukita et al., 1963; Hoshino et al., 1966; Ukita, 1967), and serious local intoxications with methylmercury have occurred in Minamata City and in the Agano River area. Therefore, the use of organomercurial pesticides has been banned and the release of methyl-, ethylmercury as well as inorganic mercury into rivers and bays is strongly restricted in our country.

In 1964, when we began this series of investigations, it was difficult to formulate general concepts of the relationships between the structures of mercury compounds and their behavior in living organisms, because although a number of reports of the studies concerning the *in vivo* distribution of several types of mercury compounds were encountered at that time (Prickett et al., 1950; Swenson, 1952; Aikawa et al., 1955; Aikawa and Fitz, 1956; Friberg et al., 1957; Friberg, 1959; Swenson et al., 1959a, 1959b; Miller et al., 1960, 1961; Suzuki et al., 1961; Sebe and Itsuno, 1962; Ulvarson, 1962; Suzuki et al., 1963; Bidstrup, 1964; Fukuchi et al., 1964; Gage, 1964; Sadakane, 1964; Suzuki et al., 1964; Urakubo et al., 1967), they had been carried out under different experimental conditions: species of animals, as well as routes of administration and periods of observation. We therefore attempted a comparative study on the distribution of several types of mercury compounds, such as mercuric chloride, phenylmercuric chloride, ethylmercuric chloride, n-butylmercuric chloride, and S-ethylmercuric cysteine in rats under controlled conditions (Takeda et al., 1968a).

Fate of Mercury Compounds in Experimental Animals

First, the excretion of mercury after subcutaneous injection of ^{203}Hg-labeled mercury compounds in the urine and feces of rats was evaluated. As is seen in Fig. 1, it was found that in general alkyl mercury compounds such as ethylmercuric chloride, ethylmercuric cysteine, and butylmercuric chloride were excreted more slowly than the inorganic or phenylmercuric compound.

In Fig. 2, the distribution of mercury in the blood, liver, muscle, and kidney of the rats, to which the mercury compounds were subcutaneously administered, is shown. The figure indicates that in the case of inorganic mercury, mercury which was accumulated in the liver, muscle, and blood rapidly disappeared but in the kidney it was retained up to at least 8 days after administration. The accumulation pattern for phenylmercuric chloride was similar to that of inorganic mercury, but the patterns for ethyl- and butylmercury were quite different, indicating high concentrations and a long-lasting accumulation of mercury in the blood. In such cases, mercury which was accumulated in the liver and muscle did not disappear as rapidly as the cases of inorganic or phenylmercury compounds. Thus, alkylmercury compounds were found to be excreted more slowly and retained in higher concentrations for a longer period of time in the body than were inorganic or phenylmercury compounds.

As alkylmercury compounds showed specifically high accumulation of mercury in the blood, a further investigation on distribution of mercury in the blood of rats was carried out (Takeda et al., 1968b). Figure 3 shows the distribution of mercury in blood constituents of rats which received several types of mercury compounds. The blood was separated into plasma, stroma, and stroma-free hemolysate of erythrocytes. As can be noted in the figure, almost all mercury was found in erythrocytes 1 day after administration, and was retained for many days in the blood of rats receiving alkylmercury compounds. Contrary to this, when inorganic mercury was administered, the majority of mercury in the blood was distributed in the plasma, and 4 days after administration some part of the mercury was transferred inside the erythrocytes. In the case of phenylmercury compounds, the type of time-dependent distribution pattern of mercury in the blood was between the two types just mentioned.

In the next experiment, the uptake *in vitro* of ethylmercury by erythrocytes was investigated. As is shown in Fig. 4, ^{203}Hg-labeled ethylmercury was added to normal rat blood to a concentration of 100 μg Hg/ml and the distribution of mercury in the stroma-free hemolysate was plotted against time. The data indicates that within 30 minutes almost all the mercury added was concentrated inside the membrane of the erythrocytes.

It was thought of interest, therefore, to investigate the fate of ethylmercury in the erythrocytes; that is, in what kind of component, and how the mercury is retained in the erythrocytes. The erythrocytes which consumed the labeled ethylmercury were disrupted and the stroma-free hemolysate was subjected to gel filtration on Sephadex G-25 and G-100. As is shown in Fig. 5, the radioactivity appeared in the fractions which contained hemoglobin. When the hemoglobin was separated into heme and globin, more than 99% of the radioactivity was found in the globin fraction.

When hemoglobin thus obtained was digested with pronase, the radioactivity liberated from the protein was parallel to the increase of ninhydrin coloration (Fig. 6), and as is seen in Fig. 7, paper chromatography tests in two different kinds of solvents indicated that R_f values of the mercury compound liberated by the digestion were identical with S-ethylmercuric cysteine.

The fate of administered ethylmercury in other organs and excreta was also investigated (Takeda and Ukita, 1970). In TABLE 1, the results of an analysis of organic and inorganic mercury compounds in urine and feces of rats which were treated with labeled ethylmercuric chloride are given. The table shows the percentage of administered mercury in urine or feces excreted in each 24 hours on the day after administration. In the urine, the content of inorganic mercury compounds was higher than that of organic mercury compounds on the 8th day, but the content of inorganic mercury compounds in the feces was much lower than that of organic mercury compounds.

In TABLE 2, the results of an analysis of the contents of organic and inorganic mercury compounds in the liver and kidney of rats treated with ethylmercury are given. In the liver, the inorganic mercury compound content was less than 10% of the total mercury compound even 8 days after administration, but in the kidney it was higher and the amount increased with time.

These results show that a part of the ethylmercury was transformed to inorganic mercury and accumulated mainly in the kidney and was excreted in the urine. But it is not obvious whether the transformation occurred in the kidney or the kidney only retained the inorganic mercury produced in the other organs. The organic mercury compounds found in the kidney were isolated and identified as ethylmercury compounds.

These experiments demonstrated that specific properties of alkylmercury compounds enter inside red blood cells, and are combined with and retained on the sulfhydryl group of cysteine residues of hemoglobin. The slow excretion, high concentration, and long-lasting

TABLE 1

Analysis of Mercury Compounds in Excreta of Rats Treated with Ethylmercuric Chloride-^{203}Hg (20 mg Hg/kg, sc)

Compound	Mercury content[a]				
Days:[a]	1	2	4	6	8
Total mercury[b] (%)	0.06 (0.04-0.08)	0.41 (0.18-0.79)	1.65 (1.41-2.02)	1.74 (1.49-1.98)	1.95 (1.88-2.00)
Urine organic mercury[c] (%)	72.7 (68.0-78.8)	11.3 (8.8 -13.8)	44.4 (33.4-62.1)	43.5 (34.0-57.8)	41.9 (32.5-55.8)
Urine inorganic mercury[c] (%)	27.3 (21.2-32.0)	88.7 (86.2-91.2)	55.6 (37.9-66.6)	56.5 (42.2-66.0)	58.1 (44.2-67.5)
Total mercury[b] (%)	0.03 (0.02-0.04)	0.17 (0.02-0.38)	0.97 (0.83-1.25)	1.09 (0.77-1.42)	1.05 (0.76-1.29)
Feces organic mercury[c] (%)	89.0 (84.8-93.7)	81.9 (79.1-87.1)	66.8 (59.5-71.1)	67.4 (65.5-69.2)	65.0 (59.5-70.8)
Feces inorganic mercury[c] (%)	11.0 (6.3 -15.2)	18.1 (12.9-20.9)	33.2 (28.9-40.5)	32.6 (30.8-34.5)	35.0 (29.2-40.5)

[a]Each value represents the mean (range) of samples from three rats. Days = days after administration.

[b]Total mercury excreted in urine of feces in each 24-hr period is expressed in percent of the administered dose of mercury.

[c]Organic and inorganic mercury are expressed in percent of the total mercury in each 24-hr period specimen.

TABLE 2

Analysis of Mercury Compounds in the Organs of Rats Treated with Ethylmercuric Chloride-^{203}Hg (20 mg Hg/kg, sc)

Organ	Days after administration	Mercury content[a]		
		Total mercury (ppm)	Organic mercury[b] (%)	Inorganic mercury[b] (%)
Liver	2	26.2 (23.0-28.8)	93.7 (92.4-94.7)	6.3 (5.3-7.6)
	8	30.1 (26.6-34.6)	93.8 (93.1-94.2)	6.2 (5.8-6.9)
Kidney	2	79.3 (61.7-88.5)	82.5 (80.1-85.7)	17.5 (14.3-19.9)
	8	108.7 (97.5-115.1)	65.9 (59.5-71.8)	34.1 (28.2-40.5)

[a] Each value represents the mean (range) of samples from three rats.

[b] Organic and inorganic mercury are expressed in percent of the total mercury in the organ.

accumulation of alkylmercury compounds in the body could be caused by this specific property of alkylmercury compounds.

The further precise distribution and accumulation of mercury compounds, especially in fetuses of pregnant animals and in the central nervous system, were studied by preparing whole body radioautograms of adult and pregnant mice, rats, as well as monkeys which were treated with ^{203}Hg-labeled mercuric chloride, ethylmercuric chloride, and phenylmercuric chloride (Ukita et al., 1967, 1969; Takahashi et al., 1971).

Figure 8 shows the distribution of radioactivity in pregnant rats 24 hours after receiving radioactive mercuric chloride and ethylmercuric chloride intravenously. Conspicuous radioactivity was found in the placenta and fetal membrane when radioactive mercuric chloride was given. On the other hand, when ethylmercuric chloride was given, a remarkable transfer of mercury through the placenta into fetuses occurred 24 hours after administration, and the fetus indicated high radioactivity in the blood and liver.

In order to see how mercury compounds distribute time-dependently in the animal body, radioautograms were prepared after several time-intervals of the intraperitoneal injection of labeled mercuric chloride and ethylmercuric chloride. In Fig. 9, the radioautograms prepared 60 minutes, 24 hours, and 4 and 8 days after the administration of mercuric chloride to rats are shown. The most widespread distribution of activity in the whole body was observed 24 to 48 hours after administration. Activity rapidly decreased after the 4th day, except in a remarkable instance where it remained in the kidney even 8 days after the injection.

In Fig. 10, similar time-dependent radioautograms were prepared from rats which received radioactive ethylmercuric chloride. In this case, high radioactivity in the blood was maintained up to 8 days after injection. A similar high activity was also observed in the liver, kidney, lung, and spleen; and the conspicuous distributions of mercury into muscle and central nervous system were also specific for this case. Radioactivity did not decrease from the whole body as rapidly as the case of inorganic mercury, even 8 days after injection.

The time-dependent concentration of radioactive mercury in the liver, kidney, and brain of rats which received radioactive ethylmercuric chloride intraperitoneally were analyzed and the results are summarized in TABLE 3.

TABLE 3

Distribution of Radioactive Mercury in Tissues of Rats which received ^{203}Hg-EtHgCl (950 μgHg/kg) by Intraperitoneal Injection

Tissue	μgHg/g wet tissues after						
	60 min	3 hr	6 hr	24 hr	2 day	4 day	8 day
Liver	3.05	4.04	4.43	4.7	5.05	3.7	3.3
Kidney	4.9	5.1	5.9	6.8	10.4	11.8	17.9
Brain	0.07	0.13	0.14	0.14	0.23	0.27	0.31

The concentration of mercury in the liver increased up to the 2nd day after injection, and decreased thereafter; that in the kidney continuously increased up to the 8th day, and that in the brain also showed a slight but steady increase to the 8th day.

In Fig. 11, whole body radioautograms prepared from cynomolgus monkeys 20 hours after intraperitoneal administration of radioactive mercuric chloride, phenylmercuric chloride, and ethylmercuric chloride are shown. In the case of the inorganic mercury compound (the top radioautogram), high radioactivity was observed in the liver, renal cortex, spleen, adrenal cortex, bone marrow, lymph nodes, mucosa of the gastrointestinal tracts, mesentery, peritoneum, and contents in the lumen of the large intestine. On the other hand, activity was low in the skeletal and heart muscles, lung, salivary gland, thymus, and pancreas. Blood and skin showed a moderate radioactivity between the above-mentioned two groups.

The radioautogram from the monkey which received ethylmercuric chloride (bottom radioautogram) revealed a remarkably different distribution of radioactivity, as distinguished from the top one. Thus, the activity was conspicuous in the central nervous system, heart and skeletal muscles, lung, tongue, mucosa of the oral cavity and throat, salivary gland, spleen, and mucosa of digestive tracts, while it was negative in the digestive lumen and secretory juices. In the brain, both radiograms revealed marked activity in choroid plexi, intracranial and extracerebral arteries, and veins and sinuses. The radioautogram prepared for ethylmercury indicated specifically remarkable radioactivity in the cerebral and cerebellar cortices, sub-cortical gray matters, various nuclei of the brain stem, and gray matter of the spinal cord and skeletal muscles but less in the cerebral and spinal white matters. However, the activity was negative in the corpus callosum and anterior commissure. The middle radioautogram from the monkey to which phenylmercuric chloride was given indicated a medium radioactivity in the central nervous system as compared with the top and bottom radiograms. In Fig. 12, the distribution of radioactivity in the heart and adjacent area is shown. The left radioautogram is for ethylmercury and the right one for inorganic mercury. The left radioautogram indicates a high specific radioactivity in the heart muscle, and in this case the activity in the blood was restricted in the corpuscles. Contrary to this, the right radioautogram shows only slight activity in the heart muscle and the activity in blood was in the plasma. In Fig. 13, a time-dependent distribution of the labeled ethylmercury in monkey is given. The top radioautogram was prepared 60 minutes after intravenous injection, and the bottom one 8 days after intraperitoneal injection of the compound. In the earlier period, the concentration of mercury is high in the blood, but after 8 days it increased remarkably in the central nervous system.

Time-dependent concentrations of ^{203}Hg in several organs of the monkey are given in TABLE 4. At 60 minutes after injection, the concentration was highest in the kidney. Those in the liver, lung, heart muscle, and arteries were also remarkable. After 8 days, the concentration in various organs tended to diminish, except in

TABLE 4

Distribution of Radioactive Mercury in Monkey Tissues 60 Minutes after Intravenous Injection and 8 Days after Intraperitoneal Injection of ^{203}Hg-EtHgCl (800 μgHg/kg/100 μCi)

Tissue	60 min		8 day	
	μg Hg/g[a]	%/g[b]	μg Hg/g[a]	%/g[b]
Kidney	6.73	0.81	8.60	0.72
Liver	6.50	0.76	3.04	0.25
Lung	5.10	0.61	0.44	0.037
Myocardium	4.62	0.55	0.81	0.067
Arteries	3.36	0.40	0.28	0.023
Intestinal wall	1.28	0.15	0.77	0.064
Muscle	1.05	0.12	0.41	0.034
Tongue	0.94	0.11	0.62	0.052
Testis	0.37	0.059	0.07	0.059

[a]Concentration of ^{203}Hg-mercury (μg Hg/g wet tissue).

[b]Percent of dose recovered (%/g wet tissue).

the kidney.

In Fig. 14, magnified radioautograms of the brain of a monkey treated with labeled ethylmercuric chloride are shown. Sixty minutes after injection (the top radiogram), radioactivity is mainly restricted in sinuses, choroid plexi, intracranial and extracerebral arteries and veins.

After 8 days, however, conspicuous radioactivity was found in the cerebrum, cerebellum, and medura, especially in their gray matter. Activity was most predominant in the gray matter of occipital lobe.

In TABLE 5, the quantitative analyses of mercury is given for the different regions of the brain of the monkey used for these radioautograms. The accumulation was highest on the 8th day rather than 60 minutes after injection.

TABLE 5

Distribution of Radioactive Mercury in the Brain of Monkeys 60 Minutes after Intravenous Injection and 8 Days after Intraperitoneal Injection of ^{203}Hg-EtHgCl (800 μg Hg/kg/100 μCi)

Portion of brain		μg Hg/g[a] after	
		60 min	8 day
Cerebellum		0.214	1.22
Cerebral cortex	Frontal lobe	0.17	1.39
	Parietal lobe	0.176	1.47
	Occipital lobe	0.213	1.68
Cerebral white matter	Frontal lobe	0.064	1.16
	Parietal lobe	0.067	1.14
	Occipital lobe	0.068	1.10
Midbrain		0.188	1.24
Corpus callosum		0.047	0.96

[a]Concentration of ^{203}Hg-mercury (μg Hg/g wet tissue).

The concentrations on the 8th day in the celebellar and cerebral cortex at occipital lobe were 1.22 and 1.68 μg per 1 g of the frozen block, respectively. The mercury compound which thus accumulated in the brain of the monkey was extracted as dithizonate and the extract was separated into inorganic and organic mercury dithizonate by use of alumina column chromatography (Ishikura and

Yokota, 1963). The results are shown in Fig. 15, indicating that approximately 97% of the mercury compound thus extracted was an ethylmercuric compound.

In the case of the monkey, the ratio of the mercury concentration in the brain to the liver was 0.42, while that for the rat was 0.094, indicating a higher distribution of ethylmercury into the brain in monkeys than in rats.

From these results it became clear that ethylmercury compounds administered to monkeys, time-dependently, accumulated as ethylmercury in the occipital lobe of the brain in especially high concentration. This specific accumulation may possibly be related to the severest, constant, and mostly advanced lesions in the visual cortex and adjacent area of the occipital lobe, as seen in human casualities intoxicated with alkylmercury compounds.

Mercury Transformation Mechanisms in Nature

Several studies concerning the methylation of inorganic mercury by microorganisms, especially by anaerobic bacteria, have been reported mainly by Swedish researchers from 1965 to 1968 (Westöö, 1966; Jensen and Jernelöv, 1967; Westöö, 1967; Westöö and Norén, 1967; Jensen and Jernelöv, 1969). Furthermore, recently in Japan, the findings of 1.86 ppm (average) of methylmercury in the hair of normal residents in Tokyo and vicinity (Ueda and Aoki, 1969), approximately 1 ppm in tuna-fish obtained from the Indian Ocean (Aoki, 1970), and specific high concentrations (37 ppm in average of nine persons) in the hair of fishermen returning from tuna-fishing were reported (Ueda and Kitamura, 1971). Prompted by these reports, we started research on the methylation of inorganic mercury in nature.

Since Wood and his co-workers reported, in 1968, a chemical transmethylation from methylcobalamin to inorganic mercury in the presence of reducing agents, we attempted to determine the exact rate of this reaction *in vitro*. It was found that the transmethylation between methylcobalamin and mercuric chloride proceeded at an unexpectedly rapid rate in a simple mixture of these two components in phosphate buffer of pH 7.0 at 37°C in the dark. Dimethylmercury appeared as an initial product and was converted to methylmercuric chloride with the elapse of time. This reaction is represented as follows:

$$2CH_3\text{-}[Co] + HgCl_2 \longrightarrow (CH_3)_2Hg$$

$$(CH_3)_2Hg + HgCl_2 \longrightarrow 2\ CH_3HgCl$$

In Fig. 16, the amount, in percent of the mercury methylated in this reaction, is plotted against time. When the molar ratio of mercuric chloride to methylcobalamin was larger than one, after 5 hours, approximately 100% of the inorganic mercury was methylated (Imura et al., 1971). This finding suggested the occurrence of a similar reaction in the animal body, where methylcobalamin is believed to be widely distributed. In the next experiment, as shown in TABLE 6, fresh livers from several kinds of animals were homogenized and incubated at 37°C for 5 hours before and after sterilization, with 2 mM concentration of mercuric chloride. On extraction and analysis of methylmercury contained in the extract, by gas chromatography, among the seven kinds of animals tested, the liver of tuna-fish indicated a remarkable increase of methylmercury. Since this increase was observed even with sterilized liver homogenate, the methylation possibly occurred chemically rather than enzymatically. When the liver of tuna-fish was substituted with muscle of the same fish, no such increase of methylmercury was observed (Pan et al., 1971).

These results suggested the chemical methylation of inorganic mercury in the animal body. Further research along this line is now in progress in our laboratory.

TABLE 6

Amount of Methylmercury Determined 5 Hours after Incubation of Mercuric Chloride with Liver-Homogenate of Various Kinds of Animals

Sample	Methylmercury found ($\mu g/g$ of wet liver)					
	Non-sterilized			Sterilized[a]		
	$-HgCl_2$	$+HgCl_2$	Δ	$-HgCl_2$	$+HgCl_2$	Δ
Yellowfin tuna (*Neothunnus albacora*)	0.013	0.155	0.142	0.014	0.121	0.107
Amber jack (*Seriola quinqueradiata*)	0.056	0.076	0.020	0.052	0.064	0.012
Mackerel (*Scomber japonicus*)	0.134	0.174	0.040	0.125	0.132	0.007
Rock fish (*Sebastodes matsubarae*)	0.395	0.380	--	0.377	0.381	0.004
Rainbow trout (*Salmo gairdneri irideuf*)	0.055	0.065	0.010	0.065	0.066	0.001
Ox	0.045	0.066	0.021	0.033	0.065	0.032
Swine	0.031	0.031	0.000	0.028	0.030	0.002
Mouse	0.570	0.557	--			
Yellowfin tuna muscle	0.011	0.012	0.001			

[a] 1 atom. 120°C, 15 min.

References

Aikawa, J. K., and R. H. Fitz: J. Clin. Invest. 25, 775 (1956).

Aikawa, J. K., A. J. Blumberg, and D. A. Catterson: Trop. Soc. Exp. Biol. Med. 89, 204 (1955).

Aoki, H.: Nihon-Eiseigaku-Zasshi 24, 556 (1970).

Bidstrup, P. L.: Toxicity of Mercury and its Compounds. Elsevier, Amsterdam (1964).

Friberg, L.: Amer. Med. Ass. Arch. Ind. Health 20, 42 (1959).

Friberg, L., E. Odeblad, and S. Forssman: Amer. Med. Ass. Arch. Ind. Health 16, 163 (1957).

Fukuchi, H., T. Takeda, A. Kamada, K. Kobayashi, and M. Aoki: Chem. Pharmacol. Bull. 12, 548 (1964).

Gage, J. C.: Brit. J. Ind. Med. 21, 548 (1964).

Hoshino, O., K. Tanzawa, Y. Hasegawa, and T. Ukita: Eiseikagaku (J. Hyg. Chem.) 12, 90 (1966).

Imura, N., E. Sukegawa, S. K. Pan, K. Nagao, J. Y. Kim, T. Kwan, and T. Ukita: Science, 172, 1248 (1971).

Ishikura, S., and K. Yokota: Chem. Pharmacol. Bull. (Tokyo) 11, 939 (1963).

Jensen, S., and A. Jernelöv: Nordforsk Biocidinformation 10, 4 (1967).

Jensen, S., and A. Jernelöv: Nature 223, 753 (1969).

Miller, V. L., P. A. Klavano, and E. Csonka: Toxicol. Appl. Pharmacol. 2, 344 (1960).

Miller, V. L., P. A. Klavano, A. C. Jestad, and E. Csonka: Toxicol. Appl. Pharmacol. 3, 459 (1961).

Pan, S. K., N. Imura, and T. Ukita: unpublished data (1971).

Prickett, C. S., E. P. Lang, and F. M. Kunze: Proc. Soc. Exp. Biol. Med. 73, 585 (1950).

Sadakane, T.: Sangyo-Igaku 6, 495 (1964).

Sebe, K., and Y. Itsuno: Nisshin-Igaku 49, 607 (1962).

Suzuki, T., T. Miyama, and H. Katsunuma: Nisshin-Igaku 48, 716 (1961).

Suzuki, T., T. Miyama, and H. Katsunuma: Japan J. Exp. Med. 33, 277 (1963).
Suzuki, T., T. Miyama, and H. Katsunuma: Japan J. Exp. Med. 34, 211 (1964).
Swenson, A.: Acta Med. Scand. 143, 365 (1952).
Swenson, A., K.-D. Lundgren, and O. Lindstrom: Amer. Med. Ass. Arch. Ind. Health 20, 432 (1959a).
Swenson, A., K.-D. Lundgren, and O. Lindstrom: Amer. Med. Ass. Arch. Ind. Health 20, 467 (1959b).
Takahashi, T., T. Kimura, Y. Sato, H. Shiraki, and T. Ukita: Eiseikagaku (J. Hyg. Chem.) 17, 93 (1971).
Takeda, T., and T. Ukita: Toxicol. Appl. Pharmacol. 17, 181 (1970).
Takeda, Y., T. Kunugi, O. Hoshino, and T. Ukita: Toxicol. Appl. Pharmacol. 13, 156 (1968a).
Takeda, Y., T. Kunugi, T. Terao, and T. Ukita: Toxicol. Appl. Pharmacol. 13, 165 (1968b).
Ueda, K., and H. Aoki: Keio-Igaku 46, 1 (1969).
Ueda, K., and M. Kitamura: personal communication (1971).
Ukita, T.: Proc. 17th Ann. Meet. Japan Med. Soc. 1, 823 (1967).
Ukita, T., O. Hoshino, and K. Tanzawa: Eiseikagaku (J. Hyg. Chem.) 9, 138 (1963).
Ukita, T., Y. Takeda, Y. Sato, and T. Takahashi: Radioisotopes 16, 31 (1967).
Ukita, T., Y. Takeda, T. Takahashi, M. Yoshikawa, Y. Sato, and H. Shiraki: Proc. 1st Symp. Drug Metabolism and Action, Pharmaceut. Soc. Japan, Tokyo, p. 32 (1969).
Ulvarson, U.: Int. Arch. Gewerbepathol. Gewerbehyg. 19, 412 (1962).
Urakubo, G., Y. Kido, and A. Hasegawa: Eiseikagaku (J. Hyg. Chem.) 13, 290 (1967).
Westöö, G.: Acta Chem. Scand. 20, 2131 (1966).
Westöö, G.: Acta Chem. Scand. 21, 1790 (1967).
Westöö, G., and K. Norén: Var Foda 19, 135 (1967).
Wood, J. M., F. Scott Kennedy, and C. G. Rosen: Nature 220, 173 (1968).

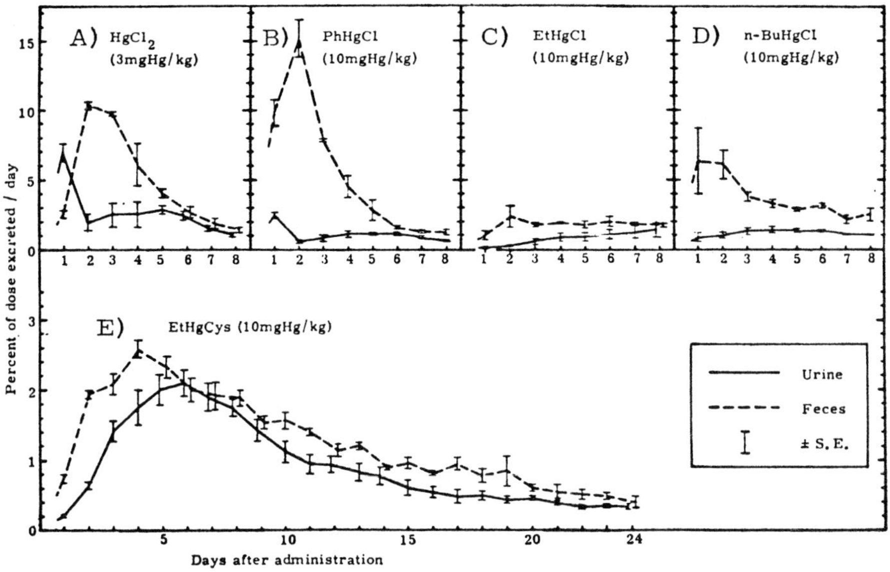

Fig. 1. Excretion of mercury after subcutaneous injection. Data for each point represent the mean of samples from three rats, and bars represent standard errors.

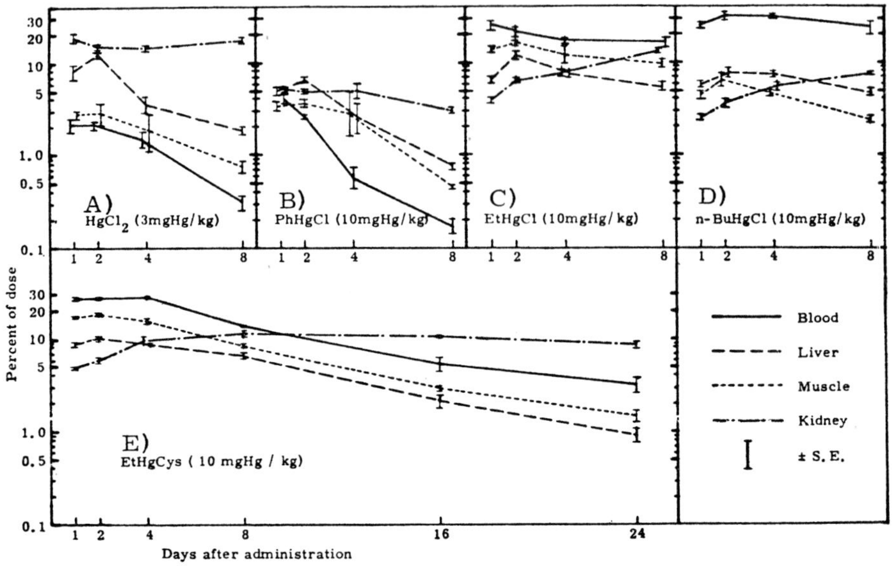

Fig. 2. Distribution of mercury after subcutaneous injection. Data for each point represent the mean of sample from three rats, and bars represent standard errors.

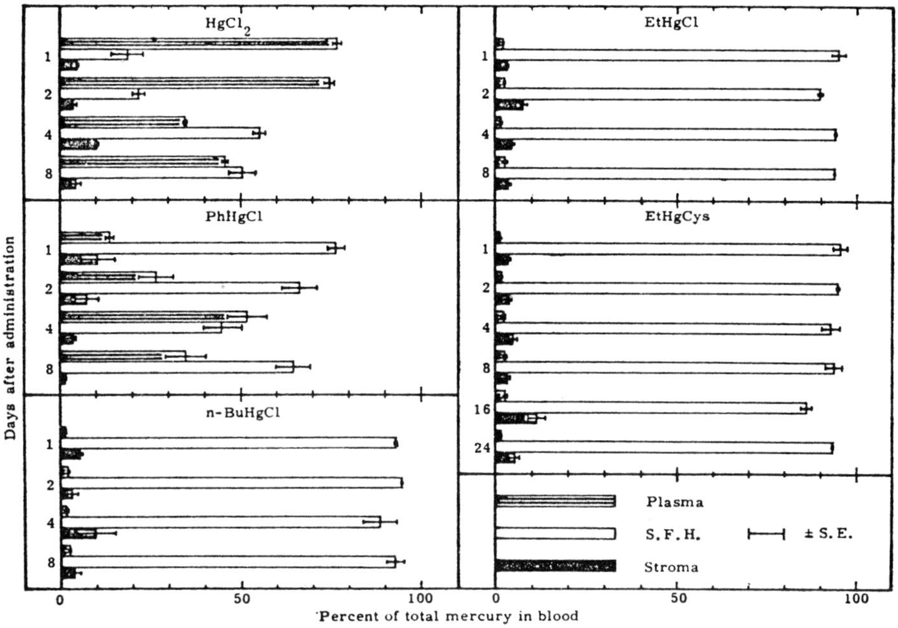

Fig. 3. Distribution of mercury in the blood constituents. Data for each point represent the mean of samples from three rats, and bars represent standard errors. S.F.H. means the stroma-free hemolysate of erythrocytes.

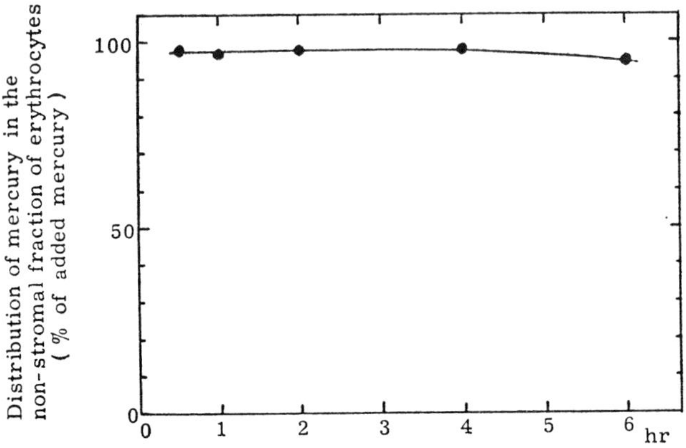

Fig. 4. Distribution of mercury in the nonstromal fraction of erythrocytes after addition of ethylmercuric chloride to the blood of the untreated rat.

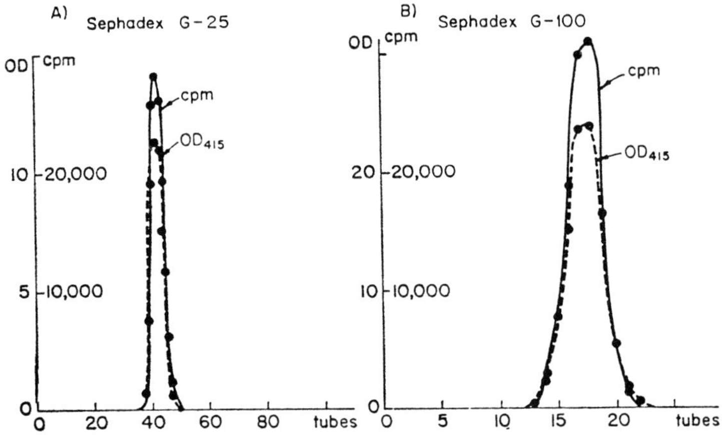

Fig. 5. Gel filtration of the stroma-free hemolysate of erythrocytes through Sephadex. (A) Column size, 2.8 x 72 cm; flow rate, 45 ml/hr, 4.3 ml/fraction. (B) Column size, 1.8 x 22 cm; flow rate, 1.5 ml/hr, 2.5 ml/fraction.

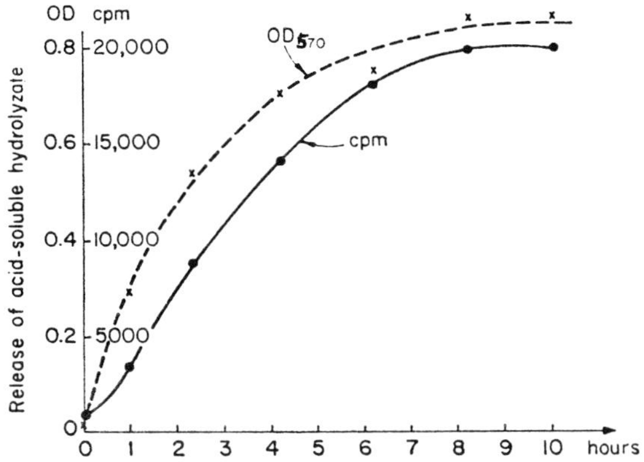

Fig. 6. Digestion with pronase of the hemoglobin of EMC-treated rats after blocking the free SH-groups.

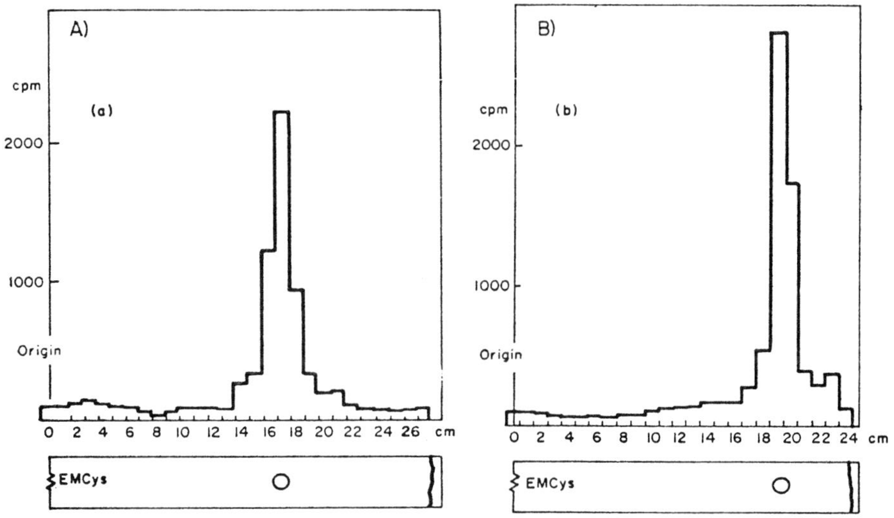

Fig. 7. Paper chromatograms of the mercury compound obtained by the enzymatic digestion of the SH-blocked hemoglobin. EMCys: S-Ethylmercuric cysteine. (A) Solvent: n-butanol-acetic acid-H_2O (4:1:5). (B) Solvent: i-propanol-conc. NH_4OH-H_2O (7:1:2).

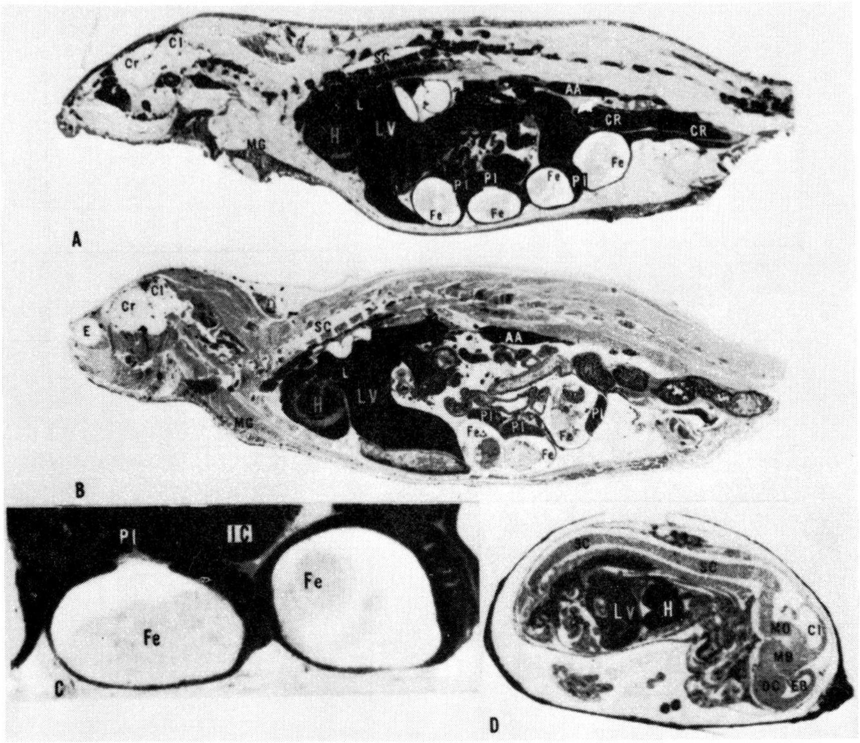

Fig. 8. Radioautograms showing distribution of radioactivity in pregnant rats 24 hours after intravenous injection of mercuric compounds (dose; 950 μg Hg/kg/ 79 μCi).

A. ^{203}Hg-mercuric chloride. B. ^{203}Hg-ethylmercuric chloride. C. Magnified radioautogram of two fetuses in A. D. Magnified radioautogram of one fetus isolated from the pregnant mother rat in B.

Abbreviations: AA, abdominal aorta; Cl, cerebellum; Cr, cerebrum; CR, contents of rectum; DC, diencephalon; E, eyeball; EB, endbrain; Fe, fetus; H, heart; IC, intestinal contents; L, lung; Lv, liver; MB, midbrain; MG, mamillary gland; MO, medulla oblongata; Pl, placenta; SC, spinal cord.

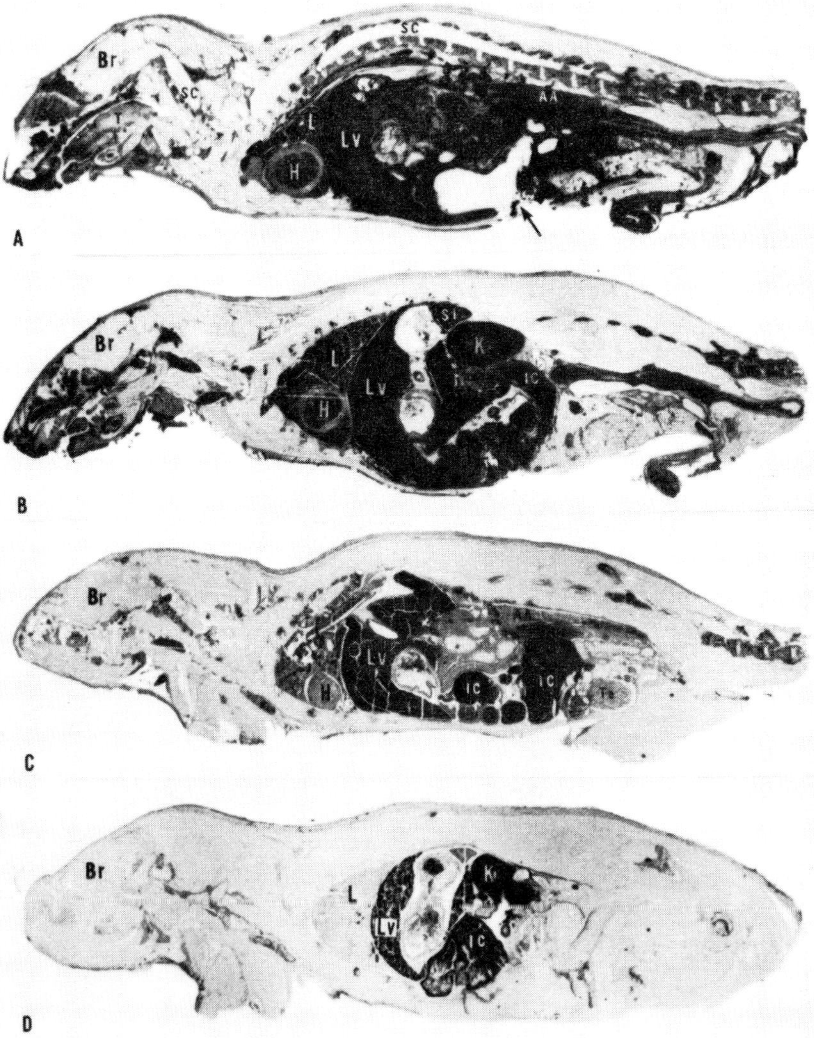

Fig. 9. Radioautograms showing distribution of radioactivity in rats after intraperitoneal injection of ^{203}Hg-mercuric chloride (dose; 950 μg Hg/kg/79 μCi).

After A, 60 min; B, 24 hr; C, 4 days; and D, 8 days, respectively.

Abbreviations: AA, abdominal aorta; Br, brain; H, heart; IC, intestinal contents; K, kidney; L, lung; Lv, liver; SC, spinal cord; Sl, spleen; T, tongue; Te, testis; Arrow, site of injection.

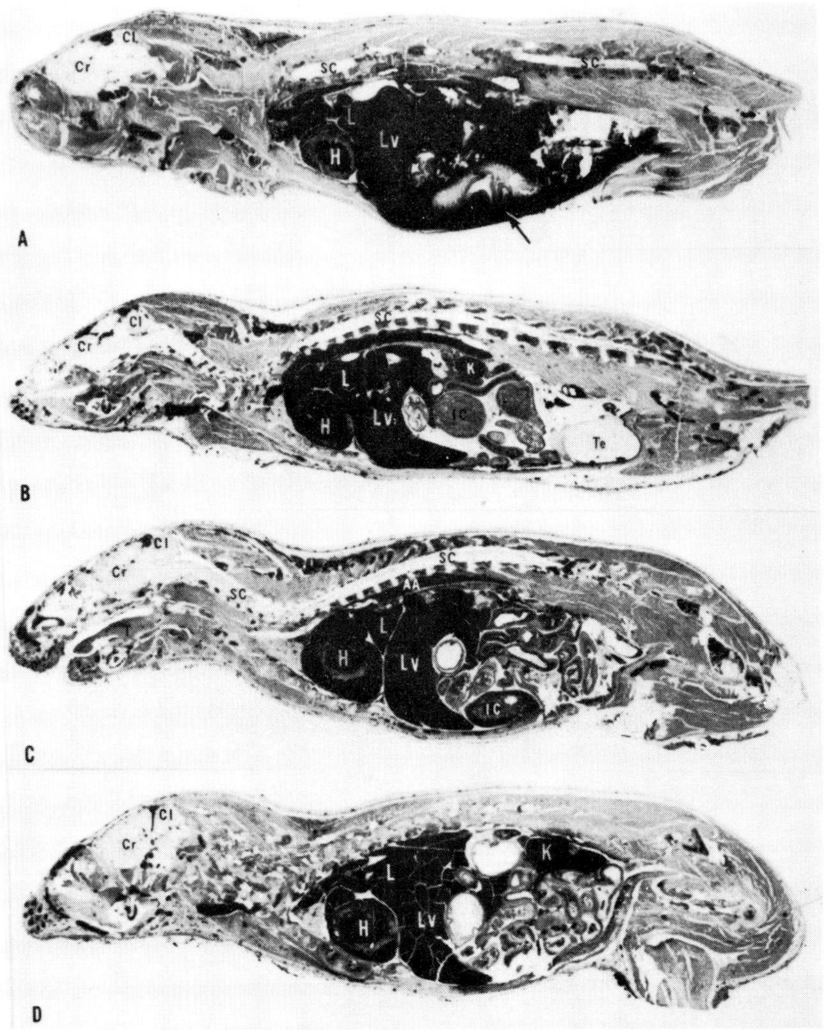

Fig. 10. Radioautograms showing distribution of radioactivity in rats after intraperitoneal injection of ^{203}Hg-ethylmercuric chloride (dose; 950 μg Hg/kg/ 79 μCi).

After A, 60 min; B, 24 hr; C, 4 days; and D, 8 days, respectively.

Abbreviations: AA, abdominal aorta; Cl, cerebellum; Cr, cerebrum; H, heart; IC, intestinal contents; K, kidney; L, lung; Lv, liver; SC, spinal cord; T, tongue; Te, testis; Arrow, site of injection.

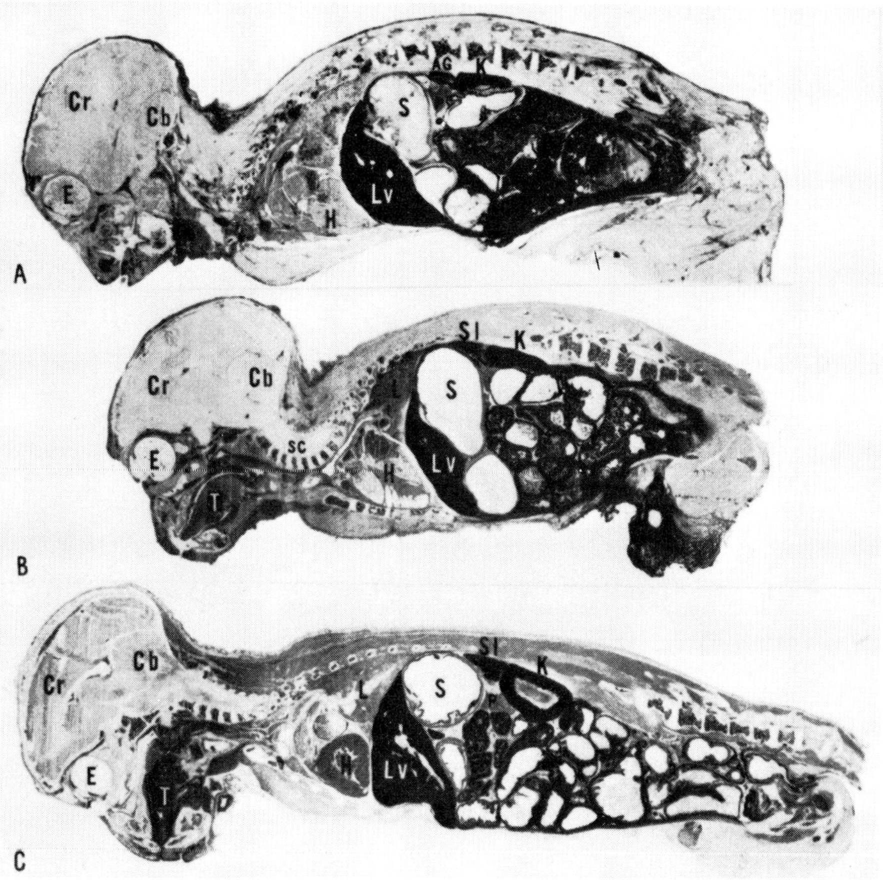

Fig. 11. Radioautograms showing distribution of radioactivity in monkeys 20 hours after intraperitoneal injection of (A) ^{203}Hg-mercuric chloride (800 μg Hg/kg/96 μCi); (B) ^{203}Hg-phenylmercuric acetate (5.1 mg Hg/kg/30.4 μCi); and (C) ^{203}Hg-ethylmercuric chloride (800 μg Hg/kg/96 μCi).

Abbreviations: Cb, cerebellum; Cr, cerebrum; E, eyeball; H, heart; L, lung; Lv, liver; K, kidney; P, pancreas; S, stomach; Sl, spleen; T, tongue; Ag, adrenal gland; SC, spinal cord.

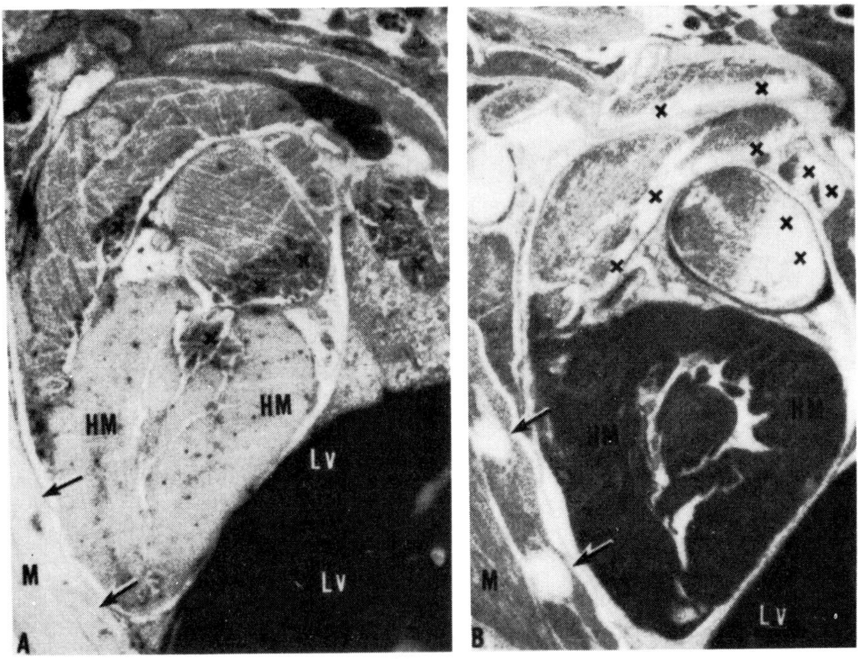

Fig. 12. Magnified heart and adjacent area of radioautograms given in Fig. 11.

A. ^{203}Hg-mercuric chloride was administered.
B. ^{203}Hg-ethylmercuric chloride was administered.

Abbreviations: HM, heart muscle; Lv, liver; M, muscle; Arrows, ribs; Crosses, plasma clots in lumen of aorta.

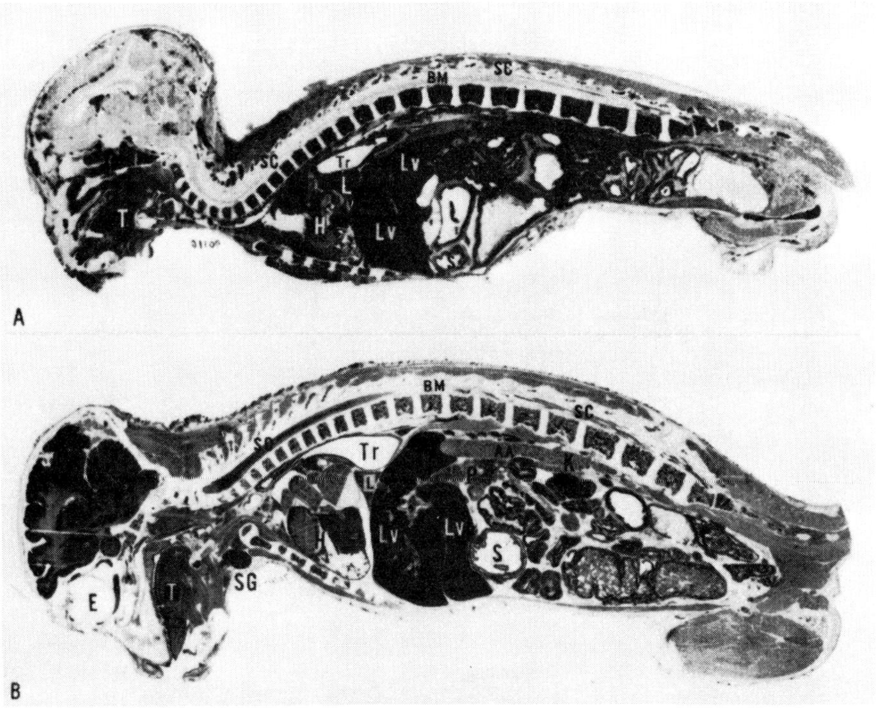

Fig. 13. Radioautograms showing distribution of radioactivity in (A) monkey (body weight 0.9 kg) 60 minutes after intravenous injection of ^{203}Hg-ethylmercuric chloride (dose; 800 μg Hg/kg/100 μCi); and (B) in monkey (body weight 1.2 kg) 8 days after intraperitoneal injection of ^{203}Hg-ethylmercuric chloride (dose; 800 μg Hg/kg/100 μCi).

Abbreviations: BM, bone marrow; H, heart; L, lung; Lv, liver; S, stomach; SC, spinal cord; T, tongue; Tr, trachea; AA, abdominal aorta; K, kidney; P, pancreas; SG, salivary gland.

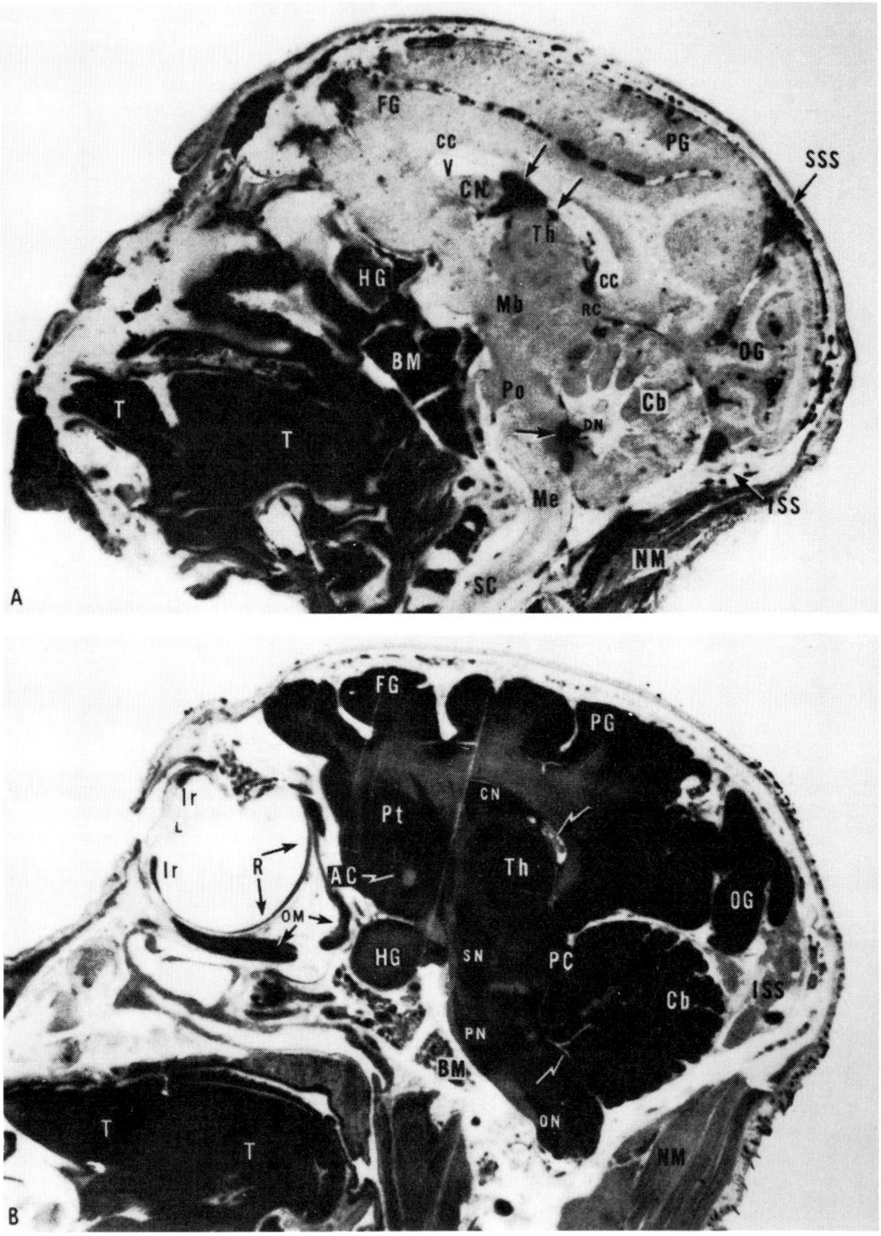

Fig. 14. Caption on following page

Fig. 14. Magnified radioautograms of brain and adjacent areas as in Fig. 13A (A) and Fig. 13B (B).

Abbreviations: AC, anterior commissure; BM, bone marrow; Cb, cerebellum; CC, corpus callosum; CN, caudate nucleus; DN, dentate nucleus; FG, frontal gyrus; HG, hypophyseal gland; Ir, iris; ISS, inferior saggital sinus; L, lense; Mb, midbrain; Me, medulla oblongata; NM, neck muscle; OG, occipital gyrus; OM, ocular muscle; ON, olivary nucleus; PC, posterior colliqulus; PG, parietal gyrus; PN, pons nuclei; Po, pons; Pt, putamen; R, retina; RC, rostral colliqulus; SC, spinal cord; SN, substantia nigra; SSS, superior saggital sinus; T, tongue; Th, thalamus; V, ventricle; Arrows, choroid plexi.

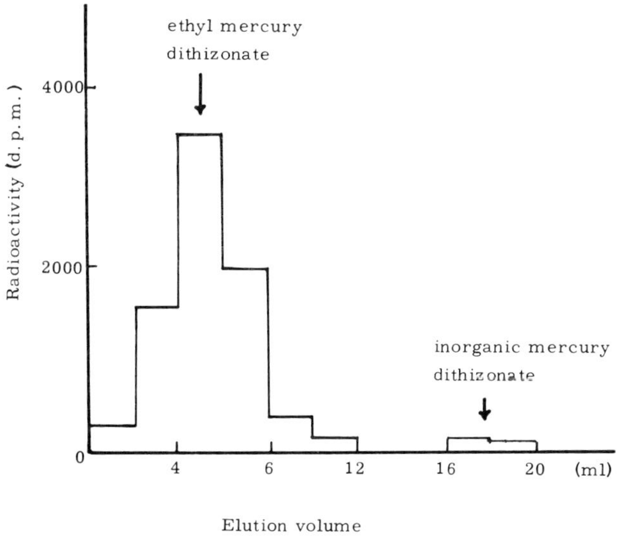

Fig. 15. Chromatographic separation of organic and inorganic mercury dithizonates from the extract of the brain of monkey 8 days after intraperitoneal injection of ^{203}Hg-ethylmercuric chloride.

The column was successively eluted with 10 ml of CCl_4, 5 ml of CCl_4-$CHCl_3$ (95:5), and 10 ml of CCl_4-$CHCl_3$ (1:1).

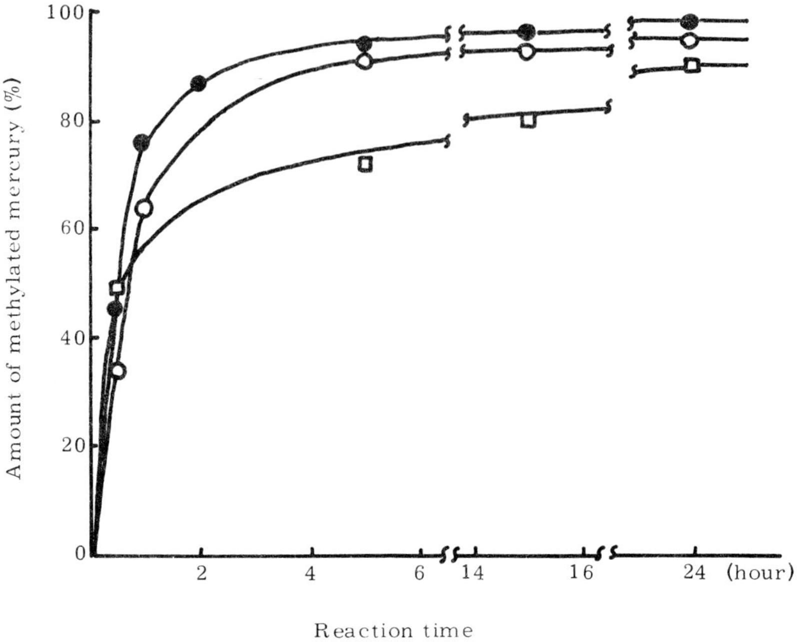

Fig. 16. Formation of methylated mercury from mercuric chloride in the presence of methylcobalamin.

Reaction condition: methylcobalamin (0.1 μmole) was treated with 0.05 (□), 0.1 (○), and 0.5 (●) mole of mercuric chloride at 37°C in 1 ml of 0.2 M potassium phosphate buffer (pH 7.0) in the dark.

HUMAN TOXICOLOGY OF MERCURY

Leonard J. Goldwater, M.D.

Duke University Medical Center
Durham, North Carolina

Introduction

The first definite allusion to mercury is found in the writings of Aristotle who referred to it as liquid silver. This was in the fourth century B.C. Hippocrates, the leading physician of ancient Greece (known as the Father of Medicine), used mercury sulfide in some of his prescriptions for ointments, thus giving evidence that he did not consider it poisonous when used externally. On the other hand, Galen in the second century A.D. declared that all forms of mercury were so toxic that they should never be used as medicines. This view was widely adopted for more than 1500 years but was not fully accepted by the Arab physicians of medieval times (Goldwater, in press).

Toxic manifestations of both mercury and cinnabar (mercury sulfide) were described in the first century A.D. by Dioscorides, who also proposed a theory as to the mechanism of action of the metal. He believed that when quicksilver was swallowed it ate through the guts by reason of its weight (Dioscorides, 1934). Rhazes, one of the greatest of the Islamic physicians, knew better. As a result of an experiment on an ape, he held that ingested mercury would pass through the bowel unchanged and would cause only minor discomfort (Iskandar, 1959).

From the time of Dioscorides up to the present many attempts have been made to explain the basis or nature of mercury toxicity (Goldwater, in press). Currently it is generally believed that the essential mechanism is related to the strong affinity between mercury compounds and sulfhydryl (S-H) groups in biological materials, particularly in enzyme systems (Hughes, 1957; Passow et al., 1961).

Much confusion is found in the literature on mercury poisoning as a result of the failure of many writers to understand some of the basic principles of toxicology. Most important, perhaps, is the fact that any substance in overwhelming doses can cause serious or even fatal poisoning. This applies to such necessities as oxygen, sodium chloride, and water. At the same time, such potent agents as curare and strychnine have found important uses in medicine. The same applies to mercury and its compounds: they can kill and they can cure. The major controlling factors are: (a) The type of compound (chemical form), (b) size of the dose, (c) route of absorption, (d) duration of exposure, (e) synergism or antagonism of other agents, and (f) host factors (biochemical individuality).

Toxicologists who work with mammalians seem to have been slow in giving adequate recognition to an important concept that has been known to plant physiologists for at least a decade: the biphasic nature of the dose-response curve (Fig. 1). This means that very high and very low concentrations of a substance are incompatible with optimal growth and even with survival while between the extremes is a range in which growth will occur. Major applications of this principle are found in the field of micronutrients and at least theoretically in toxicology. The latter may provide an explanation of why mercury and other heavy metals suffer from a bad reputation.

Recent developments in microanalysis have made it possible to determine, with a reasonable degree of accuracy, microgram and even nanogram quantities of mercury whereas formerly, sensitivities were limited to milligrams or major fractions thereof. This means that detectable amounts of mercury in biological materials (urine, blood, tissues, plants) fell on the upper part of the dose-response curve; that is, levels which were harmful or even lethal. Smaller amounts would not be found and hence the analyses would be reported as "zero" or "absent." Detectable amounts would be associated with adverse effects and, therefore, any amount would be assumed to be dangerous. Emphasis on looking for bad effects rather than any effects may also have resulted in distorted ideas about mercury. The possibility that mercury may have essential functions in living systems is slowly gaining acceptance (Goldwater, in press).

Specific Toxicological Effects

Metallic mercury - In its liquid form, metallic mercury can enter the body by swallowing, through the skin, or by injection. For many years oral doses of quicksilver were used in treating a variety of diseases: repeated administration of a few grams for everything from Apoplexy to Worms and amounts up to a kilogram in a single dose for intestinal obstruction. There are conflicting reports in the literature as to whether these uses resulted in more benefit than harm. Starting around the year 1500 and for about 450 years thereafter, mercury mixed with a fatty base and rubbed into the skin was one of the most important measures in the treatment of syphilis. There is considerable evidence that this did more harm than good; ulcerations in the mouth, loss of teeth, and generally impaired health often were experienced by the victims of the treatment. Metallic mercury has been used as a suicidal (Vedrinne et al., 1967) and homicidal agent (Kockel, 1928). The use of mercury-weighted stomach tubes has resulted in accidents when ruptures have released the metal into diseased bowels (Hanselman and Meyer, 1962) and, through aspiration, into the lungs (Zimmerman, 1969). Mercury used as a seal in the apparatus employed for sampling blood gases occasionally escapes into the heart or a blood vessel thus producing multiple emboli (Buxton et al., 1965). This is not always followed by as dire effects as might be expected.

Mercury vapor - Because of its volatility at ordinary temperatures, spilled mercury always constitutes a potential hazard. Accidental release of hot mercury vapors can cause an acute inflammation of the lungs (pneumonitis) with necrotic changes and fatal results (Hallee, 1969). Far more common are exposures to low levels of vapor over periods of months or years, resulting at times in chronic poisoning. The classical symptoms in such cases are gingivitis, tremors, and personality disturbances (erethism). Kidney injury, manifested by proteinuria, is occasionally seen and there is almost invariably an elevation of mercury levels in blood and urine above the "normal" concentrations of 5.0 μg/100 ml and 25 μg/l, respectively. Complete recovery following termination of exposure is the rule.

Inorganic salts - Contributing to mercury's bad reputation is the erstwhile use of corrosive sublimate (mercuric chloride) for purposes of suicide. A gram or two taken by mouth will cause severe vomiting and diarrhea, bleeding from the intestinal tract, and profound injury to the kidneys. The latter can proceed to complete kidney failure and death in uremia. Cases of this type are now rarely seen. Mercurous chloride (calomel) is less soluble than the mercuric salt and hence less harmful if swallowed. It has, however, been associated with acrodynia (pink disease) in children to whom it was given as a teething powder (Warkany and Hubbard, 1948). Ammoniated mercuric chloride and mercuric oxides have been used in ointments, probably doing more harm than good. Mercury nitrate used in making felt was an important occupational health hazard but this compound has now been replaced in most countries by less harmful chemicals.

Organic Compounds

Articles still appear in which authors treat all organomercurials as a single class. From a toxicological point of view, nothing could be more misleading; organic compounds of mercury span the entire spectrum from life-saving therapeutic agents to highly lethal chemicals. This emphasizes the importance of being quite specific in any detailed discussion of mercury toxicity. Some chemists maintain that only those compounds in which there is a direct carbon-mercury bond should be called organic compounds of mercury while others include any molecule containing carbon and mercury. Toxicologically, this distinction is of more than academic interest since the nature of the bond can have important effects on toxicological behavior. Organomercurials which are useful as diuretics have been found to be unstable, readily liberating mercuric ion (Hg^{++}), and aryl compounds behave in a similar manner; alkylmercurials tend to be more stable. These differences in stability can explain at least part of the differences in toxicity among the organomercurials.

Arylmercurials - This class of compounds embraces all mercury derivatives of aromatic hydrocarbons including such groups as benzyls, tolyls, naphthyls, and many others, as well as their halogenated and other derivatives. Many hundreds of varieties of arylmercurials have been synthesized but only those containing a phenyl group are of major importance as pesticides and biocides. Phenylmercuric acetate (PMA) is the most widely used of the arylmercurials, but the benzoate, lactate, nitrate, oleate, propionate, and others have found a place in the pesticide field. There is fairly good evidence that for humans the toxicological behavior of the various phenylmercurials is similar (Ladd et al., 1964). Those which have been tested on animals have been found to have LD_{50} values in the same general range.

Extensive studies of human exposures to phenylmercurials conducted at Columbia University (Ladd et al., 1964) led to the following conclusions: (a) For human exposures there appears to be no significant difference in the toxicity of the various phenylmercurials. (b) Phenylmercurials may be absorbed through the intact skin or mucous membranes, but relatively high concentrations must be applied before measurable amounts will be absorbed through these channels. (c) Solutions of phenylmercurials in concentrations in the percent range will regularly cause second degree chemical burns of the skin (blistering), but more dilute solutions are only slightly irritating and skin sensitation rarely occurs. (d) Persons may be occupationally exposed to phenylmercurials in concentrations many times the accepted Threshold Limit Value of 0.1 mg/m^3 in air for many years without showing evidence of poisoning, even though they absorb sufficient of the compounds to cause them to excrete mercury in their urine in the milligram per liter range. (e) Poisoning due to phenylmercurials is extremely rare and chronic occupational poisoning is unknown.

Toxicologists who prefer to use experimental rather than human data point to some studies on rats which resulted in equivocal findings of kidney injury when PMA was fed at a level of 0.5 ppm for 2 years (Fitzhugh et al., 1950). Subsequent observers did not confirm this when the experiment was repeated and carried for 1 year (Wisconsin Alumni Research Foundation, 1964-66).

Alkoxyalkylmercurials - Compared with the aryl and alkyl compounds of mercury, relatively little is known about the toxicity of the alkoxys. Those which are related chemically to the mercurial diuretics are, of course, practically non-toxic (Baer and Beyer, 1966). Limited experimental work has shown that methoxyethylmercuric chloride resembles PMA in its behavior and that it is distinctly less toxic than the alkyl equivalent (Daniel and Gage, 1969).

Alkylmercurials - So great has been the number of articles on alkylmercurials published during the past year or two that everyone who reads even one scientific journal can hardly escape becoming an expert on the subject. The episodes of poisoning at Minamata and Niigata, in Iraq, Pakistan, Guatemala, and the U.S.A. have been described over and over again. First in Japan and Sweden and later in North America extensive studies of the toxicology of methylmercury and other alkyl compounds have been and are being conducted. Comprehensive summaries of the findings have been published and are continuing to appear. From these an attempt will be made to extract the most important features which apply to the effects of alkylmercurials on humans. The following statements are based on information and consensus as of the time of writing (September, 1971) (see Goldwater, in press): (a) Alkyls are distinctly more toxic than aryls, alkoxys, and inorganic compounds of mercury. (b) Fatal occupational poisoning through inhalation and non-occupational poisoning through ingestion has occurred. (c) Alkylmercurials are readily absorbed from the intestinal tract and through the skin. They can pass through the placenta and cause fetal injury. (d) The principal site of injury is the central nervous system and the damage, when it occurs, tends to be permanent. The most frequent disturbances are in vision, hearing, sensation, muscle function, and mentality. (e) Alkylmercurials show affinity for lipid tissues and when taken up are liberated slowly. The "half-life" of retention in man is about 70 days. (f) There is no known effective treatment for alkylmercury poisoning. Prolonged physical therapy may result in some restitution of function.

Methylmercury in fish - Because of the widespread concern over possible harm to humans who eat certain species of fish it seems desirable to assess the present situation. Official positions in regard to the toxicological effects of methylmercury in fish have been adopted in several countries but this does not necessarily mean that sound, scientific data to support these positions are presently available. Scientists have a responsibility to exercise their traditional role of inquiry and to challenge conclusions that appear to be inadequately supported.

There can be little doubt that methylmercury played a significant role in the Minamata tragedy (Kutsuna, ed. 1968), yet it has been claimed by at least one investigator that something in the fish other than the mercury was the responsible agent (Kiyoura, 1963). Has this alternative possibility been thoroughly explored?

It has been shown that the water of Minamata Bay was

polluted by a number of chemicals in addition to organic and inorganic mercury, specifically arsenic, thallium, manganese, selenium, and aldehydes (Hamaguchi et al., 1960). Has the possible synergistic effect of each of these been studied? There may be some significance in the observation that the most severe manifestations of mercurialism have occurred in the past in the felt hat industry where arsenic trioxide was used in addition to mercury nitrate.

Biologically methylated mercury may or may not have the same toxicological properties as the chemically synthesized compound. There is evidence on both sides of this question but there is inadequate information to justify a definite and final conclusion. Marine microorganisms have been shown to be able to detoxify mercury compounds under some conditions (Fitzgerald, 1971).

Widespread publicity has been given to a woman in New York who allegedly suffered from methylmercury poisoning as a result of eating large amounts of swordfish. Less attention has been paid to the doubts expressed by the physician who reported the case and no publicity has been given to his most recent views that ". . . conclusive documentation of the cause of Mrs. N. Y.'s symptomatology will always elude us . . ." and that ". . . there were a number of bits of information which could have solidified the diagnosis but were not present and are still not present" (Herdman, 1971). Certainly there is room for doubt in this case as there is in other situations where, in spite of deficiencies of knowledge, actions are being taken as though everything that needs to be known about the human toxicology of mercury actually is known.

Toxicological Standards

Governmental agencies, particularly those charged with the responsibility of enforcing environmental health regulations, are obliged to have standards to guide them in their actions. Standards mean numbers. Earliest of the environmental standards were the Maximum Allowable

Concentrations developed as guides for the ventilation of factories and now being incorporated into laws in the United States. Tolerances for pesticide residues were first imposed in the U.S.A. in 1938 at which time the permissible level for mercury was given as "zero." This level still remains even though the sensitivity of analytical methods has reached a point where "zero" today must be several orders of magnitude less than it was in 1938. Standards for permissible discharges of mercury compounds gaseous and fluid effluents of factories are presently being developed, based at least partially on the assumption that all such discharges may eventually be converted to methylmercury which in turn may enter food chains. Guidelines or regulations for permissible concentrations of mercury in fish have been established in several countries, one assumption in Sweden and North America being that practically all mercury in fish is in the methyl form. All of these and similar regulations relate principally to the toxicological potentialities of mercury compounds, particularly the methyl forms. All entail a number of assumptions.

The difficulties and complexities of arriving at environmental standards which are scientifically sound create a temptation to simplify. One is tempted, for example, to assume that dose-response relationships are linear functions. One is tempted to extrapolate from a small number of points on a curve. One is tempted to disparage the importance of host factors as exhibited in biochemical individuality. One is tempted to ignore the great capacity of living organism to adapt to stress, including chemical stress. Persons entering the field of environmental toxicology with backgrounds in the physical sciences, or with no scientific background at all, are more likely to yield to these temptations than those with training in the biomedical sciences.

Social sciences, including economics, cannot be excluded from discussions of toxicological standards. How do risks compare with benefits at any given level of environmental exposure? How safe or how threatened does a population feel, particularly when it is bombarded by sensationalistic statements by the news media? What about the sentimental emotionalism of zealous, misguided "friends of the environment"? How does one deal with dogmatic statements declaimed by ill-informed "experts"? The questions are many, the answers few. Therein lies the challenge to science.

References

Baer, J. E., and K. H. Beyer: Ann. Rev. Pharmacol. 6, 261 (1966).

Buxton, J. T., J. C. Hewitt, R. H. Gadsen, and G. B. Bradham: J. Amer. Med. Ass. 193, 573 (1965).

Daniel, J. W., and J. C. Gage: Biochem. J. 111, 20P (1969).

Dioscorides, P.: The Greek herbal. Englished by John Goodyer, A.D. 1655. Edited and first printed, A.D. 1933, by Robert Gunther. Oxford: At the University Press (1934).

Fitzgerald, G.: Personal communication (1971).

Fitzhugh, O. G., A. A. Nelson, E. P. Laug, and F. M. Kunze: Arch. Ind. Hyg. Occup. Med. 2, 433 (1950).

Goldwater, L. J.: History of mercury. Baltimore: York Press (in press).

Hallee, J. T.: Am. Rev. Respirat. Dis. 99, 430 (1969).

Hamaguchi, H., N. Ohta, N. Onuma, and K. Kawasaki: Nippon Kagaku Zasshi 81, 920 (1960).

Hanselman, R. C., and R. H. Meyer: Int. Abstr. Surg. 114, 207 (1962).

Herdman, R.: Personal communication (1971).

Hughes, W. L.: Ann. N.Y. Acad. Sci. 65, 454 (1957).

Iskandar, A. Z.: A study of AR-Razi's medical writings with selected texts and English translation. Unpublished Ph.D. dissertation, University of Oxford, 1959.

Kiyoura, R.: Int. J. Air Water Poll. 7, 459 (1963).

Kockel, H.: Arch. Kriminol. 83, 309 (1928).

Kutsuna, M., Ed.: Study Group of Minamata Disease. Kumamoto, Japan: Kumamoto University (1968).

Ladd, A. C., L. J. Goldwater, and M. B. Jacobs: Arch. Environ. Health 9, 43 (1964).
Passow, H., A. Rothstein, and T. W. Clarkson: Pharmacol. Rev. 13, 185 (1961).
Vedrinne, J., J. J. David, and Ch. Vitani: Bull. Med. Legale Toxicol. Med. 4, 223 (1967).
Warkany, J., and D. M. Hubbard: Lancet 254, 829 (1948).
Wisconsin Alumni Research Foundation: Reports to Great Lakes Biochemical Co., Inc. Unpublished observations (1964-66).
Zimmerman, J. E.: J. Amer. Med. Assoc. 208, 2158 (1969).

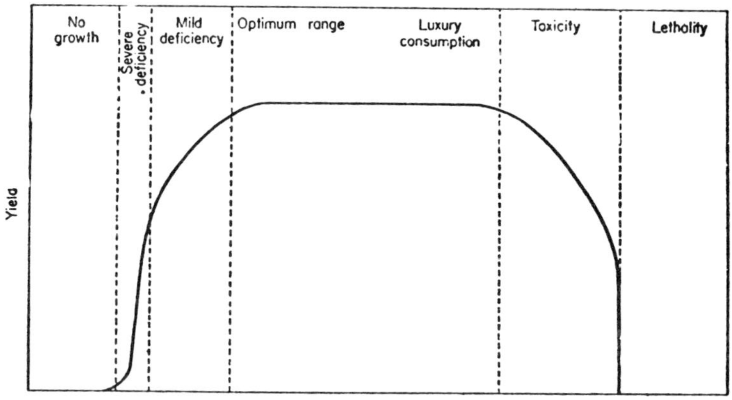

Fig. 1. Idealized diagram of growth of an organism as a function of the concentration of an essential nutrient. (P. F. Smith, Ann. Rev. Plant Physiol. 13, 81, 1962.)

RESIDUE ANALYSIS OF ORGANOMERCURY FUNGICIDES SPRAYED ON RICE PLANTS

Kazuo Fukunaga, Yutaka Tsukano, and Jun Kanazawa

National Institute of Agricultural Sciences
Nishigahara, Kita-ku, Tokyo

Introduction

Rice blast is the most destructive disease of rice in Japan. In attempts to control this disease, a number of foliar fungicides had been tested, but no successful results had been obtained until the early 1950's. Rice crops had, therefore, been seriously damaged when infested severely by the blast fungus.

It was found in 1952 that phenylmercury acetate, which had been used almost exclusively for seed treatment, gave excellent control of blast when applied to rice plant leaves, together with lime.

Because of a food shortage, raising crop yields was of utmost importance in Japan at that time. Potential chronic toxicity of mercury compounds was overwhelmed by the excellent effectiveness of this compound, and it was put to practical use the following year.

In addition, various arylmercury compounds, such as phenylmercury iodide, phenylmercury chloride, and phenylmercury dinaphthylmethanedisulfonate (PMF) were put on the market one after another. These mercury fungicides became widely used throughout the country within a short period, and have played an important role in rice plant protection as essential foliar fungicides, and contributed greatly to maintaining high levels of rice yields.

In addition, these mercury fungicides were also successfully put to use for controlling various diseases of fruit trees and vegetables. The total annual consumption, therefore, increased quickly to 200 tons (metallic mercury) by the late 1950's.

TABLE 1 shows the organomercury fungicide consumptions in Japan. The first five arylmercury compounds were used for rice plant protection. The alkyl- and alkoxyalkylmercury compounds listed were for seed and soil treatments.

TABLE 1

Organomercury Fungicides used for Crop Protection

Compound	Abbreviation
phenylmercury acetate	PMA
phenylmercury chloride	PMC
phenylmercury iodide	PMI
phenylmercury dinaphthylmethanedisulfonate	PMF
phenylmercury p-toluenesulfonamide	PMTS
ethylmercury p-toluenesulfonamide	EMTS
ethylmercury phosphate	EMP
methoxyethylmercury chloride	MEMC
methylmercury chloride	MMC
methylmercury iodide	MMI

Organomercury compounds are, however, far from ideal fungicides. The potential toxicity and possible human poisoning which might result from food contamination with these chemicals is of great concern among chemists and toxicologists as well as administrators.

In 1956, Tomizawa first reported that when ^{203}Hg-labeled phenylmercury acetate was applied to rice plant leaves around the heading stage, the labeled mercury was absorbed by the leaves, translocated in the plants, and finally accumulated in rice grains. The mercury accumulation in rice grains attracted much attention from both scientists and administrators when a symposium was held on the action and fate of mercury fungicides the following year at the annual meeting of the Phytopathological Society of Japan.

In this situation, the Ministry of Agriculture and Forestry organized a committee for the investigation of the mercury residue problem. We present the results of the studies conducted by the committee members, with some additional data obtained later.

Translocation of Mercury in Rice Plants

Figure 1 is a radioautogram presented by Tomizawa in 1956 as the first observation on mercury translocation. As shown in this figure, ^{203}Hg was accumulated in the head of rice when ^{203}Hg-labeled phenylmercury acetate was applied to rice plants at the heading stage. The penetration and translocation of labeled mercury were also observed at all stages of plant growth when the labeled fungicide was sprayed to the plant. An example is shown in Fig. 2 which was obtained 12 days after labeled phenylmercury acetate was sprayed to the seedling. Mercury translocation is shown by the presence of radioactivity in the roots and two newest leaves which developed after treatment. Mercury seems to have been translocated from sprayed leaves to roots and then into newly developed leaves.

Mercury translocation was also studied in relation to blast disease. Accumulation of labeled mercury was observed in tissues surrounding the blast-infected lesion of a rice plant leaf when the roots of the plant were dipped in an aqueous solution of labeled phenylmercury acetate.

Penetration of phenylmercury acetate into rice plant leaves was shown to be affected by the formulation. Thus, the largest amount of the fungicide penetrated when applied as emulsion, while the lowest amount was obtained with a dust formulation.

The penetration and accumulation of organomercury fungicides into rice grains is governed by the type of

TABLE 2

Mercury Residues in Rice Grains from Plants Treated with several ^{203}Hg-labeled Organomercury Compounds[a]

Growth stage of rice plant	Fungicide	Content of mercury (ppm)		
		Husk	Raw rice	Polished rice
Tillering stage (Aug. 5)	PMA	0.219	0.106	0.042
	PMC	0.148	0.102	-
	PMF	0.168	0.106	0.034
	PMTS	0.236	0.102	0.066
	EMTS	0.139	0.068	0.042
	MEMC	0.318	0.202	0.106
Flowering stage (Aug. 31)	PMA	0.482	0.330	0.230
	PMC	1.842	1.420	-
	PMF	0.618	0.634	0.190
	PMTS	0.926	0.710	0.582
	EMTS	1.198	0.690	0.570
	MEMC	0.230	0.182	0.146

[a] From Tomizawa, 1966.

compound, as shown in TABLE 2 (Tomizawa, 1966). Six compounds were sprayed as 50 ppm emulsions. Considerable differences were observed in mercury residue in raw rice when sprayed at the flowering stage. Phenylmercury chloride gave the highest level. This table also shows that the treatment at the flowering stage gave higher mercury levels in rice than the same treatment at the tillering stage. This difference was presumed to be attributable to the direct deposition onto the head and the active translocation of plant materials from leaves to grains.

Also by tracer technique, Araki et al. (1965) found that about 10% of the phenylmercury acetate sprayed deposited on rice plants at the booting stage, about half the deposit penetrated into plant tissues, and about 5% of the penetrated mercury was accumulated in rice grains.

Mercury in Plant Tissues

As evidence accumulated on penetration and translocation of organomercury fungicides, it became necessary to reveal in what form the mercury existed in rice plants.

For this purpose, rice plant leaves treated with ^{203}Hg-labeled phenylmercury acetate were extracted with a buffer solution and the extract was examined by electrophoresis and paper chromatography. The labeled mercury reacted with proteins, which suggested association of mercury with proteins in plant tissues. Furthermore, no phenylmercury compound was liberated upon addition of inorganic mercury, or detected when the extract was hydrolyzed with alkali.

It was presumed from these phenomena that the mercury atom of phenylmercury acetate existed in rice plant tissue in association with proteins, but not in the original form of phenylmercury acetate (Moriya et al., 1965).

Distribution of mercury in the rice plant leaf tissues was then examined by fractional ultracentrifugation. The leaves were obtained 2 weeks after spray application. Of all leaf cell components, the soluble protein fraction contained mercury at the highest concentration of 26 ppm, followed by the mitochondrial fraction at 13 ppm. The fibrous tissues fraction contained 60 to 70% of the total mercury in leaf tissues though the concentration was only 4 ppm (Moriya et al., 1965).

Mercury in Rice Grains

In view of the evidence of mercury accumulation in rice obtained by tracer studies, it was necessary and important to determine the amount of mercury in field-grown rice.

Since the official A.O.A.C. method gave unsatisfactory recoveries of mercury from rice grains, it was improved by replacing the catalyzer and adding a volatilization-preventing condenser. In this way, mercury in rice, fruits, and tea leaves at 0.05 to 1.5 ppm levels could be determined with recoveries of more than 80%.

This improved method was then applied to rice samples from plants dusted with phenylmercury acetate under various conditions at local agricultural experiment stations. Part of the data is shown in TABLE 3. As seen in this table, mercury contents were higher in grains from plants treated at later stages. Distribution of mercury levels obtained on all 55 samples is given in Fig. 3, together with results of spray treatments.

Unpolished rice grains from treated plants contained 0.05 to 0.60 ppm mercury, with one exception of 1.31 ppm, and the average content was 0.20 ppm. Samples containing 0.1 to 0.2 ppm mercury accounted for 40% of the total samples. The average mercury content in untreated samples was 0.04 ppm.

Distribution of mercury in rice grains is shown in TABLE 4. The content of mercury in rice grains was highest in bran, followed by husk and albumen (polished rice). About 40% of the mercury in raw rice grains was removed by customary polishing. Mercury in rice grains was found to be hardly extractable with water, but about 38% of the mercury could be extracted with 0.2% potassium hydroxide solution. It was, therefore, presumed that a

TABLE 3

Relationship of Mercury Residues in Raw Rice Grains and Treating Conditions with Phenylmercury Acetate-Dusts[a]

Location (variety)		Mercury contents of dust (%)	Application			Hg found ppm
			Hg, g/ha	Date and growth stage[a]	No.	
Tohoku	A	0.25	50	7-20(T), 8-8(J)	2	0.16
(Norin 41)	B	0.25	100	7-20(T), 8-8(J)	2	0.20
	C	0.25	50	8-16(H)	1	0.17
	D	0.25	100	8-16(H)	1	0.23
	E	0.25	50	8-23(F)	1	0.19
	F	0.25	100	8-23(F)	1	0.40
	G	0.25	50	7-20(H), 8-23(F)	2	0.16
	H	0.25	100	7-20(H), 8-23(F)	2	0.40
	I	Untreated				0.03
Kyushu	1	0.16	48	8-3(T), 8-11(T), 8-30(J)	3	0.09
(Jukkoku)	2	0.16	48	8-11(T), 8-30(J), 9-17(H)	3	0.16
	3	0.16	48	8-11(T), 8-30(J), 9-17(H), 9-26(F)	4	0.38
	4	Untreated				0.04
Kantotosan	1	0.16	48	7-19(T), 7-27(T), 8-28(J)	3	0.24
(Norin 25)	2	0.16	48	7-27(T), 8-27(J), 9-9(J)	3	0.56
	3	0.16	48	7-27(T), 8-27(J), 9-9(H), 9-26(F)	3	0.60
	4	Untreated				0.05

[a]Abbreviations used: T, Tillering stage; J, Jointing stage; H, Heading stage; F, Full heading stage; 7, July; 8, August; 9, September.

TABLE 4

Distribution of Mercury in Rice Grains from Plants Treated with Phenylmercury Acetate-Dusts

Sample	Husk ppm, Hg	Raw grain ppm, Hg	Polished grain ppm, Hg	Bran ppm, Hg	Yield after polishing %	Distribution of total Hg	
						in polished grain %	in bran %
Tohoku-F	1.79	0.40	0.18	1.91	91	49	51
Kantotosan-3	1.19	0.60	0.54	2.74	91	67	33
Kyushu-3	1.53	0.38	0.13	1.66	91	44	56
Nagano-B	3.77	1.31	1.04	4.99	95	80	20
Nagano-B	3.77	1.31	1.02	5.08	91	67	33

large portion of mercury residues in rice grains was conjugated with glutelin, the main protein in rice grains, as well as with cellulose. As observed with treated leaves, a spray application generally gave higher amounts of mercury residues in rice grains than a dust application (Kanazawa and Sato, 1965).

Tomizawa et al. (1966) also studied neutron activation analysis applied to mercury determination in rice. Mercury contents thus determined were mostly in a range of 0.22 to 0.69 ppm in 23 polished rice samples from dust- or suspension-treated plants, while untreated samples contained about 0.23 ppm.

Recently, gas chromatography with electron-capture detection was introduced utilizing oxidation of the extract with nitrous acid and back-extraction with aqueous cysteine. Phenylmercury compounds in rice grains from plants treated with phenylmercury fungicides could be determined with 70% recoveries, though the same method gave lower recoveries when applied to vegetables. Results are given in TABLE 5. Eleven to 17% of the total mercury in unpolished rice was proved to be in the form of phenylmercury compound(s). Traces of methylmercury compound(s) were also detected in rice from phenylmercury fungicide-treated plants.

Persistence of Organomercury Compounds in Soil and Uptake by Rice Plants

Since considerable portions of organomercury fungicides are deposited on soil in spray or dust application, these may be absorbed and translocated by rice plants through the root system.

Rice plants were grown on potted alluvial soil fortified with phenylmercury acetate at two levels, and mercury distribution in the plants was examined at harvest time. The results are shown in TABLE 6. Thus, unpolished grains harvested from soils treated at 5 and 25 kg/ha contained 0.43 and 0.63 ppm of mercury, respectively. When grown 2 years after the initial treatment, the unpolished

TABLE 5

Mercury Residues in Unpolished Rice Grains from Plants Treated with Phenylmercury Fungicides[a]

Sample	Treatment			Residual mercury (as Hg)		
	Fungicide[b]	Dose Hg-g/ha	Number of applications	Total Hg-ppb	Phenyl-Hg-ppb	Ph-Hg/Total-Hg %
Tohoku A	PMA, WP	20 ppm, 30	2	770	131	17
Tohoku B	PMA, D	0.25%, 75	2	510	79	15
Nagano A-1	PMA, D	0.25%, 100	4	840	109	13
Nagano A-2	PMI, D	0.2%, 80	4	370	59	16
Nagano B-1	PMI, WG	20 ppm, 30	4	760	93	12
Nagano C-1	PMI, WG	20 ppm, 30	4	350	39	11

[a]From Yamada, 1970.

[b]Abbreviations used: WP, wettable powder; D, dusts; WG, wettable granule.

TABLE 6

Uptake of Mercury by Rice Plant from Paddy Soil Fortified with Phenylmercury Acetate

Parts of rice plant	Mercury residues,[a] ppm		
	PMA 5 kg/ha	PMA 25 kg/ha	Control
Unpolished grain, 1965	0.43	0.63	0.05
Husks	0.20	0.22	0.17
Leaves	1.00	1.07	0.71
Stems	0.41	0.52	0.23
Stems, lower end[b]	0.67	3.32	0.23
Roots	3.73	13.89	0.39
(Soil)	1.95	13.98	0.27
Unpolished grain, 1967	0.09	0.14	0.05

[a]Dry matter basis

[b]The lowest 15-cm part.

grains contained only 0.09 and 0.14 ppm of mercury, respectively. Phenylmercury acetate seemed to gradually decompose to inorganic mercury compounds, as observed by Nose and Suzuki (1967), to become hardly absorbable by the plants (Kanazawa, 1971). These two authors reported that when paddy soil treated with phenylmercury acetate at 58.4 ppm was held flooded, the content decreased to one-tenth of the original in 25 days, though the total mercury content remained nearly constant. The degradation was slow when the soil was not flooded.

One of the important questions to be answered was whether or not previously introduced mercury, remaining in soils, affected mercury levels in succeeding rice crops.

Analysis of unpolished rice from untreated plants harvested in 1957 and 1965 gave average mercury contents of 0.041 and 0.045 ppm, respectively. This similarity suggests that mercury fungicides introduced between the two experiments had been converted into a form or forms which was not readily absorbed by the rice plants, or lost by volatilization and/or leaching.

Another study of mercury uptake and distribution was made by tracer technique. Rice plants transplanted at the seedling stage took up only 1% of the ^{203}Hg added to potted soil in the form of phenylmercury acetate.

The mercury uptake seems to be affected by the type of mercury compounds used. As seen in Fig. 4, the largest amount of absorption was observed with methylmercury chloride, while inorganic mercury sulfide was absorbed least (Kanazawa, 1971).

Conclusion

Results of these studies clearly indicated that use of organomercury fungicides should be strictly limited. Development and use of non-mercury fungicides were, therefore, encouraged and spray application of organomercury fungicides was finally banned in 1968, when sufficient amounts of non-mercuric blast control chemicals had become available.

The real form and behavior of mercury persisting in soils and environmental systems remains to be an important subject of investigation.

References

Araki, T., S. Toyota, Y. Mizusawa, and N. Suzuki: Residue analysis of organo-mercuric fungicides on rice and fruits. Japan Plant Protection Assoc, Tokyo p. 10-16 (1965).

Kanazawa, J.: Nogyo Gijutsu Kenkyusho Hokoku (Rep. Nat. Inst. Agr. Sci.) Ser. C 25, 109 (1971).

Kanazawa, J., and R. Sato: Residue analysis of organo-mercuric fungicides on rice and fruits. Japan Plant Protection Assoc., Tokyo, p. 17-25 (1965).

Moriya, S., C. Tomizawa, and M. Suwanai: Residue analysis of organo-mercuric fungicides sprayed on rice and fruits. Japan Plant Protection Assoc., Tokyo, p. 5-9 (1965).

Nose, K., and T. Suzuki: Noyaku Seisan Gijutsu 17, 29 (1967).

Tomizawa, C.: Presented 1956 Winter Meet. Kanto Division, Phytopathol. Soc. Japan, Tokyo, Dec. 15 (1956).

Tomizawa, C.: Shokuhin Eiseigaku Zasshi 7 (No. 1), 26 (1966).

Tomizawa, C., A. Kobayashi, M. Shibuya, Y. Koshimizu, and Y. Ota: Shokuhin Eiseigaku Zasshi 7 (No. 1), 33 (1966).

Yamada, T.: Nogyo Gijutsu Kenkyusho Hokoku (Rep. Nat. Inst. Agr. Sci.) Ser. C 24, 165 (1970).

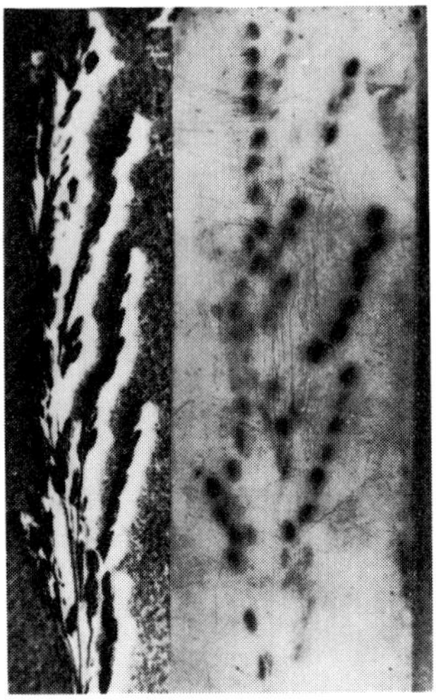

Fig. 1. Radioautogram showing the accumulation of ^{203}Hg in the head of a rice plant 11 days after dipping the top leaf with 5 ppm aqueous solution of ^{203}Hg-PMA (Tomizawa, 1956).

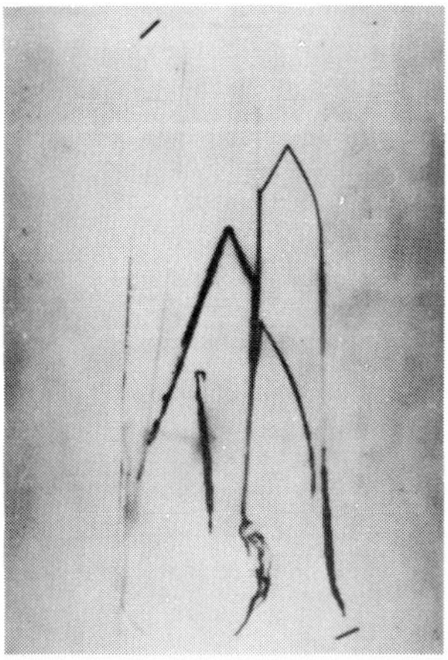

Fig. 2. Radioautogram showing the distribution of ^{203}Hg in a rice seedling 12 days after spraying with 50 ppm aqueous solution of ^{203}Hg-PMA (Tomizawa, 1966).

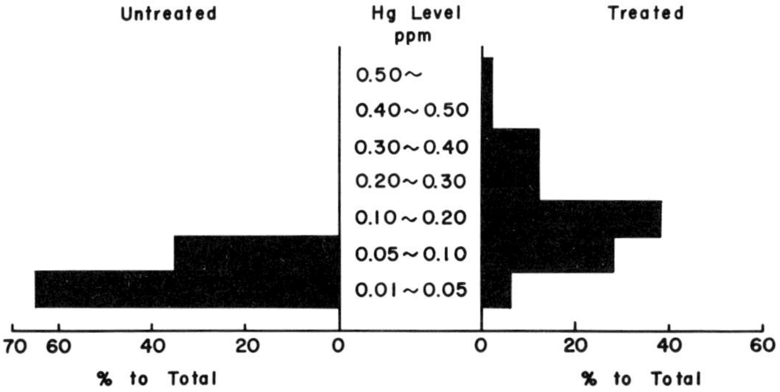

Fig. 3. Distribution of mercury residue levels in raw rice grains.

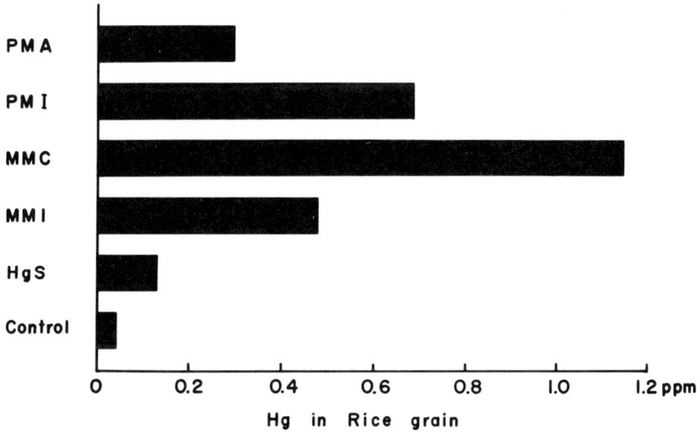

Fig. 4. Uptake of mercury from soil by rice plant in relation to type of compound.

PART III

CHLORINATED HYDROCARBON INSECTICIDES IN THE ENVIRONMENT

FACTORS RELATED TO BIOCONCENTRATION OF PESTICIDES

Eugene E. Kenaga

Ag-Organics Department, The Dow Chemical Company
Midland, Michigan

Introduction

The subject of this paper requires an understanding not only of the natural mechanisms of distribution of molecules in the world environment, but also of those properties which result in differential concentrations of pesticides in and on living organisms. The build-up of residues of DDT and its principle metabolites DDD and DDE collectively, known here as DDT-R, in the various segments of our environment, and especially in organisms in the aquatic life food-web is well documented. DDT represents a classic example of a man-made molecule which is distributed over the world in detectable amounts and in some environments in biologically significant amounts. Thus, DDT is used here as one example in the group of organochlorine insecticides which are bioconcentrated. It is also important to recognize the properties of many other economic poisons which make them relatively nonpersistent and selective in toxic action.

Bioconcentration potential, as defined here, will depend on the following factors related to the pesticide chemical (and its metabolites, conjugates, or degrates).

a. The relative amount available to the organism by contact and by ingestion. Important factors related to pesticide availability are: deposition and redistribution, adsorption, and absorption.

b. The rate of intake of the pesticide by the organism. Important factors related to intake are: concentration of the pesticide in water, food, or other environmental media of the organism; and the quantity of pesticide intake per body weight of the organism per unit of time (mg/kg per day).

c. The rate of elimination of the pesticide by the organism in terms of time and quantity. Important factors related to elimination are: the number of mechanisms for metabolism, speed of metabolism, enzymatic induction of metabolism, rate of storage in specific tissues, and ability of the organism to produce metabolites which are suitable for elimination.

The determination of bioconcentration potential of pesticides requires the identification and examination of the factors which are used for its measurement. Some of the important factors considered in this paper are the test methods used for sampling and measuring residues; the study of the distribution and accumulation of pesticide and transformation product residues in representative segments of the environment; the use of surface-area to mass ratios of environmental objects in the calculation of residues; the use of different rates of intake of food and pesticide by various organisms; the understanding of biological, physiological, and other ecological factors affecting pesticide accumulation in the environment; and the physical, chemical, and biochemical properties of pesticides.

Bioconcentration Defined

"Bioconcentration" is defined here as the amount of a pesticide residue accumulated by an organism by adsorption, and by absorption via oral or other route of entry, which results in an increased concentration of the pesticide by the organism or specific tissues. Bioconcentration is thus defined broadly because residues of compounds accumulate on the outside as well as inside of organisms whether they are plants or animals. Residues on the outside of the organisms are rarely differentiated from the total residues in and on organisms in publications reporting analytical results.

The "bioconcentration factor" is the ratio of the measured residue compared to the residue of the pesticide in the ambient air, water, or soil environment of the organism and/or the various species of food organisms consumed, as specified. Such a definition emphasizes the difficulty of putting a name to a phenomenon which has so many variable inputs, interpretations, and lack of definition in the literature.

In the emotional atmosphere of the pesticide controversy the terms "bioconcentration factor" and "biomagnification" have come to mean something undesirable. In order to evaluate the degree of undesirability, a pesticide is subjected to tests which establish definite quantitative numbers and factors to be used for comparison with other compounds. These numbers can also be used to compare amounts in various environmental components or tissues, chiefly in the food-web of organisms.

However, bioconcentration is not necessarily bad. We owe our existence to "bioconcentrations" of food and nutrients which are directed to the right place at the right time in organisms. The necessity is evident for definition of terms used to calculate bioconcentration and the standardization of test methods for comparative purposes.

DDT-R Residues and Bioconcentration Factors in Various Components of the Environment

Although DDT was applied to our environment in massive amounts in the mid-1940's, it wasn't until much later that new analytical techniques and extensive monitoring of residues produced data showing the distribution and accumulation of DDT and its metabolites, and other pesticides in various segments of the environment.

Residue monitoring data is useful for correlation with the physical and chemical properties, metabolism and toxicity of pesticides and other chemicals; the biology of organisms; the natural cycles of weather; man's activities; and the interaction of living things with soil, water, and air.

Monitoring data in the United States shows that DDT, its transformation products DDD and DDE, and dieldrin are the most commonly found organochlorine pesticides and usually in the highest concentrations (Edwards, 1970; Lichtenberg et al., 1970). Although heavy metals and other environmental contaminants may also bioconcentrate in toxic amounts in the environment, the mechanisms to study such effects are probably best illustrated at present by the organochlorine pesticides. DDT will be considered in this paper as an example illustrative of a persistent organochlorine insecticide. DDT-R residues are discussed separately from DDT, DDD, and DDE residues because much of the older analytical data does not differentiate between DDT and its individual metabolites. Regardless of the confusion in analytical techniques and residue results reported by various authors, which may not always accurately differentiate between DDT and other chlorinated or non-chlorinated environmental contaminants, it is felt that these data are useful for illustrating the principles of bioconcentration.

Edwards (1970) compiled an extensive compendium on persistent pesticides in the biota and physical environment. This compendium, although not complete, includes residue data on pesticides in soils, air, water, soil flora and fauna, aquatic organisms, plants, vertebrates, humans, and animal and human foods. From this vast amount of information he constructed a web of typical DDT-R residue values which is useful in terms of relative equilibrium data in the various segments of the environment (TABLE 1). Typical residues of DDT-R in these environments were from less than 1 ppt in water ($<1 \times 10^{-12}$) to greater than 10 ppm (1×10^{-5}) in organism. This amounts to a 10 million-fold range. No distinction is made between DDT and its metabolites in much of the older DDT-R residue data. Nevertheless, it appears that the actual maximum residues in terms of ppm DDT-R accumulated by aquatic organisms are not much greater than that held by land organisms. Those residues amount to tens or hundreds of ppm.

Based on literature residue data, Edwards calculated concentration factors of DDT-R residues in various organisms compared to those in various segments of the environment. By rearranging his data, maximum and minimum concentration factors were compared for various classes of biological organisms, based upon comparisons of the partitioning of residues with the soil, water, or air environmental components (TABLE 2). It is obvious that by far the greatest bioconcentrations occur in water, even though actual maximum residues of DDT-R in aquatic organisms are not greatly different from those in land organisms. Large bioconcentration factors (i.e., > 100) are not found for organisms feeding in or on soil.

It is concluded from the work of Meeks (1968), Edwards (1970), Hannon et al. (1970), Hamelink et al. (1971) and others concerning the distribution of DDT-R residues in various segments of the environment that:

a. High chemical, physical, and biological stability in the environment makes residues available for storage and recycling in the various environmental segments.

b. Residues of persistent compounds are widely distributed in all segments of the environment (air, water, soil, plants, and animals).

c. High residues occur on some animate or inanimate

TABLE 1

Distribution of DDT-R in the Principle Segments of the Earth's Environment

	Typical relative values[a] (ppm)	Some maximum values[b]
Air related		
Air	0.000004	0.003-56.1 x (10^{-12} $\mu g/m^3$)
Atmospheric water (rain)	0.0002	1-102 x 10^{-12} (ng/l)
Atmospheric dust	0.04	45 x 10^{-9} (μg/l)
Water related		
Fresh water	0.00001	< 1 to 109 x (10^{-12}) (ng/l)
Sea water	0.000001	< 1 x 10^{-12} (ng/l)
Plankton	0.0003	5 ppm
Aquatic vegetation (algae, vascular plants)	-	0.01-0.003 (75 by deposition) ppm
Aquatic invertebrates (shrimp, crayfish, mollusks, etc.)	0.001	0.14-151
Marine fish	0.5	0.01-136 ppm (Marine fish and Freshwater fish)
Freshwater fish	2.0	

TABLE 1 - Continued

	Typical relative values[a] (ppm)	Some maximum values[b]
Aquatic mammals	-	8.3-23.3 ppm
Predatory birds (fish eaters) (herons, eagles, pelicans)	10.0	7.1-194 ppm
Aquatic birds (herbivorous, insectivorous scavengers)	-	0.16-100 ppm 99.8 ppm
Herring gull eggs		227 ppm
Soil related		
Agricultural soils	2.0	0-131 ppm
Natural soils	?	0
Soil invertebrates (earthworms, slugs, beetles)	4.0	0-40 ppm
Plants	0.05	-
(soil treatment)	-	0.9-3.2 ppm
(foliage treatment)	-	179 (by deposition) ppm
Herbivorous and insectivorous mammals	0.5 0.05	0.01-126 ppm

TABLE 1 - Continued

	Typical relative values[a] (ppm)	Some maximum values[b]
Herbivorous and insectivorous birds	2.0	0.16-55.6 ppm
Predatory mammals	1.0	0.13-0.24 ppm
Predatory birds (owls, hawks, not including ospreys and eagles)	-	1.42-18.1 ppm
Man (adipose tissue)	6.0	< 0.01-647 ppm
Man's vegetables	0.02	0.0002-0.05 ppm
Man's meat	0.2	0.05-0.28 ppm

[a] Assigned by Edwards (1970).

[b] Adapted from tables in Edwards (1970).

TABLE 2

Bioconcentration of DDT-R Residues in Plants or Animals from its Environment[a]

Environment	Plant and animal organism	Concentration factor (DDT-R residue organism divided by residue in environment): Maximum value observed	Minimum value observed
Soil	earthworm	73	0.67
	beetles	2.81	0.31
	slugs	3.70	2.33
	crop roots	0.13	0.04
	crop foliage	0.08	-
Water	sea squirt	1,000,000[b]	200[b]
	sea hare	178,000[b]	-
	eastern oyster, clam	70,000[b]	60
	shrimp	2,800[b]	280
	crabs	144	-
	crayfish	97	17
	snails	1,480[b]	-
	plankton	16,666[b]	250
	fish	829,300[b]	5-(1450)[b]
	fish (DDD)	9,214[b]	417
	algae	33	0.34
	aquatic plants	100,000[b]	0.45
Diet	pheasant	2.91	
	woodcock	4.5	2.6
	bald eagle		
	brain	0.1	
	liver	1.9	
	fat	35.7	

[a]Adapted from Edwards (1970).

[b]DDT may be present in excess of solubility in water.

objects immediately after application due to deposition or preferential adsorption. These residues usually decline quickly afterwards.

d. High accumulation of residues occurs on or in animate and inanimate objects, compared to that found in water.

e. High residues occur on or in fish and birds, for example, which peak in concentration days, weeks, or months after application due to accumulation.

f. Accumulation and storage of residues often occur preferentially in adipose tissues of organisms and is related to lipid solubility and fat utilization by the organism.

g. The type of DDT formulation can influence the rate of release of residues into the environment, thus changing the relative distribution between environmental segments.

h. After a period of time following application of pesticide, a relatively stable distribution of residues exists in aquatic habitats between the pesticide formulation, water, particulate matter, other inanimate objects, and animate objects, which is an entirely different residue distribution than that found immediately or shortly after application.

i. Calculated (apparent) bioconcentration factors are highly susceptible to mathematical juggling depending on the test conditions and the residues in the environmental segment (such as water) used as a reference point for comparison with residues in other environmental segments.

j. No consistent pattern of maximum bioconcentration of pesticide residues occurs in various plant or animal orders which differ in magnitudes of more than 10- or 100-fold. Maximum pesticide residues may sometimes be accumulated by algae or by similar "first link" organisms in the "chain-of-life" organisms and do not necessarily result in increasing concentrations in each succeeding link of the chain.

As mentioned previously, the principle metabolites or alterations of the DDT molecule found in the environment are the persistent compounds DDD and DDE. DDD is rarely applied as a pesticide and DDE is never so used; consequently, the environmental residues are essentially derived from applications of DDT. Since each compound has different chemical, physical, and biological properties it is important to study the distribution of the metabolites of DDT as well as DDT itself. From 1966 to 1970 a number of papers were published dealing with the metabolites of DDT in various tissues, organisms, and segments of the environment. Some excellent papers reviewed were those of Hickey et al. (1966), El Sayed et al. (1967), Keith et al. (1967), Harris et al. (1969), Martin (1969); Dindal (1970), Gish (1970), Hannon et al. (1970), Johnson et al. (1970, 1971), and Findlay et al. (1971).

It may be concluded that the principle metabolites of DDT in the environment are DDD and DDE which appear to be as persistent or more so than DDT. Ratios of residues of DDT:DDD:DDE vary considerably in different segments and components of the environment. Birds, fish, and insects appear to have particularly high residues of DDE. Both DDD and DDE, like DDT, partition in favor of fat tissues and disfavor water. Plants do not appear to pick up residues systemically when grown in soil containing DDT, DDD, or DDE. Variations in the amounts of DDT in various anatomical fat parts of an individual bird may be due to differential rates of deposition, and metabolism of the fat containing DDT.

Surface Area of Organisms as Related to Accumulation of Residues of Pesticides

The bioconcentration of pesticides in organisms is in part related to the amount of residue available for sorption or ingestion. It is important to understand how to calculate or estimate the quantity of residues which may occur as the result of pesticide treatment in various segments of the environment. Residues are normally expressed as ppm or in similar terms and are therefore ratios of the weight of the pesticide residue to the object treated. Pesticides dispersed in the environment rarely result in uniform residue deposition in terms of ppm. The distribution of a pesticide in the environment and its organisms can be briefly stated to follow this sequence: application by man by use of equipment, dispersion (weather, currents of water, air, etc.), deposition (gravity), partitioning between environmental segments (water, soil, plants, air, animals), partitioning within organisms to specific tissues. Partitioning means that the pesticide will favor certain media so that the concentration will decrease in one and increase in another.

The residue, in terms of ppm of a pesticide on a given object (leaf, fruit, insect), is dependent in part on the surface area to weight (as represented by the volume) relationship of the object treated. If certain particles (algae, protozoans, bacteria, snails, sand, clay, etc.) in water are visualized as different sized spheres, then a ratio of the circular area of the sphere (at its center) to its volume (using a specific gravity of one) can be determined. A ppm relationship then may be hypothesized. Thus, particles of a radius of 1 inch, 1 cm, 1 mm, and 0.1 mm could contain 2.3, 8.4, 84, and 840 ppm, respectively, due to a 1 lb. of pesticide/A (10.4 mg/sq. ft.) treatment. Some examples of the surface area to weight relationship of animals, plants, and soil particles are shown in TABLE 3. Surface areas of objects in nature vary greatly. In general, the smaller the object, the larger the ratio of surface area to weight becomes. This varies more than 15 million-fold between man and visible particles of 10 microns in diameter. Assuming even coverage of such objects by pesticides, a large variation would occur in residue values measured as ppm.

Let us assume the pesticide is dispersed evenly over a piece of land containing a varied wooded and herbaceous habitat at a rate of 1 pound per acre. Some molecules are undoubtedly lost by volatilization to the air or drift, others fall as liquids or solids upon foliage of tall trees which screen or filter out, or adsorb part of the dosage, depending on the density of the foliage, sorption factors, etc.; others fall to lower vegetation and thus more is screened

TABLE 3

Surface Areas to Weight Relationships of Some Environmental Objects

Object	Weight kg	Surface area, square meters per gram
Animals[a]		
Man (adult)	100	.000022
(young)	10	.000048
Guinea pig (adult)	0.72	.000092
(young)	0.13	.000169
Plants		
Apple leaf	0.001	.00672
Representative Sphere Sizes[b]		
(r = 1 in.)	0.068	.000118
(r = 1 mm)	0.0000042	.003
(r = .01 mm, or 10 microns)	0.0000042×10^{-6}	300
Soil Constituents[c]		
Kaolinite clay particle		7-30
Illite		65-100
Oxide and hydroxides		100-880
Montmorillonite clay particle		600-800
Vermiculite clay particle		600-800
Organic matter		500-800

[a]Altman and Dittmer (1962).

[b]Assume specific gravity of water.

[c]Bailey and White (1964).

out before reaching the soil. Sometimes grass is the only vegetation contacted before reaching the soil. From this description it is seen that there are foliar screens and other plant and animal screens which result in different amounts of pesticide molecules reaching the soil and other parts of the habitat treated immediately after application.

As discussed by Hoerger and Kenaga (1971), the surface area-weight ratio of various animal and plant foods alone may result in large variations in ppm of a pesticide residue, even though the dosage of the pesticide is uniformly applied on an acreage basis. From TABLE 4 it is shown that the upper or maximum amount of ppm residues of a number of pesticides, on different types of plant surfaces, varies from 7-240 ppm from an application of 1 lb. per acre of pesticide. Maximum residues for vegetable and fruit foliage were 125 ppm. As an example of surface area-weight relationships, a sampling of Yellow Delicious apple leaves, averaging about 1 g/leaf and 5 sq. in./top surface area, would have a residue of about 360 ppm if dosed at 1 lb. of pesticide per acre (10.4 mg/sq. ft.). The measured maximum of 125 ppm is about 35% of the hypothetical maximum. This is reasonable considering loss of compound to the air, filtering by other intercepting foliage, and the fact that a leaf rarely presents its total upper surface area directly at a right angle to the direction of the pesticide application, thus decreasing the dosage per leaf surface basis. All pesticide residues, in terms of ppm by weight, decrease with time on living plants, mostly as the result of natural weathering, volatility and solubility factors, and to plant growth. Redistribution and degradation of the pesticide and recycling of the molecules begin immediately after application.

The Relationship of Body Weight of Animals to their Dietary Intake of Pesticides

The amount of pesticide intake of an organism is determined by the availability of the compound in the diet or

TABLE 4

Upper Limits and Typical Limits of Residues of Pesticides on Differing Categories of Plants[a]

Categories of plants	ppm Residue on the basis of a pesticide dosage of 1 lb./acre			
	Immediately after application		6 Weeks after application	
	Upper limit	Typical limit	Upper limit	Typical limit
Range grass (short)	240	125	30	5
Grass (long)	110	92	20	1-5
Leaves and leafy crops (vegetables and fruit)	125	35	20	<1
Forage crops (alfalfa, clover)	58	33	1.0	< 1
Pods containing seeds (legumes)	12	3	1.5	< 0.1
Fruit (cherries, peaches, grapes, citrus)	7	1.5	1.5	< 0.2

[a]From Hoerger and Kenaga (1971).

surrounding environment; the concentration of the compound in the diet or surrounding environment; the quantity of dietary food consumed containing the compound; and the quantity of compound penetrating the organism via the cellular barriers such as the cuticle, lungs, digestive system, gills, etc. The percent by weight of food eaten per day by various sized terrestrial animals is shown in TABLE 5 (derived from Spector, 1956). Mammals below 25 grams in weight show a tremendous increase in metabolic rate and consequent food intake. Birds follow a

TABLE 5

Ratio of Body Weight of Terrestrial Animals to their Daily Food Intake[a]

Animal	Body weight (kg)	Weight of food eaten per day (g)	% of Body weight eaten in food per day
Dairy cattle	900	12600	1.4
Beef cattle	800	12000	1.5
Horse	635	12065	1.9
Swine	227	3410	1.5
Sheep	79	2449	3.1
Man	65	715	1.1
Fox	5.7	143	2.5
Monkey	1.5-3.0	60-120	4.0
Chicken	1.82	9	5.1
White rat	0.300	15	5.0
White mouse	0.025	3.0	12.0
Hamster	0.025	3.8	15.0

[a]From Spector (1956).

similar relationship to that of mammals (Kenaga, 1971a) as shown in TABLE 6. Thus, it can be seen from TABLES 5 and 6 that the heavier animal eats more food per day than the lighter animal, but it is a smaller percentage of its body weight. If the ppm pesticide in the diet is constant, the lighter bird or mammal will have a greater intake of pesticide in terms of mg/kg per day.

To further illustrate the difference between the dietary intake of small versus large birds, it frequently occurs that small birds eat small objects and large birds eat large objects. Based on the fact that small or thin particles may have greater ppm pesticide residues than large or thick particles, the difference in mg/kg per day intake of pesticides becomes further exaggerated between

TABLE 6

Relation of Food Consumption to Body Weight[a]

Bird species	Adult weight (g)	Food eaten per day (g)	Percent of body weight eaten per day
Domestic Fowl (*Gallus domesticus*)	1800	61	3.4[b]
Common Buzzard (*Buteo buteo*)	855-900	38.5-50.0	4.5
Tawny Owl (*Strix aluco*)	442-475	22.1-23.7	5.0
Little Owl (*Athene noctua*)	164-172	9.1-9.4	5.5
Blackbird (European) (*Turdus merula*)	118	8.4	7.3
Kestrel (*Falco tinnunculus*)	200	15.4	7.7
Lapwing (*Vanellus vanellus*)	195	15.3	7.8[c]
Bobwhite (*Colinus v. virginicus*)	170	15.2	8.8[b]
Dunlin (*Erolia alpina*)	114	8.9	8.5[c]
Song Thrush (*Turdus philomelos*)	89	8.8	9.8[c]
Mourning Dove (*Zenaida macroura*)	100	11.2	11.2[b]
Chaffinch (*Fringella coelebs*)	22	2.9	13.2[b]
Robin (European) (*Erithacus rubecula*)	16	2.35	14.7
European Goldfinch (*Carduelis carduelis*)	13	2.3	17.5[b]
Great Tit (*Parus major*)	18	4.4	26.0[c]
Blue Tit (*Parus caeruleus*)	11	3.3	30.0[c]

[a]From Nice (1938).

[b]Seeds and grains.

[c]Dry weight (estimated as 40% of the dry weight of the mealworms used as food).

small birds and large birds (TABLE 7). Here, over a hundred-fold difference in mg/kg per day intake is possible in an area treated with the same pesticide dosage, depending on the size of the bird and the residues on the various sizes of food particles consumed (Kenaga, 1971a). Buhler

TABLE 7

Estimation of the Milligrams of Toxicant per Kilogram of Body Weight per Day Intake by Birds of Varying Sizes Resulting from Eating Different Foods from an Area Treated Uniformly with an Application of 1 Pound of Toxicant per Acre[a]

Illustrative examples of ppm[b] in or on different types of food eaten by birds	Mg/kg per day ingested by different sized birds		
	20 g 18%[c]	100 g 9.2%	1,000 g 3.6%
240 (sparse foliage)	43	22	9
58 (dense foliage, insects)	10	5.3	2.1
10 (seeds, fruit, large insects)	1.8	0.9	0.4

[a]From Kenaga (1971a).

[b]Pesticide residue in ppm immediately after application based on maximum values cited.

[c]Percent of body weight ingested in dry food per day.

and Shanks (1970) showed that coho salmon (Oncorhynchus kisutch) median survival time was directly proportional to body weight when fed a diet containing 200 ppm of DDT. Fish weighing 2.9 g and 15.03 g ate 1.53 and 0.63% of their weight, respectively, per day, which resulted in DDT intake of 3.07 and 1.27 mg/kg per day. Because fish are cold-blooded organisms, the percent of body weight eaten as food per day is a much smaller percentage than expected from warm-blooded mammals and birds of similar body weight.

Biological and Environmental Factors Related to the Intake and Accumulation of Pesticide Residues by Organisms

The way in which biological factors can influence the molecular transformation of pesticides and residue accumulation in the various segments of the environment are exceedingly numerous, thus only a few examples are given. Some of these factors are discussed in other sections of this paper. Because compounds like DDT are so insoluble in water it has been surprising to learn how rapidly residues in fish may accumulate. Gill surfaces are excellent for accumulating DDT from water, which then allows an exchange with blood and from the blood to fats in fish (Hamelink et al., 1971). Due to the low concentrations of DDT in water this would not seem to be a rapid exchange method until it is realized that the blood circulation time in a fish like rainbow trout, *Salmo gairdneri* (weighing about 212 g at 10°C), ranges from 48-96 seconds and has a mean value of 64 seconds (Davis, 1970). This blood circulation time appears to offer the flow system necessary for rapid transfer of DDT and other chemicals to and from fat tissues.

Grzenda et al. (1970) studied the uptake, metabolism, and elimination of DDT-R by goldfish. The ratio of the metabolites, DDD and DDE, increased in comparison with DDT with continued exposure in spite of the uniform and constant input of DDT. The change in ratio of metabolites may indicate increased ability of the fish or its microorganisms to metabolize DDT. The increased ability of organisms to detoxify chemicals often is due to enzyme induction. The interaction of DDT and dieldrin fed to rainbow trout in their diet was studied by Macek et al. (1970). The presence of dieldrin decreased, if not inhibited, the elimination of DDT, while the presence of DDT had no effect on the elimination of dieldrin. The interaction may affect the total residue accumulation in trout. Murphy (1970) found that DDE constitutes a smaller proportion of the DDT-R residues in fresh water fish than in marine fish. Marine fish can have DDT-R residues with 24-35% DDE, compared to 61-73% DDE, for the same species from coastal waters, indicating a differential conversion rate. Murphy also showed in laboratory tests that 15 parts per thousand salinity reduced the amounts of DDT, DDE, and DDD accumulated by the mosquito fish, *Gambusia affinis*, and that this salinity had a greater effect in reducing the uptake of DDT than on the uptake of DDD or DDE. This would result in a larger proportion of DDD and DDE than DDT in fish from brackish water.

Keith (1970) related the differences in fat storage metabolism to the health of birds exposed to insecticides, particularly DDT. Birds of the same species living in the same pond area vary greatly in fat content. Since DDT-R residues are associated with lipid tissues in birds, the impact of stresses which cause birds to utilize their fats, thus releasing DDT-R in toxic amounts into their bodies, is very critical to birds. The feeding habits of birds are important. Larger predator species such as hawks, pelicans, and Western grebes may not feed regularly and may lose as much as 35% of their weight during periods of self-induced starvation. This is in contrast to small insectivorous and omnivorous species which feed almost constantly. Some birds do not eat during migration, nest-building, or incubation times. Keith (1970) has concluded: "The biological vulnerability to insecticides of an individual bird, a species, or a species population is a function of interacting relationships that include: (a) the behavior of insecticides in the environment, (b) the life requirements or ecological niche of the species, (c) the incidence of the ecological coalescence of individual birds with the environmental contaminations, (d) the degree of residue accumulation in individual birds, (e) the innate behavior of the species in relation to such physiological processes as lipid turnover, (f) the population dynamics of the species, and (g) variation in these relationships brought about by differences in environmental factors such as food availability, weather, and habitat conditions. These relationships produce an extremely complex situation." Keith's conclusions apply equally to almost any species of animal or plant.

The Relationship of the Physical and Chemical Properties of Pesticides to Bioconcentration

Accumulation of pesticide residues occurs when a molecule is selectively directed or attracted toward specific surfaces or into specific cells or cell products from a reservoir and is accumulated faster than it is released. Such accumulation occurs with natural substances such as the essential mineral elements, amino acids, fat, and other building blocks used for life processes. Animals and plants contain a variety of organic and inorganic, polar and non-polar, ionic and non-ionic, high and low molecular weight, and high and low water-soluble molecules. This variety of properties for each molecule merely serves to illustrate the complex nature and relationship of the common molecules that are necessary in everyday life processes. Each of these molecules has physical, chemical, and biological properties directly related to its function and movement and to other surrounding molecules. Pesticide molecules are designed to interfere with certain normal life processes by physically, chemically, or otherwise blocking specific essential molecular functions. In order to block life functions, a pesticide molecule must have the ability to move to or be transported in a life system toward the proper blocking site. A pesticide will often go to a surface or into a cell in a manner similar to the essential molecule whose natural physical and chemical properties it most closely duplicates. Since every molecule is different and since a man-synthesized molecule may imitate a natural molecule only in part, the living organism may not be capable of disposing of the "useless" molecules and thus may accumulate them. Some of the important physical and chemical characteristics of pesticides are molecular weight, molecular size, specific gravity, solubility, latent heat of solution, polarity, acidic dissociation constants, partitioning coefficients, volatility, latent heat of vaporization, light and photolysis reactions, heat stability and molecular transformation rates. Many of these factors relate to adsorption and absorption of molecules to and in various components of the environments.

Adsorption, absorption, desorption and solution exert a profound influence on the distribution and partitioning of compounds applied to the environment. "Adsorption" is the accumulation of the pesticide at an interface such as a solute to solid (i.e., on the surface of a solid at the clay colloid-water or organo-colloid water interface). This may represent the first step in bioaccumulation. "Absorption" is the movement of solutes from one place to another through an interface of a two-component mixture, including penetration into plant or animal cells and microorganisms. "Desorption" is the releasing of sorbed molecules. Once a compound is applied to the environment, a dynamic equilibrium is established between its adsorbed and solution phase which is known as a partitioning effect. Guidelines for the use of such properties in evaluating the safety of pesticides for use in the wildlife environment are given by Kenaga (1968, 1971b). Discussion of the interpretation of data concerning such properties will be limited here to solubility and partitioning coefficients.

Solubility - The solubility of a compound in various commercial solvents is important in formulating it for proprietary use, ease of handling, safety, distribution, and many other purposes related to human activities. Many of these solvents are not synthesized in nature, but are put there by man, sometimes in large quantities (CCl_4, o-dichlorobenzene, methylene chloride, etc.). Many solvents are synthesized in nature and are concentrated by man (olive oil, kerosene, benzene, etc.). Many more additional types of organic solvents occur in nature at very low concentrations. Water, of course, occurs as the principal solvent in the world and all compounds must react to it accordingly. Because of this, water solubility must be known in order to predict the distribution of the compound in various segments of the environment.

Water solubility is useful for calculating partitioning

coefficients, saturation dosages for bodies of water, maximum equilibrium solubility for volatility calculations, adsorption equations for particulate matter, and many other factors related to reactions with water. Almost without exception, organisms metabolize pesticides to more polar compounds, which makes them much more water soluble.

The true solubility of DDT in water appears to be approximately 1.2 ppb at 25°C, being much less soluble than in comparison with the solubility in other solvents. The other solvents listed in TABLE 8 solubilize DDT at 8% or more (i.e., 80,000,000 ppb or more), which helps illustrate the great difference in the solubility of DDT in nearly everything, in preference to water. This is one of the main factors related to its accumulation in polar tissues in animal life. Unfortunately, very little information is available on the solubility of the two major metabolites of DDT; however, DDD appears slightly less soluble and DDE slightly more soluble in non-polar solvents than DDT. Since DDT has often been applied at concentrations in water far higher than its water solubility limits, the excess DDT deposits on particles or objects in the water or it volatilizes out of the water. In essence, the insoluble DDT may remain as a reservoir to maintain an equilibrium concentration (not necessarily saturated) in water for an extended period of time.

In nature there is rarely any water approaching chemically pure water, the greatest impurity often being halides, principally in the form of sodium chloride, as found in sea water. A great many organic and inorganic molecules are found in small or trace amounts and may change the solubility and stability of DDT in natural water to some degree compared to the 1.2 ppb published level for pure water. Wershaw et al. (1969) suggests that sodium humate, a major natural organic constituent of soil, lowers the surface tension of water much like sodium lauryl sulfate (a strong anionic surfactant) suggesting that sodium humate also might help solubilize compounds which are highly water insoluble. Tests with DDT show that a

TABLE 8

Solubility of DDT, DDD, and DDE in Various Solvents

Solvent	Solubility g/100 ml solvent (w/v)		
	DDT	DDD	DDE
Olive oil[a]	9.6	7.5	
Olive oil (37°C)[c]	10.5	8.0	21.1
Hexane (25°C)[b]	3.9		
Fuel oil no. 1 (27-30°C)[d]	Ca 8		
Mineral oil (15°C)[f]	5.0		
Kerosene[d]	4-8		
Diesel oil (20°C)[f]	6.5		
Xylene[d]	56-62		
Benzene[d]	77-83		
Dioxane	91-100		
Acetone (27-30°C)[d]	50-55		
Acetone[a]	21	50	
Cyclohexanone[d]	100-120		
Methylene chloride[d]	84-91		
Carbon tetrachloride[d]	46-48		
o-Dichlorbenzene[d]	63-71		
Benzyl benzoate[d]	39-41		
Ethyl alcohol (95%)[d]	Ca 1.5		
Water (2°C)[e]	5.9[g]		
Water (18°C)[e]	100[g]		
(25°C)[e]	1.2[h]		
(25°C)[e]	37.4[g]		
(37.5°C)[e]	4.5[g]		

[a]Riemschneider (1948).

[b]Hadaway et al. (1970).

[c]Metcalf (1948).

[d]Jones et al. (1945).

[e]Gunther et al. (1968).

[f]Hartung and King (1970).

[g]ppb

[h]Sieved through 40 micron membrane for microcrystals--probably most accurate measurement.

0.5% sodium humate concentration increases solubility in water at least 20 times. Wershaw et al. (1969) showed increased DDT solubility up to 4 ppm. Apparent water solubility of DDT is confusingly high due to the occurrence of micro-particulate DDT which can be filtered from water through a membrane containing 0.45 micron micropores. The solubilization of DDT, a non-polar compound by sodium humate; and the sorption of 2, 4, 5-T, an anionic compound, by humic acid are two varied types of solubility interactions which might be expected when any organic pesticide is applied to a natural organic soil-water system.

Water is the solvent in which most pesticides are least soluble. In contrast with DDT which has a solubility differential between the best known organic solvent and water solvency of about 10^{13}, 2, 4-D has a much smaller differential of about 10^5 and dichlobenil and diuron of about 10^6. However, 2, 4-D is almost as soluble in the non-polar solvents n-heptane, CCl_4, and diesel oil as it is in water, which is also in great contrast to DDT. Thus, it may be judged that DDT is more selectively solubilized in various components in aqueous environments than many pesticides (Freed, 1966).

Partitioning coefficients and distribution between various solvents and environmental surfaces - Although DDT is soluble at about 1 ppb in water, it is rarely found at this concentration in field tests within 1 day to 1 week after application, even when a reservoir of DDT remains available in the water environment to be solubilized. Edwards (1970), in his tabular review of pesticide residues of DDT occurring in various water sources, shows a maximum of 0.9×10^{-3} mg/l (0.9 ppb) and averages below 1×10^{-6} mg/l. The solubility of DDT in organic solvents such as acetone, alcohol, chloroform, etc. is several million times that of water. In layman's terms, DDT doesn't like water and tends to escape it as best it can. There are many alternative choices in natural waters. Since DDT is quite fat soluble, and since most plant and animal surfaces have various lipids, they present good sorptive surfaces for solubilizing DDT or partitioning in favor of nonaqueous surfaces.

As a purely physical phenomenon it has been shown that DDT partitions between water and oil somewhat in proportion to the ratio of its solubility in each. Hartung et al. (1970) showed that concentrations of 0.5% petroleum oils in sediments which are present in the bottom of petroleum polluted streams are able to act as partitioners and concentrate non-polar substances from water. For DDT, the partitioning coefficient between mineral oil and water in laboratory tests, and also between sedimented oils and water in the Detroit River, was about 10^6 when the water and cosolvents contained 7 ppt to 30 ppm of DDT (TABLE 9). Voerman (1969) determined the distribution ratio (partitioning coefficients) for some chlorinated hydrocarbons between hexane and water. The value for DDT was 91,000 in favor of hexane. Seba et al. (1969) have found surface slicks (organic films) in the marine environment to be efficient concentrators of persistent chlorinated pesticides. Sea water containing undetectable or less than 1^{-12} (1 ppt) of such pesticides have surface slicks containing as high as 3.46 ppb DDT, 9.25 ppb DDE, and 0.04 ppb of other pesticides which is greater than 10^5 concentration over that in water (TABLE 9). Slicks are also known to contain a 1500-fold concentration of particulate carbon and a 5-fold concentration of dissolved organic carbon. Other particulate matter, including wind-borne dust, may contribute to the source of pesticides in slicks. Additionally, there are other organics such as acetone, butyraldehyde, and 2-butanone present in sea water slicks at concentrations of 26-40 ppb which may act as cosolvents for DDT in water or the slicks. Of these organics, only acetone appeared to concentrate in oil slicks and this to less than 3-fold.

The accumulations of residues of DDT (and DDT-R) as reported by Seba et al. (1969), Edwards (1970), and Hartung et al. (1970) are situations where a great reservoir of DDT in water (represented by a river, lake, or ocean) is present as an almost inexhaustible supply of DDT water solution for partitioning into the much smaller volume of solvent in water (oil slicks, oil sediments, fish, etc.).

TABLE 9

Partitioning Coefficients for DDT from Water into Petroleum Oils and Marine Surface Slicks

Sorbant chemical	DDT concentration at start	Relative partitioning coefficient over water	Reference
Mineral oil	10-30 ppm[a]	1,300,000	Hartung and King (1970)
Sedimented mineral oil	1.24 ppb[b]	1,080,000	Hartung and King (1970)
Sedimented oil	7 ppt 40 ppt	1,450,000[c] 8,310,000	Hartung and King (1970)
Surface slicks on marine water (containing acetone, butyraldehyde, 2-butanone, carbon)	<1 ppt	100,000	Seba and Corcoran (1969)

[a]DDT dissolved in mineral oil, and stirred during test period.

[b]DDT dissolved in ethanol, and put in flowing water--reached equilibrium in flowing water containing 1.2 ppb DDT in about 12 days.

[c]Calculated apparent coefficient from sedimented oil in Detroit River containing an average of 7 ppt (7×1^{-12}) in filtered water and 40 ppt in unfiltered water.

This allows maximum partitioning of DDT-R between water and other better solubilizing fluids for DDT which are present in water. These examples represent extreme cases where considerable time may elapse for partitioning to occur. Solubility is not the only criterion for measuring partitioning coefficients based on the way the term has been used in the literature. Since DDT is very water-insoluble it will readily deposit or sorb on nearly any surface of objects in water. In an attempt to measure the uptake of DDT by various species of phytoplankton (diatoms), Keil et al. (1969) and Cox (1970) measured the relative concentrations of DDT placed in water versus concentrations of DDT on the phytoplankton after various periods of time in laboratory tests (TABLE 10). Under these conditions it is seen that a considerable residue of DDT may occur in several minutes, amounting to a calculated 10^5 increase in residues over that in the water. Such a rapid accumulation of residue is not likely to be due to solubility but to adsorption. The phytoplankton represent fine particulate matter with an immense surface-to-weight ratio, and since the residue is calculated on a weight basis, the surface area of phytoplankton in a limited volume of water (and DDT) is a major determinant factor in the magnitude of the residue on phytoplankton. No doubt, under test tube conditions, with a given concentration of DDT in water, one could put a constant weight of glass beads in such a test tube and obtain greatly varying ppm of DDT merely by varying the diameter of the glass beads.

The presence of emulsifiers, wetting agents, etc. in formulations may also affect the solubility and partitioning coefficients of pesticides. Vance et al. (1969) studied the "biological concentration of pesticides by algae" using laboratory cultures. When DDT was applied in acetone at concentrations of 1 ppm (far above its water solubility) four species of algae were noted to accumulate from 134-270 ppm. No evidence was presented to show whether the algae were absorbing the DDT or merely adsorbing it. In such a test, the presence of acetone as a DDT solubilizer may greatly modify (reduce) the partitioning coefficient.

TABLE 10

Partitioning Coefficients for DDT from Water onto Phytoplankton

Sorbant organism	Temp. (C)	Time exposed	DDT concentration in water or media at start[a]	Relative partitioning coefficient	Reference
Syracosphaera carteraea (a coccolithophorid)	18	few min.	2.0-2.6 ppt[c]	260,000	Cox (1970)
Amphidinium carteri (a dinoflagellate)	18	few min.	2.4-3.3 ppt[c]	150,000	
Thalassiossera fluviatilis (a centric diatom)	18	few min.	0.5-1.0 ppt[c]	170,000	
Cylindrotheca closterium (a diatom)	?	21 days[b]	100 ppb[d]	265	Keil et al., (1969)

[a]ppt = 10^{-12} wet weight g/g water.

[b]In 21 days, 8.8% of the DDT was converted to DDE, but none to DDD.

[c]Ethanol used as cosolvent for DDT.

[d]Acetone used as cosolvent for DDT.

From these discussions it is shown that the magnitude of residue accumulation and the distribution ratio varies according to the interrelationship of the given environmental situation and that the relative size of the DDT reservoir, the relative solubility in various solvents, the relative adsorption on various objects and the surface area of the adsorbant in relationship to its weight, are main criteria for determining partitioning coefficients. On this basis, the bioconcentration factor using partitioning coefficients can become a "numbers game," since by varying any one of these parameters in laboratory tests the bioconcentration factor with DDT can be varied from several hundred to several million.

The Need for Standardization of Test Methods for Criteria Used in the Determination of Bioconcentration Potential and Bioconcentration Factors

The preceding information in this paper illustrates the variable nature of data used for the calculation of bioconcentration and bioconcentration factors. Since these two criteria are the products of many variables, even small errors or misinterpretations in the actual values used can cause large errors in the final determinations. Without standardization of the test methods small errors are inevitable. Some of the variables (by no means all-inclusive) which are often ignored or overlooked and are important to consider in the evaluation of bioconcentration potential are as follows:

a. The environmental component (water, soil, diet, animal, plant) used as a reference point for comparison with residues held by the animal or plant organism is one of the chief reasons for variation in the bioconcentration factor. This reference point must be specified. Residues in water are generally chosen for the reference in aquatic studies, and those in soil in terrestrial studies.

b. Residues of pesticides in the whole plant or animal body or of special tissues may be chosen as a basis for comparison with the residue of the reference environmental component (i.e., water, soil, or air). Fat tissues, for example, are often chosen as the residue basis for DDT-R in animals, frequently without mentioning the residue level in the body or the percent fat in the animal. In the calculation of bioconcentration factors it is not correct, for example, to use the residues retained by algae for comparison with residues in the fat tissue of birds (instead of in whole bird tissues), without specifying such comparisons.

c. It is important to specify the time at which residue data are taken (shortly after or long after application of the pesticide), since residue values at equilibrium are often much different than residue values taken immediately after application. Furthermore, residue data should be obtained at specified sequential times after application of the pesticide. Without such data, the relative rise and fall of residue levels held by various objects, organisms, or tissues in the environment cannot be studied and used for calculation of rates of partitioning and accumulation of the pesticide.

d. The residues in soil and plant or animal tissues should be stated in terms of wet or dry weight. Many reports of data do not take this factor into consideration. Residues in soil and tissues of plants and animals are variously measured and reported.

e. The presence or absence of particulate matter, soluble salts, etc. in water should be reported when measuring pesticide residues in water since adsorption and other factors may have a greater effect on the measurement than solubility of the pesticide in water.

f. Measurement of ppm residue in bottom muds or sediments or in soil must be accompanied by the depth of sampling and preferably by a weight per volume measurement (specific gravity). Since many pesticides are sorbed on the surface layer of mud, ppm measurements are meaningless without such accompanying information.

g. Analytical procedures have improved greatly in the past few years. Older residue data may often be suspect, especially when not identified adequately by test methods other than by gas chromatographic analysis. DDT data may or may not include DDD and/or DDE, or other metabolites, or overlap polychlorinated biphenyls or other environmental contaminants, in the gas chromatographic spectrum. The analytical method description should specify the limits of the interpretations to be placed upon such results. Some analytical procedures result in alteration of residues during measurement, sometimes creating new chemical structures not found in the original pesticide residue sample.

h. Variation in residue values on objects of different surface-area to mass ratios or in different depths of water or soil are often overlooked or miscalculated.

i. Measurement of metabolites of the pesticide, particularly persistent metabolites, is not often well documented in the literature and is sometimes a more important consideration than the originally applied pesticide.

j. Variation in the food intake capacity of various organisms is an important consideration in calculating the total pesticide residues consumed.

k. The physical, chemical, and biochemical properties of pesticides are not sufficiently determined by standardized, sensitive test methods or, if determined, are not sufficiently used in explaining or predicting interaction with various environmental components.

l. Many laboratory test results are used to predict the effects of pesticides without taking into account the natural biological and ecological factors affecting pesticide bioconcentration in the environment.

m. Bioconcentration of residues within specific tissues or organisms is most often related to time of exposure and partitioning due to sorption and solubility. Chain of life residues of pesticides do not necessarily result in increasing concentrations in terms of ppm in each succeeding link of the "food chain." Residues accumulated by algae, for example, may be as high as those accumulated by fish or birds.

Additional comments on variables in residue analyses and the need for standardization of biological and analytical test methods are found in the publications of Gunther

(1962) and Lisk (1970).

Summary

Molecules such as DDT and its stable transformation products DDD or DDE, have been shown to persist and become bioconcentrated. The degree of bioconcentration of DDT and other materials depends on first, the temporary equilibrium of residue concentration reached initially by adsorption on organisms in competition with the air, water, and soil segments of the environment; and second, the redistribution of these adsorbed pesticides in organisms and their tissues by ingestion, absorption, metabolism, partitioning, storage, and elimination. Some of the important criteria for selecting test methods to measure bioconcentration discussed are the physical, chemical, and biochemical properties of the pesticide and its transformation products; surface weight relationship of treated environmental objects; consumption rates of various organisms; and other biological and ecological responses of various organisms to pesticides as used in practice. Bioconcentration factors are highly variable for a compound such as DDT due to great variation in the parameters and test methods used as a basis for the measurement and calculation of the factor. Standardization of environmental test methods are needed for comparative purposes. Serious ecological problems may arise from bioconcentration when the degree of partitioning of a substance or its transformation products from water results in translocation to and storage in critical tissues of the organisms. Bioconcentration in many species can result in high residues caused by constant exposure to widely distributed toxicants, and an intake exceeding the capacities of the organism to metabolize and eliminate them. Some of the important properties of pesticides resulting in high bioaccumulation are low water solubility; high fat solubility; high partitioning coefficients from water to environmental components; and high stability under various hydrolytic, light, heat, and microbiological conditions.

References

Altman, P. L., and D. S. Dittmer: Biological Handbook. Fed. Amer. Soc. Exp. Biol. Washington, D.C. p. 344, 366 (1962).

Bailey, G. W., and J. L. White: J. Agr. Food Chem. 12, 324 (1964).

Buhler, D. R., and W. E. Shanks: J. Fish. Res. Board Can. 27, 347 (1970).

Cox, J. L.: Bull. Environ. Contam. Toxicol. 5, 218 (1970).

Davis, J. C.: J. Fish. Res. Board Can. 27, 1860 (1970).

Dindal, D. L.: J. Wildl. Manag. 34, 74 (1970).

Edwards, C. A.: Persistent pesticides in the environment. CRC Press. The Chemical Rubber Co. Cleveland, Ohio. p. 1-78 (1970).

El Sayed, E. I., J. B. Graves, and F. L. Bonner: J. Agr. Food Chem. 15, 1014 (1967).

Findlay, G. M., and A. S. W. de Freitas: Nature 229 (5279), 63 (1971).

Freed, V. H.: Chemistry of Herbicides. Soil Science of America. ASA Special Publication No. 8. Madison, Wisconsin. p. 25-43 (1966).

Gish, C. D.: Pesticide Monit. J. 3, 241 (1970).

Grzenda, A. R., D. F. Paris, and W. J. Taylor: Trans. Amer. Fish Soc. 99, 385 (1970).

Gunther, F. A.: Adv. Pest Control Res. 5, 191 (1962).

Gunther, F. A., W. E. Westlake, and P. S. Jaglan: Residue Rev. 20, 1 (1968).

Hadaway, A. B., F. Barlow, and C. R. Turner: Bull. Entomol. Res. 60, 17 (1970).

Hamelink, J. L., R. C. Waybrant, and R. C. Ball: Trans. Amer. Fish Soc. 100, 207 (1971).

Hannon, M. R., Y. A. Greichus, R. L. Applegate, and A. C. Fox: Trans. Amer. Fish Soc. 99, 496 (1970).

Harris, C. R., and W. W. Sans: Pesticide Monit. J. 3, 182 (1969).

Hartung, R., and G. W. King: Environ. Sci. Technol. 4, 407 (1970).

Hickey, J. J., J. A. Keith, and F. B. Coon: *J. Appl. Ecol.* *3* (Suppl.) (1966).

Hoerger, F., and E. E. Kenaga: Pesticide residues on plants: correlation of representative data as a basis for estimation of their magnitude in the environment. *Environmental Quality and Safety.* Academic Press, N.Y. In press (1971).

Johnson, B. T., C. R. Saunders, H. O. Sanders, and R. S. Campbell: *J. Fish. Board Can.* *28*, 705 (1971).

Johnson, D. W., and S. Lew: *Pesticide Monit. J.* *4*, 57 (1970).

Jones, H. A., H. T. Fluno, and A. B. Hendricks: *J. Econ. Entomol.* *38*, 207 (1945).

Keil, J. E., and L. E. Priester: *Bull. Environ. Contam. Toxicol.* *4*, 169 (1969).

Keith, J. O.: Variations in the biological vulnerability of birds to insecticides. *The biological impact of pesticides in the environment.* J. W. Gillett, Ed. *Environ. Health Series No. 1.* Oregon State University. Corvallis, Oregon (1970).

Keith, J. O., D. B. Knapp, and W. C. Johnson: Kinetics of pesticides in natural marsh ecosystems. *Ann. Prog. Rep., Calendar year 1967.* *Denver Wildlife Research Center.* USDI, Denver, Colorado (1967).

Kenaga, E. E.: *Down to Earth* *23*, 11 (1968).

Kenaga, E. E.: Factors to be considered in the evaluation of the toxicity of pesticides in the environment. *Environmental Quality and Safety.* Academic Press, N.Y. In press (1971a).

Kenaga, E. E.: Guidelines for environmental study of pesticides: Determination of bioconcentration potential. *Symp. Guidelines for Environmental Studies of Pesticides.* *162nd Nat. Meet. Amer. Chem. Soc.* Washington, D.C. Sept. 12-17 (1971b).

Lichtenberg, J. J., J. W. Eichelberger, R. C. Dressman, and J. E. Longbottom: *Pesticide Monit. J.* *4*, 71 (1970).

Lisk, D. L.: *Science* *170* (3958), 589 (1970).

Macek, K. J., C. R. Rodgers, D. L. Stalling, and S. Korn: Trans. Amer. Fish Soc. 99, 689 (1970).

Martin, W. M.: Pesticide Monit. J. 3, 102 (1969).

Meeks, R. C.: J. Wildl. Manage. 32, 376 (1968).

Metcalf, R. L.: The mode of action of organic insecticides. Chemical-Biological Coordination Center Review No. 1. National Research Council. Washington, D.C. (1948).

Murphy, P. G.: Bull. Environ. Contam. Toxicol. 5, 405 (1970).

Nice, M. M.: Bird-Banding 9, 1 (1938).

Riemschneider, R.: Die Pharmazie 9, 675 (1948).

Seba, D. B., and E. F. Corcoran: Pesticide Monit. J. 3, 190 (1969).

Spector, W. S.: Handbook of biological data. W. B. Saunders Co., Philadelphia. p. 196 (1956).

Vance, B. D., and W. Drummond: J. Amer. Waterworks Assoc. 61, 360 (1969).

Voerman, S.: Bull. Environ. Contam. Toxicol. 4, 64 (1969).

Wershaw, R. L., P. J. Burcar, and M. C. Goldberg: Environ. Sci. Technol. 3, 271 (1969).

BHC RESIDUES IN THE ENVIRONMENT

Ryo Tatsukawa, Tadaaki Wakimoto, and Tsunehiko Ogawa

College of Agriculture, Ehime University
Matsuyama, Ehime, Japan

Introduction

The organochlorine insecticides have been widely used for the past 20 years in Japan. While environmental concern has been focused on food contamination for some time, the investigation of pesticides in terms of environmental pollution, from both geochemical and ecological viewpoints, has started only recently.

The purpose of this paper is to examine the fate and distribution, in the environment, of BHC as used in paddy fields. We will also analyze the factors which determine the distribution of BHC in the environment.

BHC was chosen for study for the following reasons: (a) BHC was the most widely used organochlorine pesticide in Japan. (b) The degradation and fate of BHC under water-logged conditions was unknown, yet was supposed to be different from upland conditions. (c) The technical BHC used in Japan contains several isomers (TABLE 1). We wanted to know how each isomer behaved in different environments.

Besides the above reasons, BHC is a synthetic compound which is persistent in nature, and it can be detected in very minute amounts. These properties of BHC make it useful as a geochemical tracer.

The nature and fate of BHC in the environment might also be applicable for determining the fate of other organochlorine pollutants, such as DDT, PCBs, and chlorinated paraffins.

TABLE 1

Properties of Technical BHC

Isomer	Composition (%)	Solubility in water[a] (ppm, 25°C)	Vapor pressure[b] (mm Hg, 20°C)
α	68-78	1.6	2×10^{-2}
β	9	0.7	5×10^{-3}
γ	13-15	7.9	3×10^{-2}
δ	8	21.3	2×10^{-2}

[a]From Kanazawa, 1971.

[b]From Slade, 1945.

Statistical Surveys

The annual input of BHC to the environment on a prefectural basis was estimated by the sales statistics of BHC (Anonymous, 1963-71). The summaries of these calculations are shown in Fig. 1 and 2. The greatest use in grams of gamma-BHC per hectare of arable land was in the western part of Japan. Production and use of BHC increased annually to the peak year 1968, when 890 grams of gamma-BHC was used (per hectare) on arable land. In Fukuoka and Saga Prefectures of North Kyushu more than 4 kg of gamma-BHC per hectare was used.

In 1970 the use of BHC decreased sharply, although in eastern Japan BHC was used slightly more than in previous years.

As BHC had been sprayed mainly to paddy fields, the statistical data are not calculated on the basis of all arable land but on the basis of the paddy field acreage. As areas of paddy soils and upland soils are nearly even, the amount of gamma-BHC used per hectare may be about twice as large as in paddy fields.

Analysis of BHC Residues

During the course of the study, the methods of analysis progressed. At present, we analyze the organochlorine compounds by the methods shown in Fig. 3. According to the nature of the samples and the purposes of the study, the methods used are different in specific detail.

We have conducted environmental surveys since 1968. A summary of this work follows. Since residue levels of BHC in soil, water, air, and various biota are roughly proportional to the amounts used, the western part of Japan would be expected to be more contaminated with BHC than the eastern part of the country. BHC residues in paddy soils are also higher in the western part of Japan than in the eastern part, but the over-all concentrations of BHC in paddy field soils are not as high in spite of intensive spraying. BHC is degraded faster under water-logged conditions than under upland conditions. Also, the higher the soil temperature, the more rapid the degradation of BHC. Among the BHC isomers, the gamma isomer is most rapidly degraded, followed by the alpha, beta, and delta forms. Consequently, soils following rice harvesting have high levels of the beta isomer.

Although accumulation of BHC in soils is higher in western Japan, residue rates (residual gamma-BHC/total gamma-BHC used per the same unit area) are higher in eastern Japan, probably because of its colder climate.

BHC has not been as widely used in upland soils, but degradation under these conditions is very slow, and therefore the residue levels in upland soils are approximately the same as in paddy soils.

BHC levels in fresh surface waters fluctuate widely, being high in the growing season and low in the winter. The isomer composition of BHC residues in the environment are near to technical BHC just after spraying, but gradually changes with time. The isomer composition of BHC in waters is generally highest with alpha, followed by the beta, gamma, and delta forms. Marine waters surrounding Japan also contain BHC at levels up to 1 ppb.

BHC sprayed on paddy fields vaporizes rather rapidly. This causes atmospheric and rain pollution in wide areas. Urban air usually contains 0.1-0.01 $\mu g/m^3$ of BHC. Both beta- and delta-BHC are more stable in the atmosphere than alpha- and gamma-BHC. Less than 1% of the BHC used precipitates again to the land surface via rain water.

Biota in the environment also accumulate BHC. Beta-BHC, among the isomers, is the most persistent and most concentrated in animal tissue. Adipose tissue of average Japanese people contains 10-20 ppm of total BHC, of which beta-BHC comprises more than 95%. BHC in the air and drinking water comprises less than 1-2%, and almost all BHC reaching man comes from food, in particular, beef, cows milk, butter, and cheese.

In April 1971, the use of BHC was essentially banned. Therefore, BHC in our environment should diminish with time. However, BHC in man and in cows is expected to persist for a long time.

Accumulation and Fate of BHC in the Environment

Estimated concentrations, accumulation amounts, and isomer composition of BHC are shown in TABLE 2. The fate of BHC in the environment is schematically shown in Fig. 4.

The total load of BHC in the environment in Japan is estimated to be 2,000-7,000 tons, of which more than 90% is in soils, especially in arable soils. Approximately half of the total consists of the beta isomer, and gamma-BHC occupies only a small portion, probably less than 10% of the total.

However, soils are not the only sites of BHC accumulation. It also is found on the treated plants. By comparison with soils, the accumulation of BHC in biological materials is not as great.

Air and fresh surface water are not the terminal

TABLE 2

Residue Distribution of BHC in the Environment (after harvesting rice, 1968-1970)

	Estimated concentration (ppm)	Amounts of accumulation (ton)	Isomer composition
Soil			
Paddy	0.2-1.0	800-4,000	$\beta > \alpha > \delta > \gamma$
Upland	0.1-0.5	400-2,000	
Others	0.01	500	
Total Soil		1,700-6,500	
Water, fresh surface	0.0001-0.001	10-100	$\alpha > \beta > \gamma > \delta$
Air	0.00001-0.0001	5-50	
Biological materials			
Rice, whole	1-10	40-400	
Livestock	1-5	3-15	$\beta > \alpha > \gamma > \delta$
Man	2-5	10-25	
Others	?	?	
Total Biota		ca. 50-500	
Grand Total		ca. 1,800-7,000	

Input of BHC into the Environment:

Year	1970	13,000 ton
	1969	24,000
	1968	33,000
	1967	28,000

environmental sites of BHC, but rather the transport mechanisms. The atmosphere, in particular, plays an important role in the fate of BHC in the environment, both directly by drift, and indirectly as vapor from the surface of waters, soil, and plants. A large proportion of BHC used eventually reaches the atmosphere. In spite of constant supplies of BHC to the air, the estimated amount of BHC in the air is very small. In addition, BHC in the atmosphere is thought to undergo rapid degradation by sunlight.

Since the earth is a closed system, degradation products of BHC, and as yet unknown compounds derived from BHC, will eventually reach the land surface. It is essential that we further study the fate and biological effects of BHC and its degradation products in the atmosphere in order to clarify and reduce the load of BHC in the environment, and to guarantee the safety of our planet.

References

Anonymous: Noyaku Yoran (Yearbook of Pesticides). Japan Plant Protection Assoc., Tokyo (1963-1971).
Kanazawa, J.: Kagaku, Tokyo, 41, 384 (1971).
Slade, R.: Chem. Ind., 314 (1945).

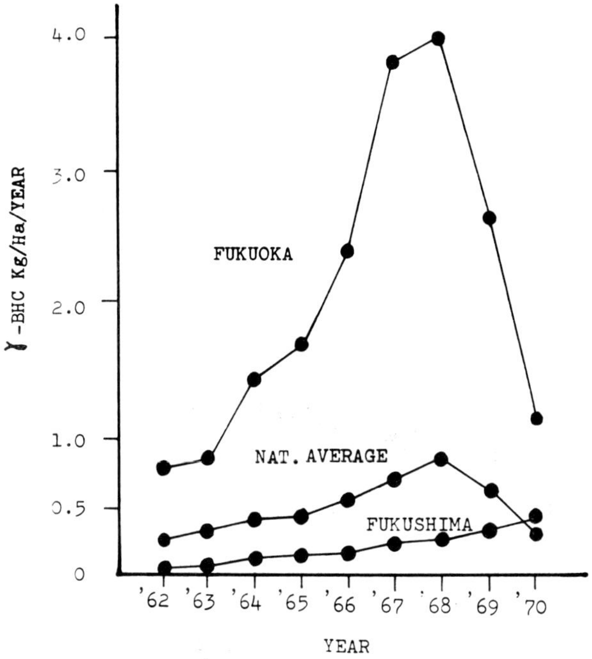

Fig. 1. Annual sales of γ-BHC in Japan.

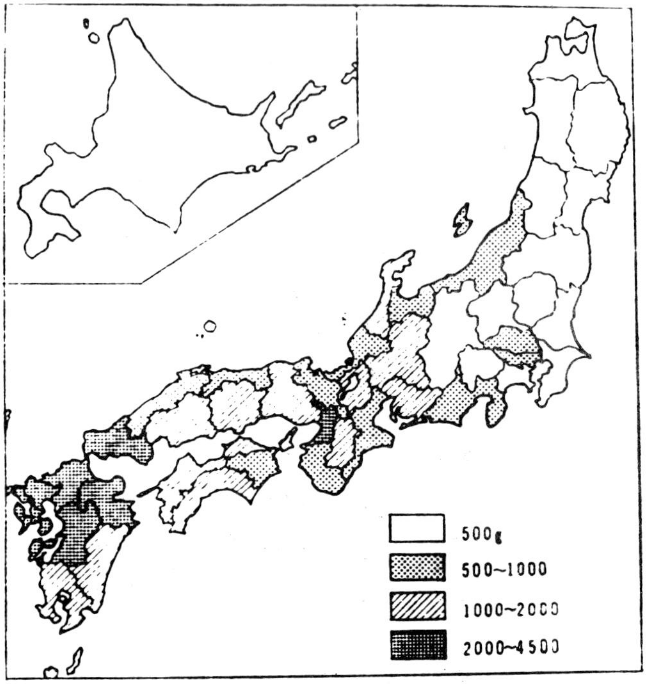

Fig. 2. γ-BHC sales of prefectures per hectare of arable land (pesticide year 1968).

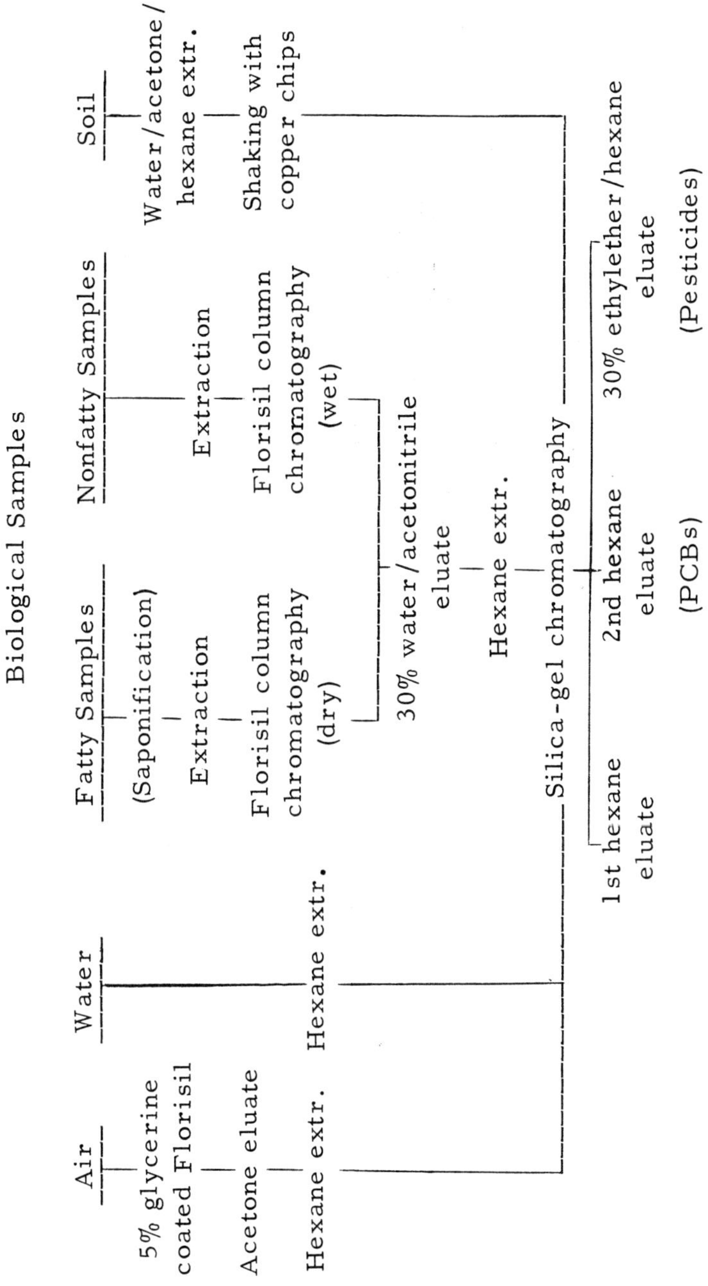

Fig. 3. Outline of method for organochlorine residues in the environmental samples.

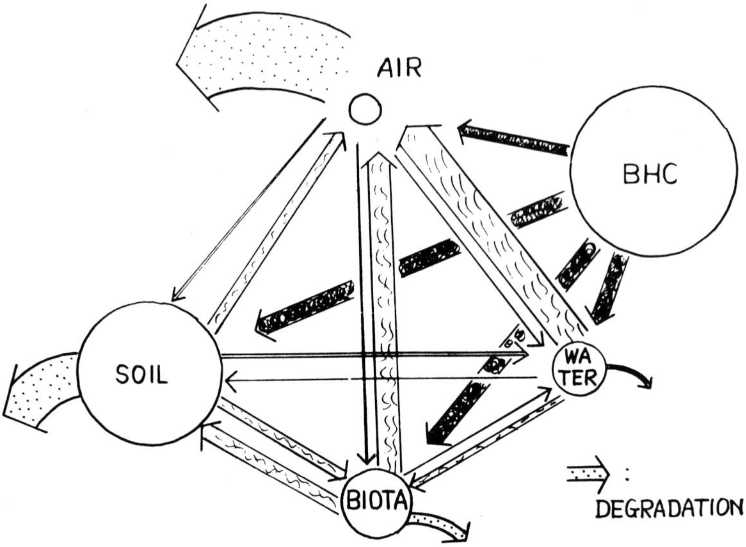

Fig. 4. The fate of BHC in the environment.

CONTAMINATION OF MILK WITH CHLORINATED HYDROCARBON PESTICIDES

Hiroya Tanabe

National Institute of Hygienic Sciences
Tokyo, Japan

Introduction

Because of the stability of most chlorinated hydrocarbon pesticides, contamination of the environment with these compounds has caused anxiety in highly industrialized countries. Contamination of cows milk with DDT has caused some trouble in several western countries. In Japan, also, contamination of animal foods with these pesticides has been noted from time to time. In 1965, collaborative studies were initiated in order to determine accurate and precise methods for detecting residues of these pesticides in cows milk as well as dairy products.

In 1966, members of the Hygienic Institute of Kochi Prefecture reported results of a survey on the contamination of many kinds of foods with chlorinated hydrocarbon pesticides. At that time, no unusual contamination was noted. The government, in 1969, planned a nation-wide survey on the contamination of animal foods. This was in response to international news information concerning the contamination of the environment with these compounds--especially DDT. At the same time, Dr. Ueda of the Hygienic Institute of Kochi Prefecture, reported his findings regarding the serious contamination of several kinds of animal foods with beta-BHC to the government. A national survey was begun on the basis of his report. It was a cooperative effort with six local governmental hygienic institutes and the National Institute of Hygienic Sciences. Also, studies on the toxicity of beta-BHC were started at the National Institute employing monkeys and mice. Additionally, the pesticide manufacturers in Japan stopped producing BHC and DDT.

Detection of Chlorinated Hydrocarbons in Cows Milk

The principal foods of the survey were milk and dairy products due to their nutritional significance. In addition, Hokkaido and Nagasaki Prefectures' Hygienic Institutes cooperated in this project so that remote areas of Japan were included as well. Studies to trace the source of contamination of cows milk were also carried out by several of the above cooperators, as well as a research group organized by the Ministry of Agriculture and Forestry. The method employed for the detection and determination of chlorinated hydrocarbon pesticides was similar to that described in the U.S. Food and Drug Administration's Pesticides Analytical Manual. However, modified points in the analytical procedures were as follows: Fat content was determined by the Gerber method; n-hexane was substituted in almost all procedures, except extraction in cases of heavy emulsions, for petroleum ether; and Florisil was always kept at 130°C before use (Fig. 1). Beta-BHC, in cases of especially high contamination, was identified by GC-Mass. method.

Many samples of raw milk obtained in the field, as well as milk from the market, were analyzed. Also, changes in contamination levels were followed employing several specified cows.

At the beginning of the survey, the origin of beta-BHC contamination of cows milk was revealed (TABLES 1 and 2) to be rice plant straw, the main feed for cows in Japan. The BHC had been applied to rice to suppress damage due to *Chilo suppressalis*. In addition, the technical grade of BHC was used on rice instead of the purified gamma-BHC, which was more than twice as expensive.

TABLE 1

Chlorinated Hydrocarbon Pesticide[a] Residues in Cow Fodders

Samples	α-BHC	β-BHC	γ-BHC	δ-BHC	DDT	DDE	DDD
Rice straw	3.494	5.200	1.936	1.810	0.508	0.015	0.101
Corn grass	1.156	0.047	0.061	0.052	nd	nd	nd
Wheat straw	0.044	nd[b]	0.238	nd	0.014	0.006	nd
Wheat bran	0.075	0.035	0.102	0.035	tr	tr	nd
Broad bean pad	0.193	tr[b]	0.010	tr	nd	nd	nd
Hay pellet	0.087	0.054	0.031	0.023	0.022	tr	nd
Bean cake	0.019	0.069	0.009	0.015	0.019	tr	tr
Beer cake	0.033	tr	0.022	0.023	nd	nd	nd
Beet	0.035	0.011	0.018	0.013	0.094	0.011	0.030
Assorted feed	0.027	tr	0.019	0.008	nd	nd	nd
Hay	0.011	0.023	0.006	0.007	nd	nd	nd
Pressed barley	0.013	0.018	0.002	0.009	nd	nd	nd
Molasses	0.029	tr	0.008	0.004	nd	nd	nd
Soy sauce cake	0.019	tr	0.010	nd	nd	nd	nd
Bean-curd refuse	0.007	tr	0.002	nd	nd	nd	nd
Soy bean pad	0.002	tr	0.002	nd	nd	nd	nd

[a] In none of the cases were residues of aldrin, dieldrin, and endrin found to be present, except 0.07 ppm of dieldrin in beet. All values are expressed in ppm.

[b] nd = not detected; tr = trace (< 0.001)

TABLE 2

BHC Residues in Fodders and Raw Milk[a]

Samples	Location[b]	α-BHC	β-BHC	γ-BHC	δ-BHC	Total
Rice straw	A	0.418	0.307	0.108	0.174	0.997
	B	0.156	0.115	0.072	0.081	0.424
Wheat bran	A	0.032	tr	0.069	tr	0.101
	B	0.009	tr	0.002	tr	0.011
Green grass	A	0.065	0.036	0.077	0.023	0.201
	B	0.006	0.003	0.003	0.002	0.014
Assorted feeds	A	0.037	0.012	0.017	0.004	0.070
	B	0.028	0.004	0.015	0.010	0.057
Raw milk	A	0.172	0.191	0.007	0.018	0.388
	B	0.034	0.060	0.002	0.003	0.099

[a] tr = trace (< 0.001). All values are expressed in ppm.

[b] A = a location where the raw milk was highly contaminated with BHC, and
B = a location where the raw milk was less contaminated with BHC

The results with milk are shown in Fig. 2, 3, 4, and 5. Generally, contamination of BHC from different areas almost paralleled pesticide use. Contamination levels of beta-BHC in the north-eastern part of Japan including Hokkaido, Tohoku, Kanto, and the north-eastern part of the Chubu area were not as high in 1969, or in early 1970. However, the levels in other parts of Japan (i.e., the south-western portion) were considerably higher, especially in both the Osaka and Nagasaki Prefectures, at the beginning of 1970.

After the disclosure of the cause of contamination, the government made every effort to lower the beta-BHC level in cows milk. At first, the BHC applied to growing rice plants was strictly regulated, and later its use was prohibited. They also educated the dairy farmers not to feed rice straw. As a result of these measures, the beta-BHC levels in cows milk lowered little by little.

In general, in highly contaminated areas residue reduction in the early stage was easy. However, later it took much more time to lower the levels. The result of one of the studies illustrating the changes of residues with time was carried out at the Okayama Dairy Institute, where feed could be controlled (TABLE 3).

Tentative ADI for beta-BHC was estimated to be 0.05 mg/kg per day according to the data for subacute toxicity in monkey and Chronic (1 year) toxicity in mouse, obtained at the Department of Toxicology of the National Institute of Hygienic Sciences. From this value a tentative guideline for beta-BHC levels in cows milk was established as 0.2 ppm. One liter of milk containing this residue level would contain an equivalent of a little less than one-tenth of ADI dose of beta-BHC.

Fortunately, by 1971 the beta-BHC levels in cows milk have been lowered to less than 0.2 ppm, even in the south-western part of Japan.

TABLE 3

Chlorinated Hydrocarbon Pesticide[a] Residues in Raw Milk of Cow Fed on Fodders without Rice Straws

		α-BHC	β-BHC	γ-BHC	δ-BHC	DDT	DDE	DDD
1970	1	0.065	0.069	0.002	tr	0.004	0.004	0.002
	2	0.024	0.023	0.003	tr	0.002	0.003	0.001
	5	0.035	0.038	0.002	0.002	0.007	0.005	0.008
	9	0.024	0.023	0.002	0.002	0.004	0.005	tr
	10	0.011	0.022	0.001	0.003	0.004	0.003	nd[b]
1971	1	0.004	0.003	tr[b]	0.001	tr	0.002	tr
	3	0.010	0.015	0.002	0.002	nd	0.005	nd

[a]Throughout, no aldrin and endrin residues were found to be present in any of the samples. All values are expressed in ppm.

[b]nd = not detected; tr = trace (<0.001)

BHC Residues in Human Milk

It is quite natural that some members of the survey group were interested in the contamination levels of human milk. During the cows milk survey, data on human milk contamination with beta-BHC were reported, and the urgent need for a wider survey discussed.

During the latter half of 1970, the participants of the survey group analyzed several kinds of meat and fish and found that, as a rule, beef was much more contaminated with beta-BHC than any other animal foods (except milk). Therefore, it was assumed that efforts to lower the residue levels of cows milk would result in lowering that of beef and then human milk as well. When the contamination level of cows milk was lowered to about one-tenth of that present at the beginning of the survey, the government organized a new survey group to examine residue levels in human milk. Hygienic institutes of up to 24 out of 43 prefectures in Japan participated in this survey. Human milk was sampled (70-100 ml) from each mother. The mothers who participated in this study were requested to answer a number of questions.

The first question concerned where they lived and what they did for a living; i.e., did they live in an urban area and had never engaged in agriculture, or had they resided in the country and engaged in agriculture? The second question was for mothers who lived in the country and had engaged in agriculture. They were asked whether or not they spread much pesticides around their houses, the frequency of using pesticides by themselves per year, and the kinds of pesticides used.

They were also questioned as to kinds of water they had used (well water or city water), the amount and kinds of meat and fish they usually consumed (beef, pork, chicken, freshwater fish, or marine fish), and their daily intake of cows milk (specifically, the amount of butter or margarine they ate). Finally, they were asked what kinds of vegetables they liked, leafy vegetables or root crops; and whether or not they smoked.

The methods of residue analysis of human milk were similar to those used in analyzing cows milk. However, since each sample amount obtained was very small, the weight of the evaporation residue of the n-hexane extract was regarded as fat content.

The preliminary results obtained are shown in Fig. 6, 7, and 8. Just as with cows milk, contamination levels of beta-BHC were considerably higher in the south-western part of Japan. In such areas, beta-BHC levels in milk from rural mothers who had engaged in agriculture were, despite their more frequent opportunities to be in contact with pesticides, lower than those from urban mothers who had never engaged in agriculture. According to statistical studies on the relationships between beta-BHC levels in human milk and location of dwelling and types of foods consumed, only two items appear significant: the daily intake of cows milk and the fondness of the individual for beef (Fig. 9 and 10). It thus became clear that the main contamination route of human milk (and body) was the oral intake of these foods. Also, the data may be partially explained by the fact that rural people usually prefer more vegetable foods to animal ones in comparison with urban people.

The previously mentioned assumption that efforts to lower the contamination levels in cows milk might bring desirable results to human milk was also thus scientifically supported.

A larger scale survey on pesticide residues in human milk is beginning, with hygienic institutes in almost all prefectures in Western Japan, as well as many in the north-eastern district, participating.

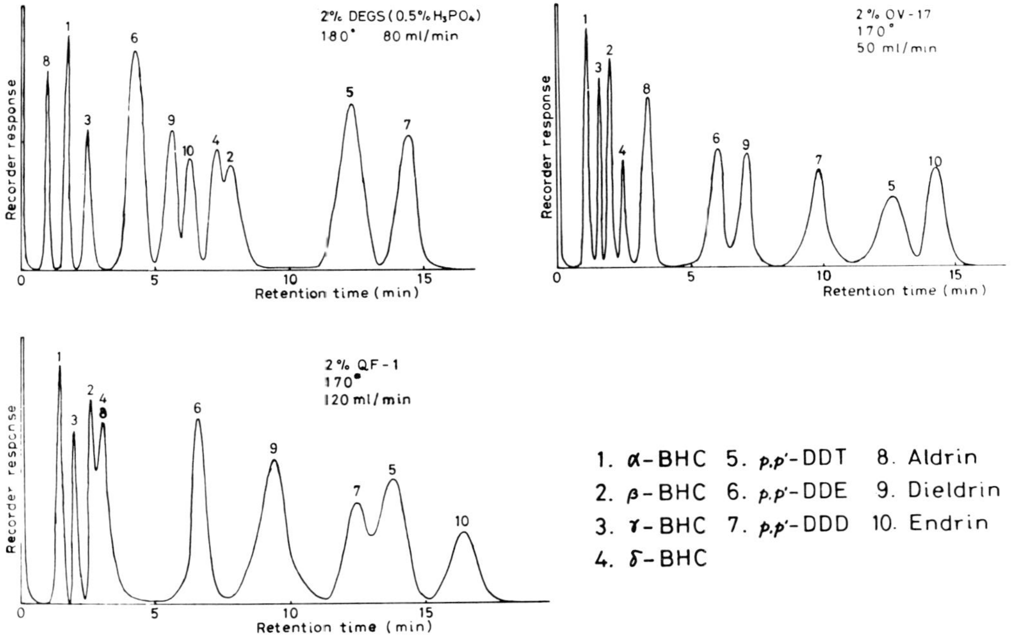

Fig. 1. Gas chromatogram of chlorinated hydrocarbon pesticides (as detected by an electron capture detector).

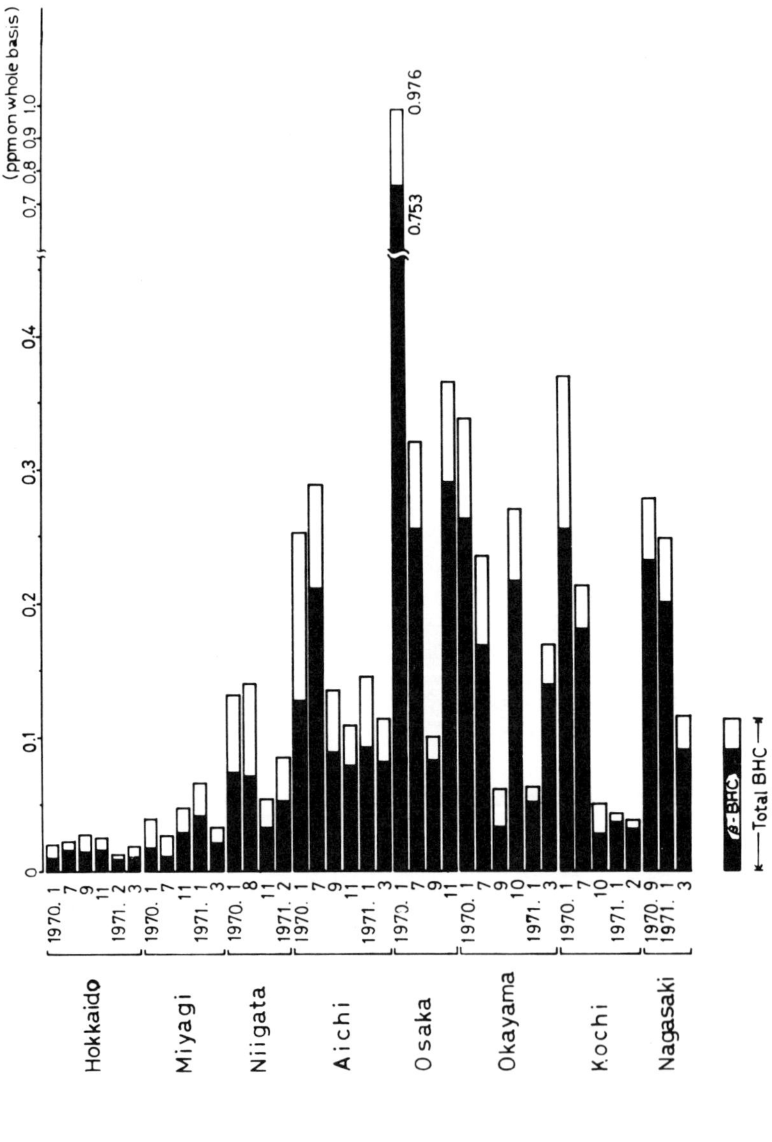

Fig. 2. Variation in BHC residues in raw milk by location and season.

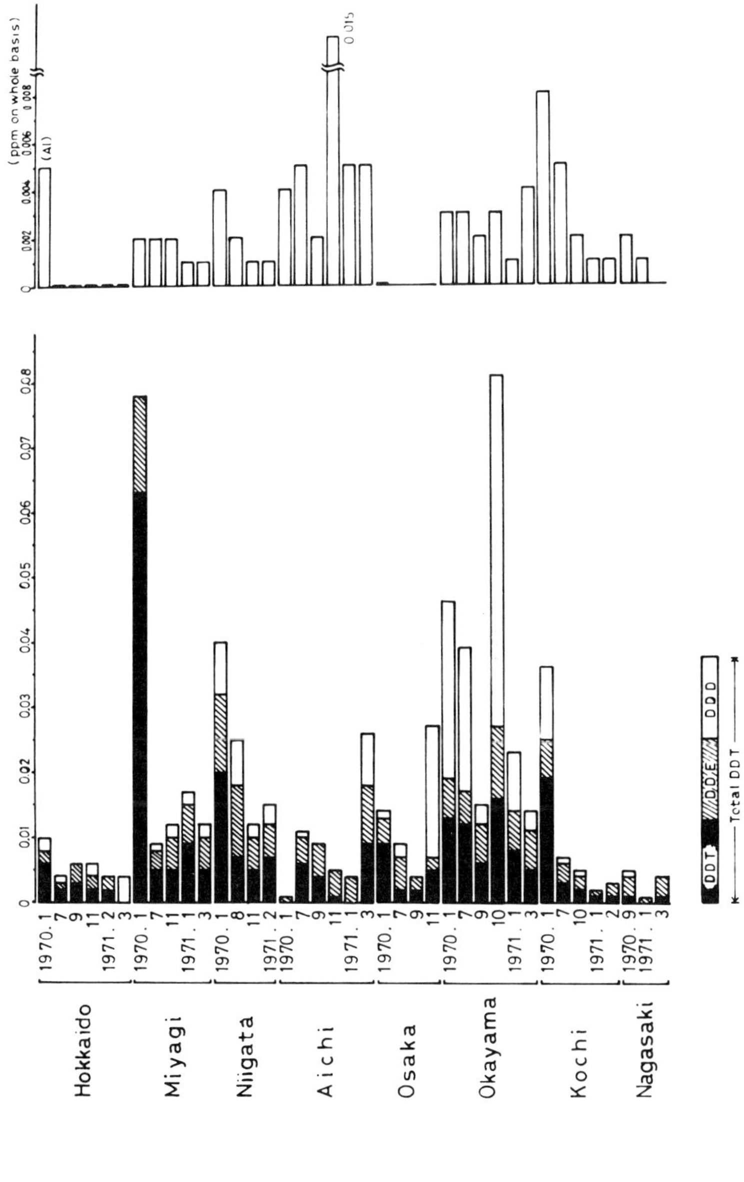

Fig. 3. Variation in DDT and dieldrin residues in raw milk by location and season.

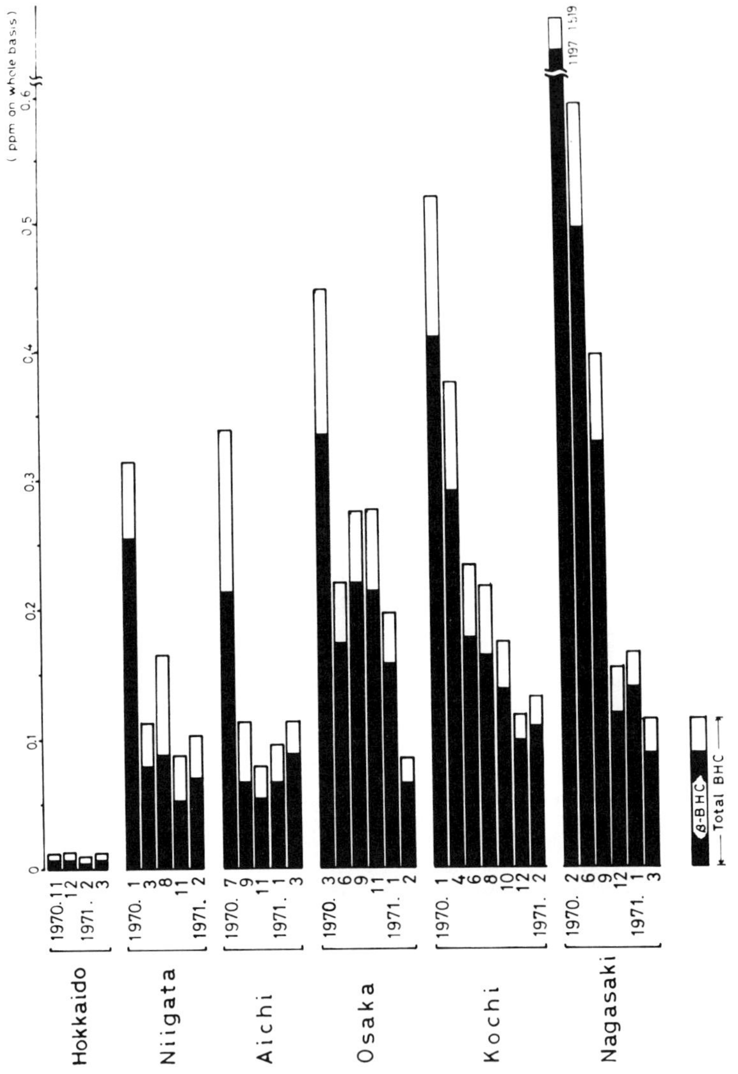

Fig. 4. Variation in BHC residues in market milk by location and season.

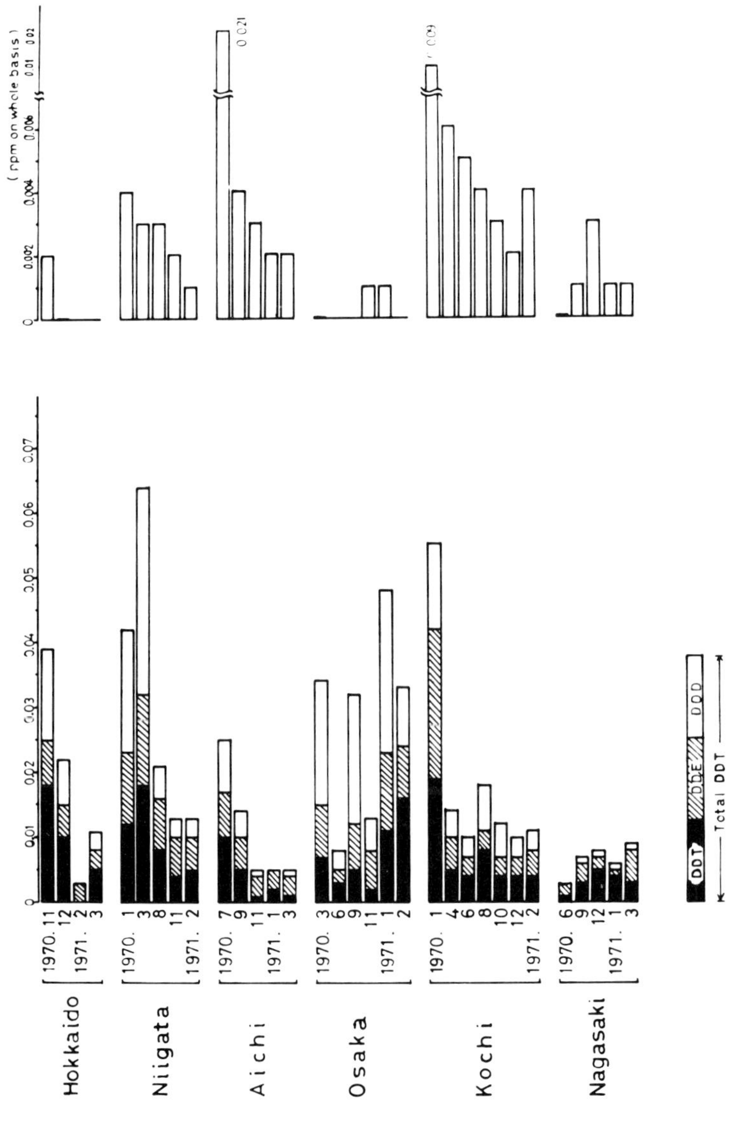

Fig. 5. Variation in DDT and dieldrin residues in market milk by location and season.

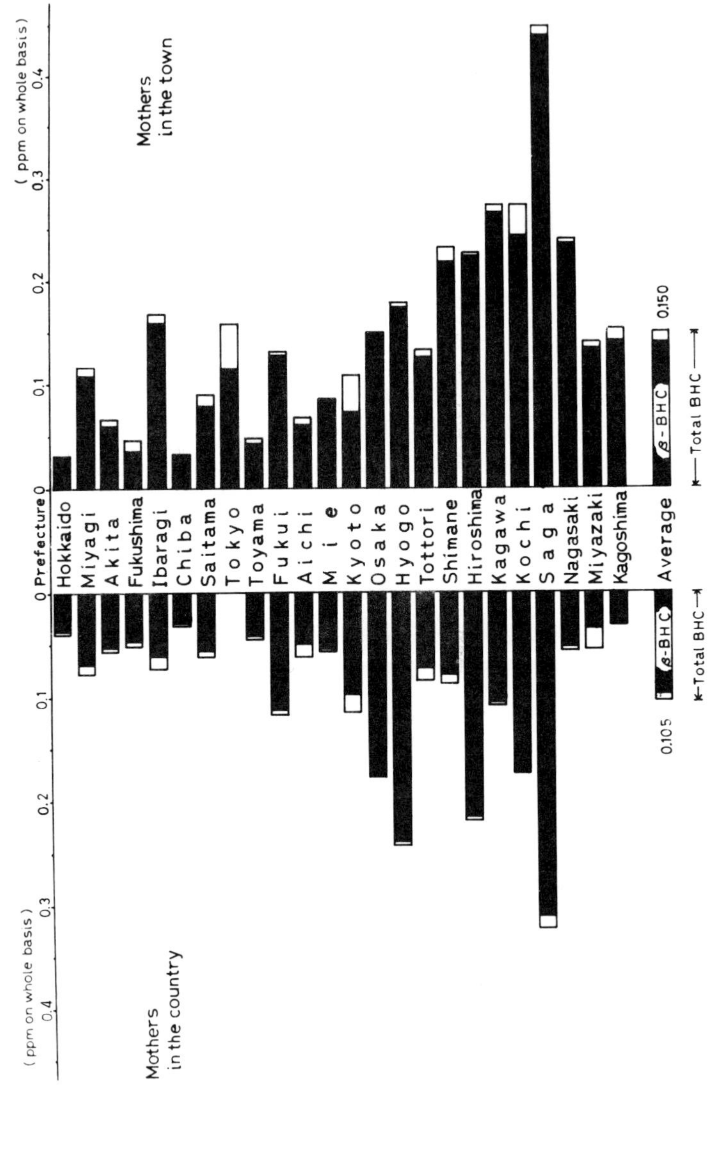

Fig. 6. BHC residues in human milk.

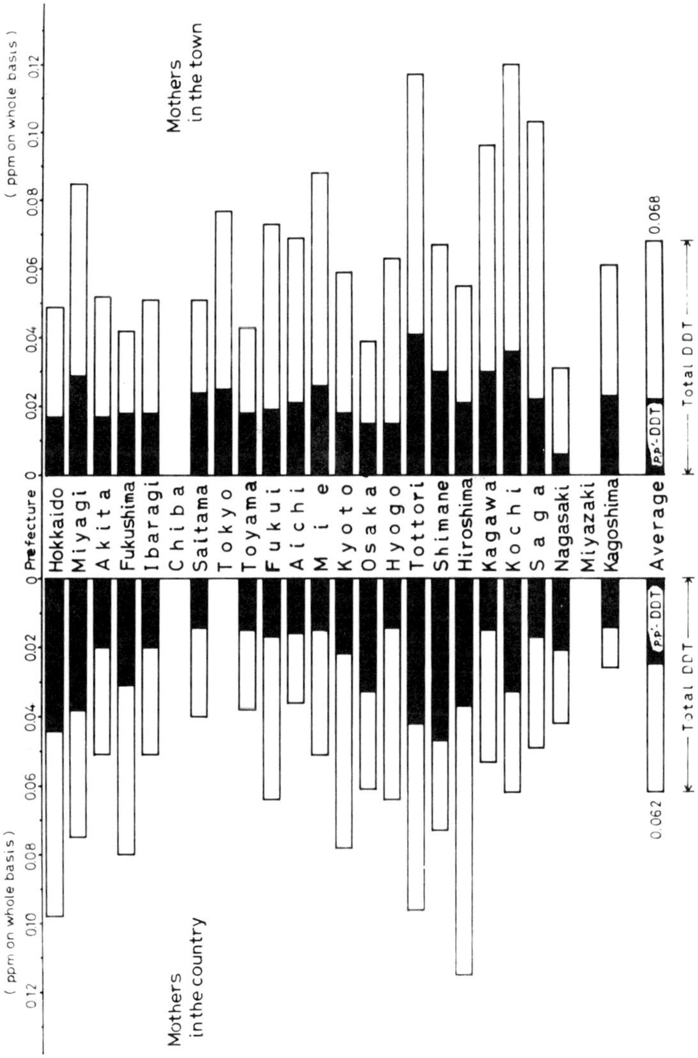

Fig. 7. DDT residues in human milk.

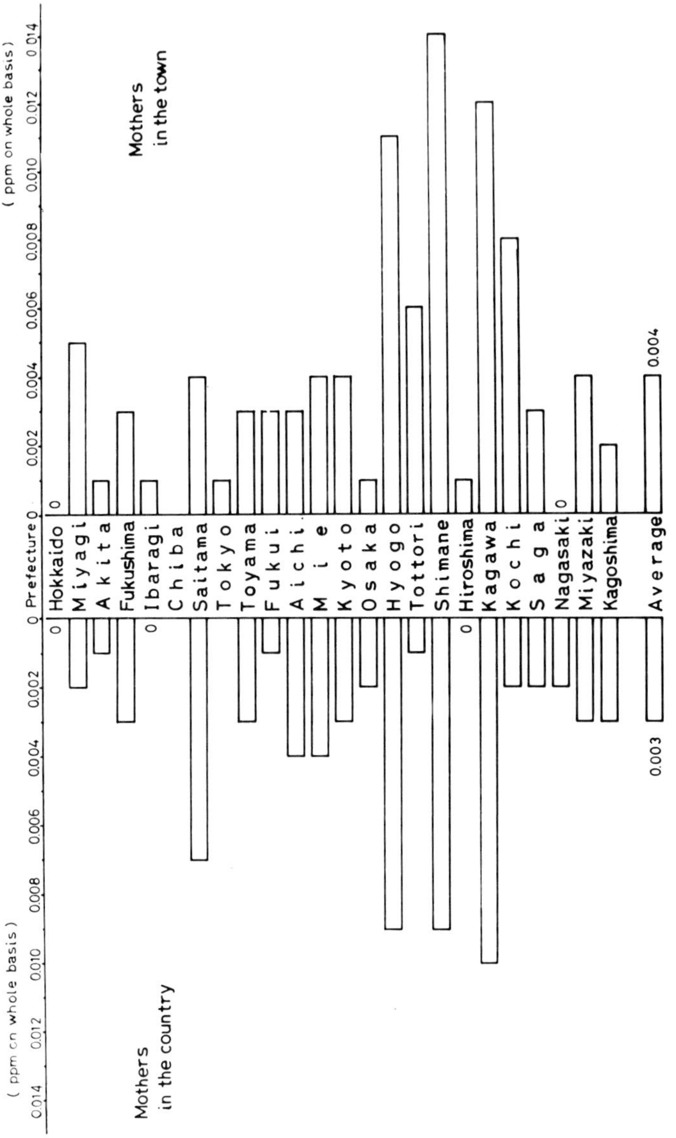

Fig. 8. Dieldrin residues in human milk.

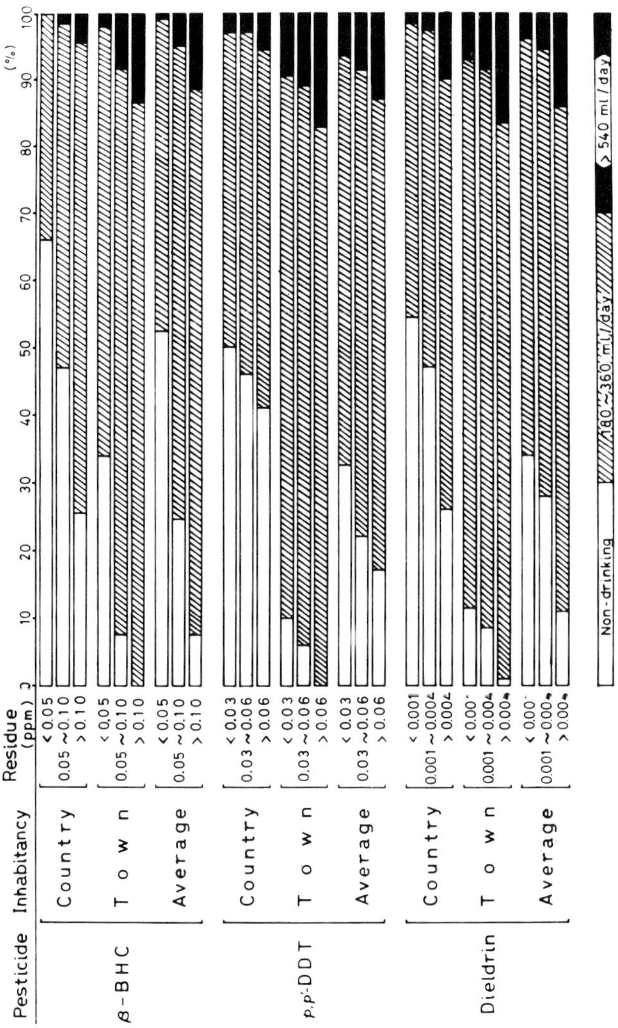

Fig. 9. Relationship between chlorinated hydrocarbon pesticide residues in human milk and intake of cows milk.

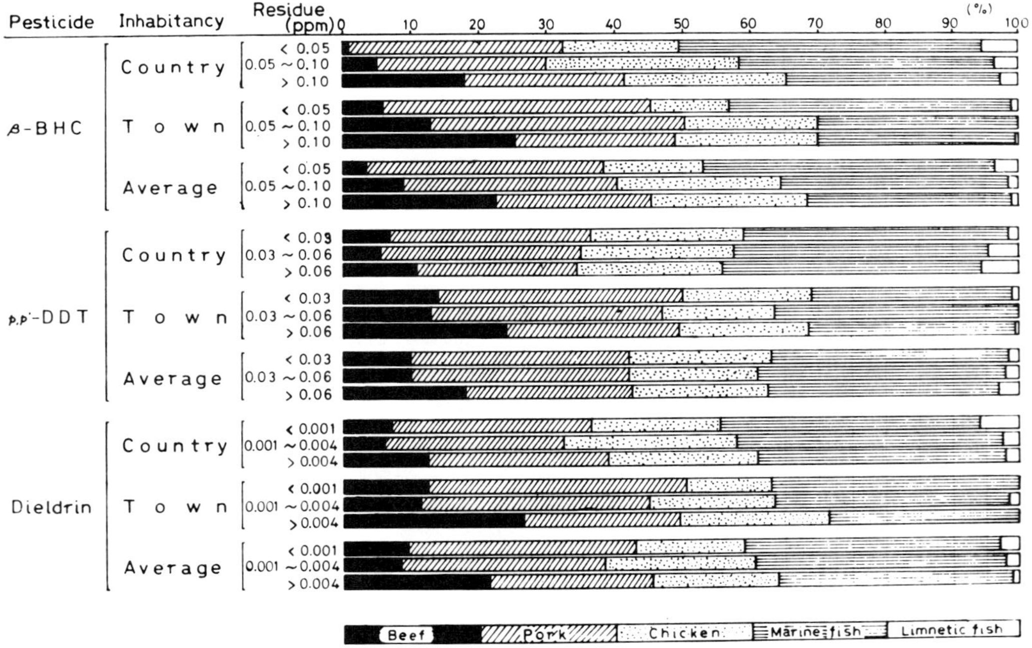

Fig. 10. Relationship between chlorinated hydrocarbon pesticide residues in human milk and intake of meat and fish.

LONG-DISTANCE TRANSPORT OF PESTICIDES

Henry Hurtig

Research Coordinator
Environmental Quality Research Branch
Canada Department of Agriculture, Ottawa, Canada

Introduction

Agriculture and health agencies are the main users of pesticides and in many countries they have the major responsibility for research. Until the late 1960's, the major concern of research scientists and regulatory authorities has been to ensure that these new chemical tools are used to abate the menace of insect-borne diseases of man and to produce and protect food and fibre with minimal hazard to the user. At the same time emphasis was placed on the development of a technology for use and surveillance to ensure no hazard to health results from consumption of food produced in this manner. We have evolved a sophisticated inter-disciplinary, social and economic philosophy for coping with these parameters of safety. In most countries pesticide residue control in food is recognized as an accepted requirement and routine monitoring systems either exist or are being developed. This current general agreement on management of this problem did not come about fortuitously. It is the result of men of goodwill from the several scientific disciplines, agricultural production agencies, health, industry, and governments working together to develop the necessary background knowledge. Experience in these labors can be spanned by the average length of professional career of those present today, since no useful precedents were available.

Governments are confronted with the political necessity of selecting goals they wish to achieve, deciding upon their relative importance and the best means of achieving them within a reasonable time. We have today a fairly sophisticated set of criteria for limiting the residues of pesticides in the food of man and his immediate environment. In the past decade we moved into an additional arena of concern about air, water, and soil pollution. In particular, due to the propensity of some pesticides of finding their way into food chains of non-target organisms, we are now involved with the necessity for defining a new set of parameters for safety.

A notable example demonstrating objective collection and evaluation of facts required to approach this subject is the international collaborative OECD-sponsored exploratory program on wildlife tissue residues (Holden, 1970). A collaborative check sample analysis program of the type carried out in Canada is required to validate analytical procedures for residue determination. The monographs and publications of the FAO/WHO Joint Meeting on Pesticide Residues (1965-1970) indicate the type of critical reviews required.

The available data suggest that in the chain of events surrounding the ecological fate of pesticides, the ultimate step in global pathways of distribution involves long-distance transport associated with low-altitude fallout. The sea is considered to be the ultimate sink, and hence this valuable global resource is assumed to be threatened. The objects of most studies for significance of these observations have been the fish-eating sea birds.

The ultimate significance of these observations depends upon an assessment of studies carried out by qualified investigators on dose-effects relationships involved in the specific concentrations of particular individual pesticides now being detected in monitoring operations. The length of time of exposure to the concentrations involved appears to be of more importance in food chains of lower animals. In the meantime the various disciplines concerned with the welfare of agriculture must direct part of their resources to understanding the mechanisms of regional and long-distance transport of pesticides. From these future studies, hopefully we can arrive at an agreed restriction of use of persistent non-biodegradable pesticides, supervise the management of their use and type of application, but not necessarily ban a whole chemical class of pesticides due to the undesirable properties of one or more.

Persistence of Pesticides

There are many routes by which pesticides will reach water. The first, and probably the one receiving most attention today from the standpoint of possible reduction of hazard, is the direct injection into the atmosphere during the application process. A large body of literature involving engineering, physics, meteorology as applied to plant protection indicates the scope of attempts current today to reduce drift away from the target for which estimates vary from 10 to 70% of emitted pesticide.

Of more recent interest is the propensity and mechanisms by which pesticides in soil are recycled to the other sectors of the environment. A brief review of persistence in soil is appropriate here. If one consults five literature sources on the subject, five different forecasts of persistence will result. But for our purpose today, the summary developed by Kearney and his colleagues (Kearney et al., 1969) is useful in terms of orders of magnitude. Caution must be used in generalizing here also, since many new pesticides degrade or change to end products which also persist. But, persistence in itself is not a hazardous property.

The generalizations by these authors, shown in TABLE 1, are for persistence in soil and many are subject to qualification as to soil type, method of incorporation, climate, etc.

TABLE 1

General Degrees of Persistence of Pesticides

Pesticide	Persistence
Chlorinated hydrocarbon insecticides	- more than 18 months
Urea, triazine and picloram herbicides	- 18 months
Benzoic acid and amide herbicides	- 12 months
Phenoxy, toluidine, and nitrile herbicides	- 6 months
Carbamate and aliphatic herbicides	- 3 months
Phosphate insecticides[a]	- 3 months
Chlordane	- 5 years
DDT	- 4 years
BHC,[b] dieldrin	- 3 years
Heptachlor, aldrin	- 2 years
Diazinon	- 12 weeks
Disulfoton	- 4 weeks
Phorate	- 2 weeks
Malathion, parathion	- 1 week
Propazine, picloram	- 18 months
Simazine	- 12 months
Atrazine, monuron	- 10 months
Fenuron, divron	- 8 months
Linuron	- 4 months
Prometryne	- 3 months
2, 3, 6-TBA	- 12 months
Bensulfide	- 10 months
Diphenamide	- 6 months
Amiben	- 3 months

TABLE 1 - Continued

Pesticide	Persistence
CDAA, Dicamba	- 2 months
Trifluralin	- 6 months
2,4,5-T	- 5 months
Dichlobenil	- 4 months
MCPA	- 3 months
2,4-D	- 1 month
TCA	- 12 months
Dalapon, CIPC	- 8 months
CDEC	- 6 months
IPC, EPTC	- 4 months
Barban	- 2 months

[a]Many exceptions among new phosphate insecticides under cold soil conditions, with persistence extending up to and in excess of a crop-year.

[b]Unlikely that lindane persistence is more than 1 year.

Mechanisms of Pesticide Loss from Soil

Residues in soil have varying significance, depending upon the concentration, availability and nature of the soil residue, agronomic practices, soil type and genesis, and the several physical and chemical factors involved (Lichtenstein, 1966; Lichtenstein et al., 1962, 1971; Harris, 1966; Harris et al., 1966; Harris and Sans, 1969; Kearney et al., 1969). The major phenomena are considered to be volatilization, photodecomposition, chemical decomposition, adsorption leaching, dilution, mechanical removal (erosion), uptake by plants, microbial decomposition, and co-distillation.

The abiotic and biotic components of the biosphere have been outlined in simplified form by Robinson (1969). A more detailed outline of possible post-application sources and sinks of a single pesticide in the atmosphere near a treated area is presented by Hage (1971) in Fig. 1. This diagrammatic conceptual analysis emphasizes the important principle of recycling this type of persistent pollutant, and the complexity of the systems approach to mathematical modelling for predicting the behaviour of a pesticide whose persistence may even only span a single crop year.

Translocation of Organochlorine Insecticides in Aquatic Environments

Trophic Levels in Aquatic-Terrestrial Systems

In the soil-water and atmosphere-water sequence of transfer of pesticide residues, the magnification phenomenon in aquatic food chains has been singled out by a series of critical events. The potential hazard to the oxygen supply of the world has even been attributed to presumed effects of DDT on plankton (Wurster, 1968). Some suggested concepts of "biomagnification" in aquatic food chains in water require revision in the light of recent studies in ponds involving organochlorine insecticides and Caesium-137. It is evident from these that there has been gross oversimplification visualized and inadequate attention paid to materials balance studies of the chemicals behaviour in the total system.

The fate of dieldrin in a pond ecosystem suggests that mechanisms other than food chain at lower levels may be equally important in the transfer process to the atmosphere or higher animals (Rosenberg, 1970). Sufficient dieldrin was applied in July 1967 to produce an approximate concentration of 1 ppb in water. Dieldrin levels in mud, water, vegetation, and invertebrates were monitored by gas chromatographic methods until the autumn of 1968. The residues were undetectable in mud, water, and vegetation by spring of 1968, but persisted at very low levels in the non-transient invertebrate population throughout the remainder of the study. While analysis of all samples is not yet complete due to the difficulty of identifying immature stages of certain invertebrates, the data available now discourage an over-simplified model suggesting dieldrin is concentrated in lower trophic food-chain levels.

At the lowest level in this food chain, dieldrin concentrations in plankton (Daphnia) decreased from a high in excess of 360 ppb to less than 40 ppb in 3 weeks, and to less than 10 ppb by the following spring. In the next series of studies by Rosenberg, the same decreases versus time accumulated by other invertebrates in this aquatic food chain are repeated until the final levels are reached of less than 10 ppb. Components of these steps in the food chain involve the snails Lymnea stagnalis, Lymnea elodes, and Heligoma trivolvis at the level above plankton and in another step the chironomids (several genera, at least 14). The same phenomena are observed in the predatory leeches of the Erpobdellidae and Glossiphoniidae groups, and the Corixids and Notonectids (water-boatmen and backswimmers). While high levels were accumulated in the initial population samples, within weeks the final levels in samples of remaining populations were less than 10 ppb. The levels of dieldrin in water, mud, emergent vegetation, and submergent vegetation all dropped dramatically in less than 2 months.

At the Whiteshell Manitoba Forest Reserve Research Station of Atomic Energy of Canada, Guthrie (1970) has been studying the ecological consequences of radioactive pollution of water from nuclear power plants. His results of following concentrations in aquatic food chains are similar to those of Rosenberg. In a controlled artificial pond system contaminated with known concentrations of radiolabelled Caesium, instead of the expected increasing concentration in a producer-primary consumer-predator system in chronological order, the levels measured decreased an order of magnitude in each time scale and trophic level examined.

To emphasize again that experience gained with one

pesticide cannot serve as a predictive model for chemically related compounds one should refer to the recent work of Metcalf and his colleagues (Kapoor et al., 1970). In a laboratory scale model aquatic-terrestrial system using radiolabeled compounds, the difference in metabolic pathways and degradability in this ecosystem of methoxychlor as compared with DDT has been demonstrated. The metabolic pathway differences in mammals for these two organochlorine insecticides has also been clarified.

The concentration in a food chain with fish at the top for DDT (TABLE 2) confirms concern for approximately 100,000-fold concentration from water but a concentration

TABLE 2

Distribution of Metabolites of DDT from Model Ecosystem[a]

	Concentration (ppm)			
DDT	H_2O	Physa	Culex	Gambusia
p,p'-DDT	0.00022	22.9	8.9	54.2
p,p'-DDE	0.00026	7.6	1.8	18.6
p,p'-TDE	0.00012	1.6	0.4	5.3
polar metabolites	0.0032	0.98	1.5	0.8

[a]From Kapoor et al. (1970).

of three orders of magnitude only for methoxychlor (TABLE 3).

Regional Differences: Soil-Water-Fish
Long Distance Transport

Shellfish, bivalves, and predatory fish are useful in monitoring operations for detecting the overt contamination of water from pesticide atmospheric drift and for the slower transport of pesticides with erosion and drainage from treated lands and forests. Governments and plant

TABLE 3

Distribution of Metabolites of Methoxychlor from Model Ecosystem[a]

	Concentration (ppm)			
Methoxychlor	H_2O	Physa	Culex	Gambusia
p, p'-methoxychlor	0.00011	13.2	-	0.17
mono-OH-methoxychlor	0.0013	1.0	-	-
di-OH-methoxychlor	0.00003	-	-	-
polar metabolites	0.00125	0.8	-	0.16

[a]From Kapoor et al. (1970).

protection authorities anxious to limit water contamination would be well advised to examine each use of the pesticide concerned and its method of application in relation to the agronomic practices, soil type, and geology of the region. Some water drainage systems lead to almost completely closed final large marine or freshwater sinks and peculiar food-chain systems for fish. Examples of these in relation to unacceptable levels of DDT in fish and fish-feeding birds are Clear Lake, California, Lake Michigan, and the Baltic Sea. The peculiar hydrological features of Lake Michigan are, in part, responsible for the DDT pollution situation defined there, since this large lake flushes possibly only once in a hundred years. One would guess a similar situation is responsible for problems in the Baltic.

A study was initiated recently in Ontario to determine sources and severity of pesticide contamination of the Great Lakes. In interpreting the results to date, it is necessary to relate the incidence of pesticide residues to the land use pattern, hydrology, and geology of the areas involved. The land use pattern in Ontario shows concentration of the major agricultural production zones following the pattern of heaviest population density from the St. Lawrence River Valley to Lake Ontario, Lake Erie, and the triangular southwestern segment of Ontario bordering on these two Great Lakes and Lake Huron. The main recreational areas near the large cities of Toronto and Hamilton are in the rocky lake-strewn Laurentian Shield geological area to the north, and the area west of Ottawa. The most productive vegetable-producing areas are in drained peat-type marshland--the Holand Marsh near Toronto draining into Lake Simcoe, and the area from London to Windsor draining into Lake Erie. In this southwestern region the light soils are the most productive tobacco-growing areas of Canada. They also drain into Lake Erie. The interesting feature of the drainage of marshland into Lake Erie is that the area is diked, since it is below the level of Lake Erie. Drainage canals collect water which is then pumped up into the lake.

One of the main drainage systems for the tobacco-growing area is Big Creek in Norfolk County, a shallow, slow-moving stream, bottom covered with sedimentation from the light soil of the area. DDT was still used for tobacco cutworm control in this area in 1971, since potential economical replacements had not yet been properly evaluated on a field scale for acceptability by the end of 1970. Surprisingly, the DDT residues in the tobacco soil are low compared with those in orchard and vegetable soils (TABLE 4). Water samples in this drainage system average only 16 parts per (American) trillion, mud 0.066 ppm, and fish 0.85 ppm of DDT. This represents a concentration factor of 4120X from water to mud, and 53,000X in water to fish (Miles and Harris, 1971).

Compared with this drainage into Lake Erie, a different situation was found in the drainage canals of the Erieau Marsh vegetable-producing area. DDT use was discontinued here in 1968, but 1969 measurements of soil residues indicated high levels (38.2 ppm) remaining. Data summarized in TABLE 5 illustrate the high degree of retention of residue by the high organic content soil

TABLE 4

DDT Residues in Big Creek, Norfolk County, Ontario[a]
(tobacco-growing area drainage system)

Sample	Total DDT			
Soil (1969)	3.4 ppm			
Water (1970) average	16. $pp10^{12}$	) 4120X	)	
Mud (1970) average	0.066 ppm	)	)	53,000X
Fish (1970) average	0.85 ppm		)	

[a]From Miles and Harris (1971).

TABLE 5

DDT Residues in Erieau Drainage Ditch,
Kent County, Ontario[a]
(vegetable-growing area, marsh drainage system)

Sample	Total DDT			
Soil (1969)	38.2 ppm			
Water (1970) average	43.0 $pp10^{12}$	) 5130X	)	
Mud (1970) average	0.22 ppm	)	)	17,600X
Fish (1970) (one only) (scavenger)	0.76 ppm		)	

[a]From Miles and Harris (1971).

when compared with residues in water, sediment, and a fish (Miles and Harris, 1971).

Equally indicative are the results of an extensive survey of residues of DDT and dieldrin in fish in Ontario from 1968-1969 (Frank, 1971). The most interesting feature of the data summarized in TABLE 6 is the wisdom in removing permission to use DDT in the northern recreational areas for blackfly, mosquito, and forest insect control. The Muskoka Lakes is an area of beautiful summer homes

TABLE 6

Pesticide Residues[a] in Fish, Ontario, 1968-1969

Location	Number analyses	In muscle tissue (ppm) DDT	DDD	DDE	Dieldrin	Fat[b] (%)	In fat[b] (ppm) DDT	DDD	DDE	Dieldrin	DDT (%)
Trent River - Bay of Quinte	329		.507		.004	1.75		29.0		0.23	32
Holland Marsh and Lake Simcoe	312		.682		.023	2.77		24.6		0.83	27
Muskoka Lakes and Rivers	519		7.97		.030	3.60		221.4		0.83	63
Great Lakes	404		.750		.025	4.25		17.6		0.59	44
Ottawa River	57		.118		.009	3.22		3.66		0.28	36

[a] Based on extractable fats.

and vacation resorts, located on a thin layer of forest soil on the rocky Laurentian Shield, bordering on myriads of small streams and lakes. Not only do fish from this area contain 10 times more DDT than those found in the Great Lakes, but the much higher percentage of DDT present in their tissues as unchanged DDT (63%) indicates the extremely fast movement of DDT used on land to these water bodies. Both extremely thin soil cover and pesticide drift are probably responsible for this undesirable situation.

Computer Generated Curves for DDT in Aquatic Ecosystems

Reference has already been made to our future necessity for a predictive capability on pesticide behaviour in the environment. As yet, very little has been done in mathematical modelling to provide this long look ahead, but the possibilities have been demonstrated. Biomathematicians have analyzed three years of accumulated data on the fate of ring-radiolabeled DDT in a small pond. Computer models were devised similar to those used to study the transfer of Caesium-137 in the lichen-caribou-Eskimo food chain. The transfers of DDT in this pond ecosystem, as predicted by the computer generated curves, was verified by the experimental results (Eberhardt, 1970, cited by Peterele, 1970).

The rapid removal of DDT from unfiltered water originating in a treated marsh warns that transfers of DDT dissolved in water can be quite different from DDT associated with matter suspended in water. In the next major step the DDT residues in sago pondweed (*Potamogeton pectinatus*) were revealed by the computer generated curve to follow a two-compartment model. No residues were detected in these small water plants after 3 years. The first stage of residue reduction was rapid reflecting transfer from water; the second stage appears as a slower uptake which probably involves DDT actually being taken into the plant rather than adsorbed on its surfaces.

Computer generated curves for DDT accumulation in crayfish (Orconectes immunis) also demonstrate properties of the two-compartment model and no detectable residues were found after 3 years.

This type of predictive modelling can be useful in providing descriptive curves based on factorial treatment of increments of increasing levels of pesticide application and the monitoring of resulting residues. The data can be applied to a simulated system with various frequencies of inflow of pesticide to the system. However, before principles for forecasting pesticide behaviour can be developed, a vast array of experimental data must be generated which is reliably representative of pesticide behaviour under different climatic conditions, soil types, and crop management procedures. The resulting mathematical models next must be tested under field conditions.

Atmospheric Long-Distance Transport of Pesticides

While there is a large mass of continuing literature on the long overdue improvement of pesticide application equipment, there has been little research carried out on the atmospheric transport aspects. However, two forces are operating today which should make possible a better understanding of the phenomena involved, and hopefully remove them from the realm of highly speculative "ecological forecasting." The first is the availability of recently developed information on the long-distance travel of atmospheric industrial pollutants, and modelling that is now possible with the declassification of information developed by defense agencies on the long-distance travel and fallout of nuclear materials, chemical warfare agents, and simulants for them. The second force is the worldwide interest in global and regional transport of pollutants as exemplified in programs now being formulated for discussion and possible action at the 1972 UN Conference in Stockholm on Human Environment.

An attempt at rationalization of current observations of DDT and PCB residues in the environment in comparison with latitudinal fallout of Strontium-90 from the stratosphere has been made by Appleby (1969). Reports in the literature of residues in atmosphere, rivers, human food intake, human tissues, soil, fish, birds, birds' eggs, other wildlife, plankton, sewage, fish, and vegetable oils have been plotted in comparison with the recorded latitudinal fallout of Strontium-90. The main thrust of his thesis is that the occurrence of DDT residues reflects the main intensity of man's use of DDT, his garbage (including the Antarctic), and his abuses which should be discontinued, e.g., sheep-dips, textile treatments, certain seed-dressings, forest insect control, recreational area biting-fly control, etc. He also argues that no DDT residues will occur in Antarctic snow and ice pack at locations and depths free of possibility of human contamination, while Strontium-90 residues are found in these sites. We shall reserve judgment on this thesis and compare it in a few moments with recent observations on the long-distance transport of carbon monoxide.

A closer look at consequences of large scale aerial spray operations involving a replacement for DDT hopefully will become available from studies recently initiated in New Brunswick. Fenitrothion is now used on a large scale to attempt to prevent coniferous forest devastation by the spruce budworm. Monitoring of fenitrothion in air in towns and cities close to and at distances up to 40 miles away from the treated area was carried out in 1969 (Yule et al., 1971); as examples, two cities--Grand Falls and St. John (N.B.)--are approximately 40 miles away from the closest treated area boundaries. The daily records of analysis of air samples, measured as "P" collected as particulates and vapour, is shown as μg/day. At all sampling sites operated from 21 May to 28 June (spanning the treatment period and pre- and post-treatment periods) recoveries from air of both particulate and vapour forms of insecticide were recorded. Particulate recovery was much higher with cumulative totals in excess of 100 μg/day at distant collection sites. This pesticide was presumed to be short-lived and quickly biodegradable. But recent experience suggests this whole question must be more intensively studied, since in addition to these data on transport from site of application, rather unexpected persistence on coniferous foliage has been encountered extending to several months.

Long-Distance Movement of Non-Pesticidal Materials: Models for Atmospheric Transport

To obtain some concept of growth rates and travel from continuous pesticide emissions in a large-scale aerial control operation it is necessary to refer to information developed in relation to other problems. Comparisons based on observations of growth rates of instantaneous source clouds in the atmosphere have been made for smoke puff sources, lycopodium spores, gases, Be tracers, fluorescent tracers, and nuclear clouds (Fig. 2) (Hage, 1971). This type of microscale and mesoscale transport concerns the growth and movement of an airborne pesticide cloud from the moment of release to the atmosphere until the cloud has expanded horizontally to the dimensions of the order of hundreds of kilometers. The ordinate in this figure is the lateral standard deviation of the concentration distribution and the abscissa is travel time following release.

These models illustrate both the importance of space and time scale in the turbulent diffusion process and the great variability that occurs at any particular scale.

A practical application of applying synoptic scale local and regional meteorology in relation to long-distance transport is provided by analysis of meteorological records in relation to migration tracts of black rust in Europe (Zadoks, 1965, cited in Mukammal and McKay, 1971) (Fig. 3). Equations developed for spores having dimensions of 5μ by 3μ predict flight ranges of up to 12,400 km, but the complexity of atmospheric motions invalidates general application.

Another example of long-distance chemical transport can be found in a recent study of the concentrations of carbon monoxide measured over the Greenland ice-cap from July 23 to August 3, 1967 (Fig. 4). This can now be compared with the airmass trajectories from densely populated industrial areas to the north-eastern hemisphere for the same period, based on geostrophic winds at 700 mb (Fig. 5) (Robinson and Robins, 1969). By extrapolating from this type of data analysis it is not difficult to accept that there are additional plausible explanations for DDT and dieldrin residues reported in polar bears in the Canadian arctic (i.e., other than minor use by man locally or in the garbage of food imported to the area).

Recommendations - As Developed by the Canada Committee on Agrometeorology

It is of special interest that in Canada concern over lack of information on long-distance travel of pesticides was not prompted by DDT or dieldrin, but from the constant large rate of use of 2,4-D and MCPA herbicides in the huge cereal monocultures of the prairie provinces of Alberta, Saskatchewan, and Manitoba. While the consumption in Alberta suggests an average of between 6 and 8 million acres treated per year, in Saskatchewan the estimates would be closer to 13 million acres. A conservative estimate would be 1.3 million ounces per year of 2,4-D and related compounds injected into the atmosphere.

For these and other reasons an interdisciplinary and interagency project has been established in western Canada to develop meteorological, physical, engineering, and biological data to allow a better system to be developed for managing herbicide use in these monocultures.

To assist in this and related problems, the Canada Committee on Agrometeorology was asked to select scientists from the several disciplines involved to evaluate the present state of knowledge on long-distance transport of pesticides and make recommendations for the steps to be taken in a reasonable time frame. Their eight position papers have been assembled into a one-volume report and forwarded as a contribution to the special scientific and government groups concerned with global monitoring and surveillance for the 1972 Stockholm UN Conference on Human Environment. The extract, conclusions, and recommendations have been prepared.* They probably have broader application and interest in other countries and we, in Canada, hope they will prove to be useful.

The special needs for analytical instrumentation for monitoring the small amounts of pesticide involved in long-distance transport today requires collection of large samples, concentration and extraction procedures which are regarded today as unreliable. Recent innovations by Barringer Research Limited of Rexdale, Ontario, have resulted in the development of a sensor which can be installed in an aircraft to detect, analyze, and chart kinds and quantities of several pollutants in air. This sensor is known as a "correlation spectrometer." A practical demonstration of the possibilities of this type of mobile instrumentation was obtained in a recent aerial survey of plumes of nitrogen oxides pollution over Chattanooga, Tennessee. The major plume measured was related to high level nitrogen dioxide produced by a chemical factory. A mysterious smaller peak of concentration was recorded over what later proved to be a chicken farm. The abnormal nitrogen dioxide level resulted from the droppings of some 50,000 chickens. More recent modifications allow the airborne instrumentation to chart locations of concentrations of mercury in excess of a 1-nanogram background. Modification should also be feasible for organochlorine airborne pollutants. We are looking forward to and await the application of these new technologies as they become available to study the long-distance transport of pesticides.

---

*The report, "Pollution in Relation to Agricultural Pesticides: Atmospheric Transport of Pesticides" and the recommendation, can be obtained from the office of the author or from Dr. J. L. Bergsteinsson, Physics Div., Saskatchewan Research Council, Saskatoon, Sask., Canada.

References

Appleby, W. G.: Pesticide residues in our environment. Paper presented to IUCN, Delhi, 24 Nov.-1 Dec. (1969).

Frank, R.: Unpublished report. Provincial Pesticide Residue Testing Lab., Ontario Dep. Agr. Food, Guelph, Ont. (1971).

Guthrie, L.: Paper presented to Ann. Meet. Entomol. Soc. Can. and Entomol. Soc. Manitoba, 25 August 1970, Winnipeg, Man. (1970).

Hage, K. D.: Pollution in respect to Agricultural Pesticides: Atmospheric Transport of Pesticides. Can. Comm. Agrometeorol. Rep. In press (1971).

Harris, C. R.: J. Econ. Entomol. 59, 1221 (1966).

Harris, C. R., and W. W. Sans: Pesticide Monit. J. 3, 182 (1969).

Harris, C. R., W. W. Sans, and J. R. W. Miles: J. Agr. Food Chem. 14, 398 (1966).

Holden, A. V.: Pesticide Monit. J. 4, 117 (1970).

Kapoor, I. P., R. L. Metcalf, R. F. Nystrom, and G. K. Sangha: J. Agr. Food Chem. 18, 1145 (1970).

Kearney, P. C., R. G. Nash, and A. R. Isensee: Chemical Fallout. C. C. Thomas, Springfield, Ill. p. 54 (1969).

Lichtenstein, E. P. J. Econ. Entomol. 59, 985 (1966).

Lichtenstein, E. P., T. W. Ruhremann, and K. R. Schulz: J. Agr. Food Chem. 19, 718 (1971).

Lichtenstein, E. P., C. H. Mueller, G. R. Myrdal, and K. R. Schulz: J. Econ. Entomol. 55, 215 (1962).

Miles, J. R. W., and C. R. Harris: Pesticide Monit. J. In press (1971).

Mukammal, E. I., and G. A. McKay: Pollution in respect to agricultural pesticides: Atmospheric transport of Pesticides. Can. Comm. Agrometeorol. Rep. In press (1971).

Peterle, T. J.: Biological Impact of Pesticides in the Environment. Environ. Health Sci. Series No. 1. Oregon State University Press, p. 11 (1970).

Robinson, J.: Chemical Fallout. C. C. Thomas, Springfield, Ill. (1969).

Robinson, E., and R. C. Robins: J. Geophys. Res. 74, 8 (1969).

Rosenberg, D.: Paper presented to Ann. Meet. Entomol. Soc. Can. and Entomol. Soc. Manitoba, Winnipeg, Aug. 26 (1970). (Thesis in fulfillment of Ph.D. requirements, University of Alberta, 1972, in preparation.)

Wurster, C. F.: Science 159, 1474 (1968).

Yule, W. N., A. F. W. Cole, and I. Hoffman: Bull. Environ. Contam. Toxicol. 6, 289 (1971).

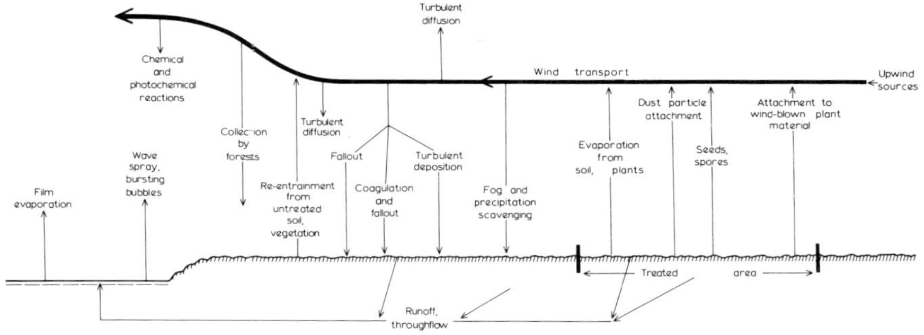

Fig. 1. Theoretical stages and pathways in transport and recycling of a single pesticide treatment to a variety of environmental sinks. (Hage, 1971)

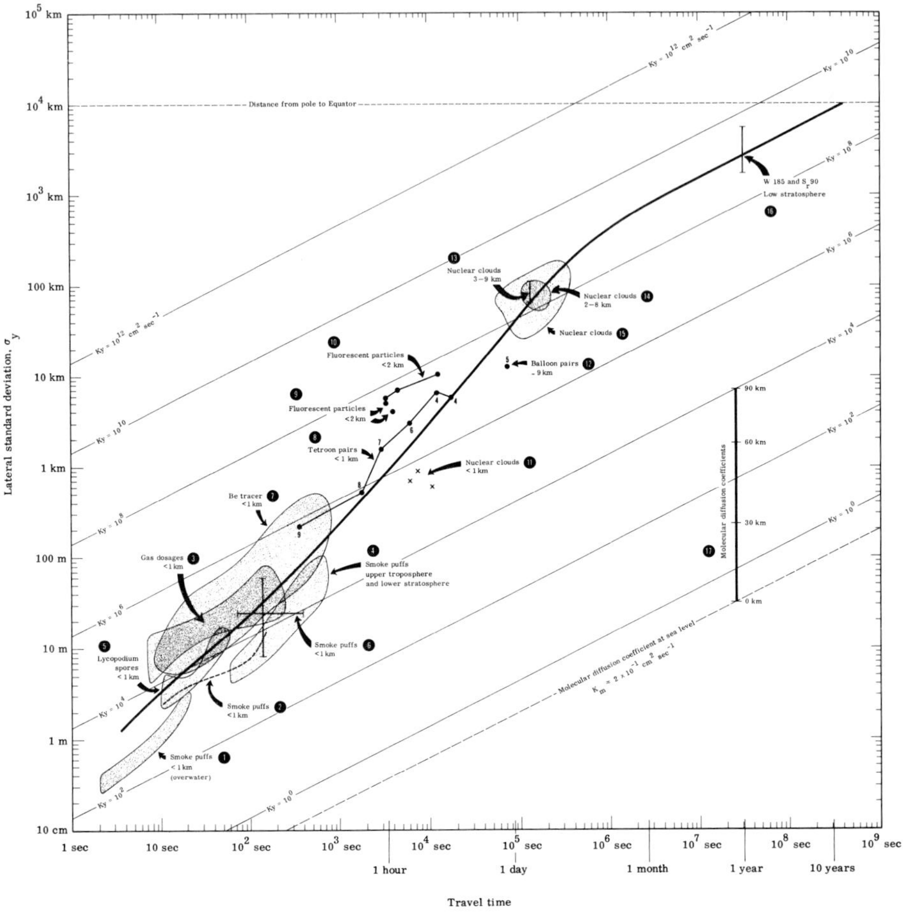

Fig. 2. Micro- and mesoscale transport of point source releases. (Hage, 1971)

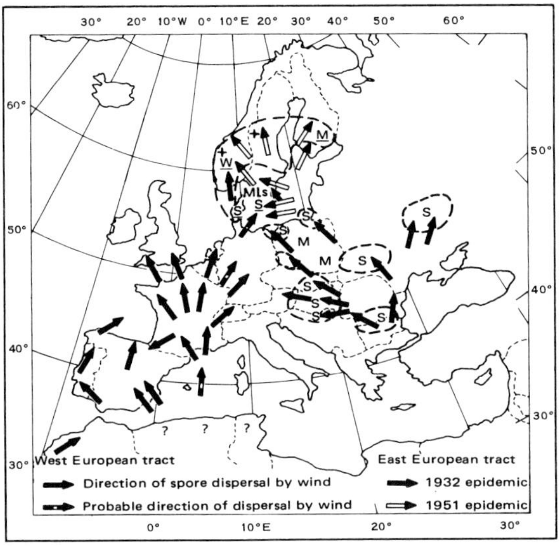

Fig. 3. Application of synoptic scale meteorology to long-distance transport of black rust spores in Europe.

Arrows over Norway (w) and that over the Danish Islands refer to the west European tract. For the eastern European tract, letters refer to 1932 when not underlined and to 1951 when underlined. Letters refer to severity S, S = severe epidemics of national importance, M, M = mild epidemics of regional importance or general epidemics with small losses. (Zadoks, 1965, cited in Mukammal and McKay, 1971)

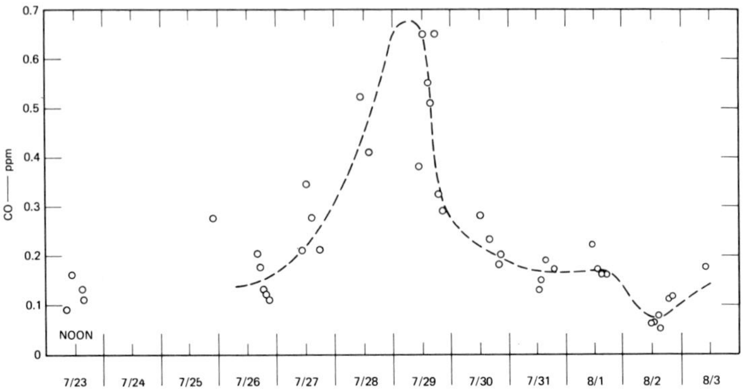

Fig. 4. Concentrations of carbon monoxide over Greenland ice-cap from July 23 to August 3, 1967. (Robinson and Robbins, 1969)

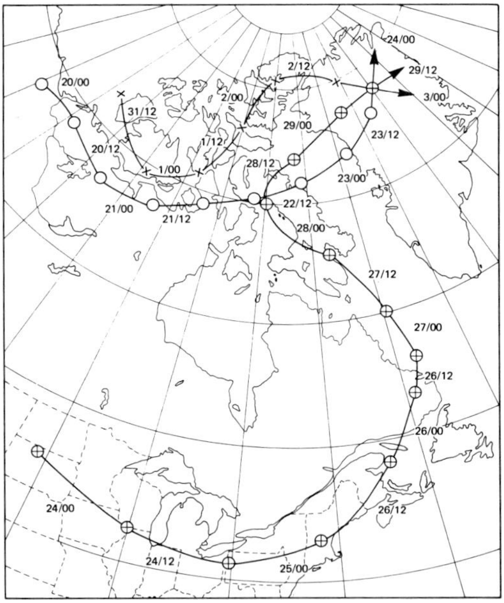

Fig. 5. Geostrophic winds at 700 mb. Airmass trajectories July 23 to August 3, 1967 in relation to CD transport to Greenland ice-cap. (Robinson and Robins, 1969)

PART IV

FUNGICIDES, HERBICIDES, ORGANOPHOSPHATES AND CARBAMATES

PHYTOTOXIC METABOLITES OF PENTACHLOROBENZYL ALCOHOL

Mitsuo Ishida

Sankyo Co., Ltd.
Agricultural Chemicals Research Laboratories
Yasucho, Yasugun, Shigaken, Japan

Introduction

The remarkable effect of 2,3,4,5,6-pentachlorobenzyl alcohol (PCBalc.) against rice blast was found in 1964 (Sumi et al., 1968) and the chemical moved into the limelight as one of the best non-mercurial rice blast control agents. Studies on its mode of action and metabolisms in rats, rice plants, and some phytopathogenic fungi were reported at the U.S.-Japan seminar on pest control held in Nikko in 1967 (Ishida et al., 1969).

The foliar sprays of PCBalc. at higher concentrations (2,000-10,000 ppm) proved to have no effect on the growth of many plant species, including varieties of rice plants (Sumi et al., 1965). Yasumura (1968), working with mulberry trees, made the initial observations of the phytotoxic effects of PCBalc. applied in soil. In April, 1968, 2 years after its introduction to the market under the trade name Blastin, some leaf malformation of tomato, melon, and cucumber plants was reported from several places in Japan, and the causal relationship between malformation and PCBalc. preparations was discussed. The most severe leaf malformation was observed following the use of compost prepared from rice straw on which PCBalc. preparations had been applied the previous year.

Identification of Phytotoxic Metabolites

An aqueous extract of a sample of strongly phytotoxic rice straw compost from Sanjo, Niigata was fractionated by extracting with ether at pH 2-3, 7-8, and 10-11, separately. Only the acidic fraction caused the same leaf malformation as noted in the compost. Therefore, the fraction was methylated with diazomethane in ether, and fractionated by preparative gas-liquid chromatography (Fig. 1A). Three fractions were isolated and identified as PCBacid, 2,3,4,6- and/or 2,3,5,6-TECBacid, and 2,4,6-TRCBacid by gas-liquid chromatography and mass spectrometry, and by using authentic samples of the most probable metabolites of PCBalc. prepared in our laboratories (TABLE 1). These two TECBacid isomers had the same behavior in gas-liquid chromatograph with a wide variety of stationary phases in packed columns and operation conditions, and also had the same fragmentation pattern of mass spectrometry. A gas chromatographic separation reported by Kirkland (1961) was tried, but it was so poor in resolving these compounds that a mixture of these TECBacids gave but a single peak. Very recently, however, a fairly good separation was achieved by a 45 meter Golay column coated with Apiezon L (Fig. 2) and studies on the identification of these benzoic acid isomers from phytotoxic soil and compost are in progress. The chromatogram of another sample of rice straw compost (Fig. 1B) shows that there are many peaks in addition to the three dominant ones shown in Fig. 1A, suggesting that phytotoxic metabolites and natural interferring materials may vary from sample to sample. The identification of

TABLE 1

Authentic Samples Prepared

Compound	Abbreviation	Isomers
2, 3, 4, 5, 6-Pentachloro-benzyl alcohol	PCBalc.	
2, 3, 4, 5, 6-Pentachloro-mandelonitrile	PCMN	(dl-)
2, 3, 4, 5, 6-Pentachloro-benzaldehyde	PCBald.	
2, 3, 4, 5, 6-Pentachloro-benzoic acid	PCBacid	
Tetrachlorobenzyl alcohol	TECBalc.	2, 3, 4, 5- 2, 3, 4, 6- 2, 3, 5, 6-
Tetrachlorobenzoic acid	TECBacid	2, 3, 4, 5- 2, 3, 4, 6- 2, 3, 5, 6-
Trichlorobenzoic acid	TRCBacid	2, 3, 4- 2, 3, 5- 2, 3, 6- 2, 4, 5- 2, 4, 6- 3, 4, 5-
Dichlorobenzoid acid		3, 4-
Hydroxychlorobenzoic acid	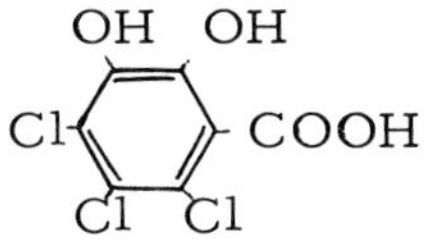	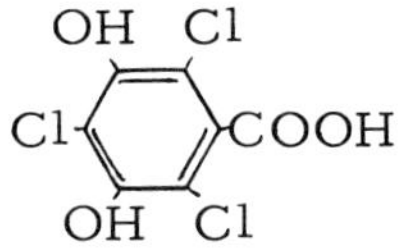
	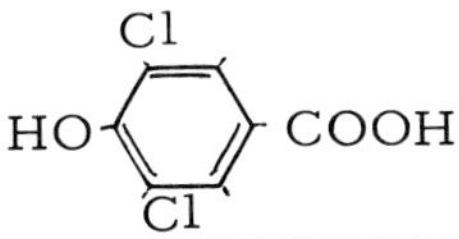	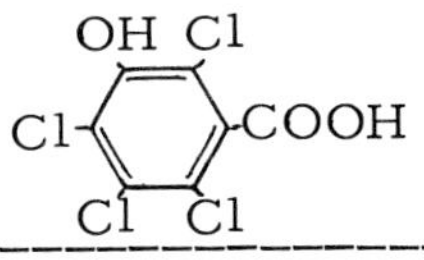

peaks only by retention times is unreliable.

Among the chlorinated benzoic acids tested, 2, 3, 5, 6-TECBacid and 2, 3, 6-TRCBacid caused the most severe leaf injury, followed by 2, 3, 4, 6-TECBacid and PCBacid (TABLE 2). The former two acids were distinguished by

TABLE 2

Phytotoxicity of Polychlorobenzoic Acids

Polychlorobenzoic acid		Phytotoxicity rating[a] (0-5)
Tri-	2, 3, 4-	0
	2, 3, 5-	2
	2, 3, 6-	5
	2, 4, 5-	3
	2, 4, 6-	2
	3, 4, 5-	0
Tetra-	2, 3, 4, 5-	1
	2, 3, 4, 6-	4
	2, 3, 5, 6-	5
Penta-	2, 3, 4, 5, 6-	4

[a]Degree of phytotoxicity and symptom vary depending upon experimental conditions such as observation periods and application methods, and plant species. These ratings are the average of several experimental results.

their strong stem-bending activity, compared with the latters, which were relatively strong in stunting and leaf-marginal curling. The four hydroxypolychlorobenzoic acids tested were substantially non-phytotoxic. Since the hypothesis for critical structure requirements of auxins proposed by Porter and Thimann (Thimann, 1963) explains quite well the strong phytotoxicity of 2, 3, 5, 6-TECBacid, a detailed comparison of auxin activities of PCBacid and

TECBacid isomers was made using Avena coleoptiles (TABLE 3). As was expected, only 2,3,5,6-TECBacid showed auxin activity, indicating that dechlorination at the para position apparently increases the phytotoxic activity.

TABLE 3

Avena Coleoptile Straight Growth Test

Compound	Growth (% to control)					
	10^{-3}	10^{-4}	10^{-5}	10^{-6}	10^{-7}	10^{-8}M
2,3,4,5-TECB acid	-	-	97	96	102	100
2,3,4,6-TECB acid	-	-	103	99	101	100
2,3,5,6-TECB acid	77	113	122	104	104	100
PCB acid	76	98	102	100	105	100
Indoleacetic acid	132	143	141	129	111	-
2,4-D	-	121	132	128	107	-

An attempt was made to avoid this activation reaction by replacing the chlorine atom with a methyl group at the para position of PCBalc. The formation of a carbon to carbon bond diminished the phytotoxicity of PCBalc. Unfortunately, however, the anti-rice blast activity of the compound was much lower than that of PCBalc. (TABLE 4).

The presence of such phytotoxic acids in the technical grade PCBalc. preparations, and in structurally related herbicides such as 2,6-dichlorobenzonitrile and 2,3,6-TRCBacid, was then investigated. The possible precursors for tri- and tetrachlorobenzoic acids were also determined by oxidizing these preparations with potassium permanganate in an alkaline condition. No reasonable amounts of the phytotoxic acids and their possible precursors were noted in these preparations.

TABLE 4

Phytotoxic Effect of Methyltetrachlorobenzyl Alcohols

Compound[a]		Tomato plants malformation rating[b]		Anti-rice blast activity rating[b]
		Exp. 1	Exp. 2	
$(o\text{-}CH_3)$	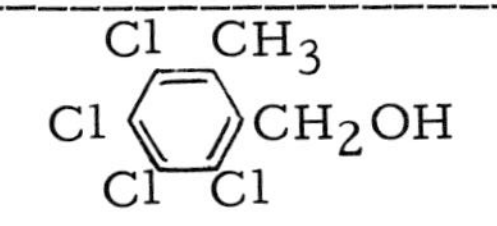	-	3	2
$(m\text{-}CH_3)$	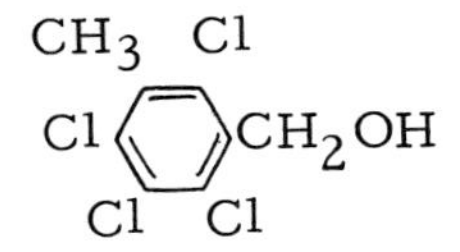	2	3-4	2
$(p\text{-}CH_3)$	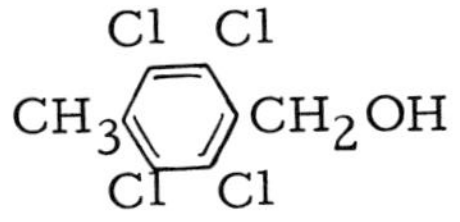	0	0 (retardative)	2
PCBalc.		2	4	5
PCBacid		5	-	-
Untreated		0	0	0

[a]Concentration in soil: 100 ppm w/w.

[b]0-5 rating system.

The Cucurbitaceae, Legminosae, and Solanaceae plants were most susceptible to penta- and tetrachlorobenzoic acids. Young melon plants were most sensitive, being affected by PCBacid at a concentration of only a few hundredths of one ppm. Gramineae and Cruciferaeae plants were very tolerant. An extensive study on the susceptibility of rice plant varieties, and on its genetical aspects were carried out by Sakurai et al. (1970). They noted that only 2 (Japanese origin) out of 281 varieties collected from the world were susceptible to PCBalc., at a concentration of 16 ppm.

Biotransformation

A preliminary experiment demonstrated that PCBalc., incorporated in non-sterilized soil, caused leaf malformation of tomato and cucumber plants at a concentration of 7.2 ppm within 2 weeks, and that a part of the alcohol was dehydrogenated to PCBacid in the soil, whereas the alcohol in heat-sterilized soil was stable and did not affect the growth of the plants. Another experiment on the metabolism of PCBalc. in a paddy field in the summer is shown in TABLE 5. The poor production of TECBacid in paddy field soil coincides with the fact that almost all the paddy field soil samples collected from various places were scarcely contaminated with TECBacid isomers (Fig. 6).

Pentachlorobenzaldehyde (PCBald.) was subject to both hydrogenation and dehydrogenation, almost at once, to yield PCBalc. and PCBacid in non-sterilized soil, respectively (TABLE 6). In addition, 2,3,4,5,6-pentachloromandelonitrile (PCMN), another PCBalc.-related rice blast control agent, was instantly decomposed to PCBald. by liberating hydrogen cyanide quantitatively, followed by further transformation of PCBald. to the corresponding alcohol and acid (TABLE 7).

Young rice plants were transplanted in soil containing PCBacid in porcelain pots and grown under the usual paddy field conditions. Four months later the soil and rice plants were found to contain the parent acid, PCBalc., and

TABLE 5

Metabolism of PCBalc. in Paddy Field[a]

Days after incorporation[b]	Concentration of metabolites ppm, dry weight basis			
	PCB alc.	PCB acid	TECB acid	
			2,3,4,6-	or 2,3,5,6-
1	3.46	0.72	ud[c]	ud
15	3.28	2.26	ud	ud
43	1.09	0.40	0.04	0.04
103	0.26	0.10	trace	trace

[a]One test plot = 13.2 m^2, 6 replicates.
[b]2.4 Kg of PCBalc./ha.
[c]Undetectable.

TABLE 6

Transformation of PCBald. in Soil

Time after incorporation[a]	Found, %			
	PCBald.	PCBalc.	PCBacid	Total
30 min.	45	7	9	61
2 days	2	25	31	58
4	2	26	53	81
6	1	22	31	53
18	0	17	48	65

[a]Initial supply was 1.8×10^{-4} moles/kg of soil.

TABLE 7

Transformation of PCMN in Soil

Time after incorporation[a]	Found, %			
	PCMN	PCBalc.	PCBacid	Total
30 min.	86	4	5	95
2 days	2	12	53	67
4	1	13	45	59
8	0	11	49	60
16	0	11	52	63

[a]Initial supply was 0.33×10^{-4} moles/kg of soil.

2,3,4,6- and/or 2,3,5,6-TECBacid (TABLE 8). Besides these, the rice plants contained trace amounts of 2,3,4,5-TECBacid.

To solve the question of whether the dechlorination reaction takes place in rice plants, young rice plants were grown in water culture solutions containing either PCBalc. or PCBacid. However, it was recognized that it was probably impossible to avoid contamination with soil microorganisms completely. A small fraction of PCBalc., absorbed in water-cultured rice plants, seemed to be dehydrogenated to PCBacid, whereas PCBacid in the plants was quite resistant to both hydrogenation to PCBalc. and dechlorination to the TECBacid isomers (TABLE 9). The fact that PCBalc. was undetectable in rice plants grown in PCBacid-containing solution may suggest that the contribution of soil microorganisms was very small, if at all.

None of the three isomers of ring substituted TECB-alc. were detected in reasonable amounts in soil, rice plants, and rice straw and compost throughout this study. However, the fact that UV-irradiated methanol solutions of PCBalc. liberate chloride ions suggests that sprayed PCBalc. might be subject to dechlorination on the surface of rice plants. Figure 3 shows that at the time when added

TABLE 8

Metabolism of PCBacid in Paddy Field Soil

Metabolite found[b] (x10^{-6} moles)	PCBacid added[a]			
	2 mg/pot (8.6x10^{-6} moles)		20 mg/pot (86x10^{-6} moles)	
	Soil	Rice plants	Soil	Rice plants
PCBacid	0.34	0.22	3.9	2.2
PCBalc.	0.57	0.09	0.57	0.09
2,3,4,5-TECB acid	UD	0.007	UD	0.035
2,3,4,6-TECB acid or 2,3,5,6-TECB acid	0.01	0.09	0.30	0.19
	0.10	0.10	0.30	0.21
Recovery (%)	14.9	6.0-6.2	7.0	3.7-3.8
	20.9-21.1		10.7-10.8	

[a]On June 10, 1968.

[b]On Nov. 10, 1968.

PCBalc. was completely photodecomposed, approximately half of the chlorine in covalent form was converted to the ionic form. The products of this photodecomposition, however, have never been investigated.

There are many reports on the persistency of 2,3,6-TRCBacid in soil. Dewey et al. (1962) demonstrated that 30-50% of Cl^{36}-2,3,6-TRCBacid incorporated in soil was biologically changed to ionic form after 1, 2, and 5 months, and no chlorine in an intermediate stage between the parent acid and ionic chlorine was found. MacRae and Alexander (1965) showed that monochlorobenzoates were quickly destroyed in soil suspension and the susceptibility of chlorobenzoates to biodegradation was related to the number of chlorines on the aromatic ring. No large peaks

TABLE 9

Metabolism of PCBalc. and PCBacid in Water Cultured Rice Plants

		Parent compound	
		PCBalc.	PCBacid
Conc. in culture solution		10 ppm	3 ppm
Period of water culture		30 days	24 days
Contents of metabolites	: PCBalc.	230 ppm	UD
	: PCBacid	7.34 ppm	554 ppm
	: TECB acids	UD	UD
	: TRCB acids	UD	UD

on the chromatogram, which might correspond to di- and monochlorobenzoic acids, were detected in this study. Calcium cyanide accelerated the biotransformation of PCBalc. in rice straw compost (TABLE 10), but did not significantly change the ratio of metabolites. Activated charcoal was effective only temporarily in preventing the formation of the phytotoxic acids.

A number of strains of microorganisms were examined for both their ability to convert PCBalc. into PCBacid, and that of PCBacid into further metabolites, hopefully non-phytotoxic ones (TABLE 11). Several strains were found to be very active in dehydrogenating PCBalc. and in the biotransformation of PCBacid, respectively. Practical application of the PCBacid metabolizing microbes in soil

TABLE 10

PCBalc. and its Metabolites in Rice Straw Compost Prepared[a] in Combination with Additives

Additive	%	x 10^{-5} moles/kg, dry weight basis				
		PCB alc.	PCB acid	TECB acid		
				2,3,4,5-	2,3,4,6- or	2,3,5,6-
None	-	2.61	13.7	0.121	0.604	0.538
$Ca(OH)_2$	5.0	2.99	14.5	0.176	0.608	0.539
$CaCN_2$	2.5	1.34	19.7	0.220	0.978	0.857
Active charcoal	0.3	8.46	7.51	0.132	0.396	0.341
Rice straw used	-	12.5	0.349	UD	UD	UD

[a]Preparation period = 3 months.

TABLE 11

Screening of Microorganisms which are capable of Transforming PCBalc. or PCBacid

Group	Substrate		Bacterium	Actino-mycete	Fungus	Yeast	Unclas-sified	Total
I[a]	No. of strain tested		145	124	335	100	75	779
	Capable	PCBalc.	4	8	16	3	0	31
		PCBacid	3	1	15	6	0	25
II[b]	No. of strain tested		229	35	142	7	0	413
	Capable	PCBalc.	2	6	19	0	0	27
		PCBacid	3	2	8	1	0	14
Total	No. of strain tested		374	159	477	107	75	1192
	Capable		12	17	58	10	0	97

[a]Strains being kept in our laboratories.

[b]Strains isolated from phytotoxic composts and soils.

fortified with PCBacid, however, was unsuccessful in reducing phytotoxicity, probably due to the strong antagonistic action of other soil microorganisms.

The possible biotransformation pathways of PCBalc. appear in Fig. 4. The presence of 2, 4, 6-TRCBacid strongly supports the view that 2, 3, 4, 6-TECBacid is also present, since the former acid can be derived only from the latter acid, assuming that there is no chance for migration of chlorine atoms. On the other hand, phytotoxic symptoms support the presence of 2, 3, 5, 6-TECBacid. It seems, as a result, that all three isomers of TECBacid are biologically produced from PCBalc. The fact that 2, 4, 6-TRCBacid is the only predominant isomer may be explained by either that its persistency is longest among the six isomers, or that the chlorine atoms on the 3 and 5 position (meta to carboxyl group) are more susceptible to biological dechlorination, compared with those at 2 and 6 (ortho) and 4 (para) position. Hence, the production of 2, 3, 4, 6-TECBacid and 2, 4, 6-TRCBacid exceeds that of the other isomers. Studies on the dechlorination reaction by isolated microorganisms are in progress.

Residues and Safety

An extensive survey on the contamination of rice chaff and straw, rice straw compost, and paddy field soil with polychlorobenzyl alcohols and polychlorobenzoic acids revealed that most of them contained PCBalc., and PCBacid and 2, 3, 4, 6- and/or 2, 3, 5, 6-TECBacid, although ratios of their amounts varied. In addition, some contained small amounts of 2, 3, 4, 5-TECBacid (TABLES 12 and 13, and Fig. 5, 6, and 7). These contaminants were expected to dissipate within several months, estimated by the data given in TABLE 5. Actually no phytotoxic damage to rice or vegetable plants was reported in 1969, except when heavily contaminated compost was carelessly used.

PCBalc. residues in unpolished rice grains were determined from 61 and 46 samples collected from dusting powder- and wettable powder-applied paddy fields,

TABLE 12

PCBalc. and its Metabolites in Phytotoxic Compost

Compost	Concentration in ppm, dry weight basis					
	PCB alc.	PCB ald.	PCB acid	TECB acid		
				2,3,4,5-	2,3,4,6- or	2,3,5,6-
I	0.15	UD	2.6	UD	1.84	2.15
II	1.6	UD	6.5	UD	3.72	4.36
III	0.02	UD	1.0	UD	0.05	0.05
IV	25.6	UD	21.1	UD	13.7	16.0

TABLE 13

PCBalc. and its Metabolites in Rice Straw

Rice straw	Dusting times[a]	Concentration in ppm, air-dry weight basis			
		PCB alc.	PCB acid	TECB acid	
				2,3,4,6- or	2,3,5,6-
I	1	18.6	1.3	0.07	0.08
II	2	0.00	0.5	0.06	0.05
III	3	1.4	1.3	0.08	0.10
IV	3	8.3	2.7	0.07	0.08

[a]4% PCBalc. dusting preparation.

respectively (Fig. 8). The average residue of all 107 samples was 0.086 ppm. PCBacid residues in unpolished rice grains were found to be approximately 0.1 ppm after two to three applications of 4% PCBalc. dusting powder preparation (TABLE 14).

TABLE 14

PCBacid Residues in Unpolished Rice Grains

Rice grain sample	Application times of PCBalc. prep.[a]	PCBacid found, ppm
I	0	0.03
II	3	0.13
III	3	0.12
IV	2	0.13
V	2	0.09

[a]4% Dusting powder preparation.

No effect of PCBalc. on the growth of bacteria and actinomycetes in soil was observed at concentrations of 100 and 1,000 ppm in soil by Sakurai et al. (1970) (TABLE 15). This has also been confirmed in our laboratories.

PCBalc. is substantially non-toxic to mammals and insects, and studies on its metabolism in rats showed that it is rapidly excreted without accumulation in organs (Ishida et al., 1969). Subacute (90 days) toxicity tests to albino rats at dosage levels of 100, 1,000, and 2,000 mg/kg per day showed no effects of the chemical on organs and blood functions, even at the highest dosage level. Acute oral toxicities of PCBacid to rats and mice (TABLE 16) was higher than those of PCBalc. Dosage levels of 10, 30, and 60 mg/kg per day were chosen for subacute (90 days) toxicity tests to albino rats. No effects were found at the highest level. The very rapid excretion of radioactivity from mice orally administered with C^{14}-PCBacid

TABLE 15

Effect of PCBalc. on Soil Microorganisms[a]

Concentration of PCBalc. incorporated	Number of microbes, x 10^6	
	Bacteria	Actinomycetes
0 ppm	10.7-11	9.5-17
100	23 -28	1.2- 2
1,000	19.8-37	6 -16

[a]Temperature, 27°C; incubation period, 1 month.
From Sakurai et al., 1970.

TABLE 16

Acute Toxicity of PCBacid

Animal	Sex	LD_{50} (mg/kg)		
		Oral	I.P.	Dermal
Mouse		415	175	2,000
		323	162	2,000
Rat		158	57	500
		175	56	500

is probably due to the high water solubility of its salt, and is favorable to its safety (TABLE 17). Rapid excretion was also demonstrated by whole body radioautography of mice, which visualizes localization and dissipation of the radioactive compound.

Production and marketing of PCBalc. preparations were discontinued in June, 1968 when its biotransformation to persistent phytotoxic acids in soil became clear. Total amounts of PCBalc. applied in the forms of dusting and wettable powder were approximately 1,570 tons and there was substantially little damage to vegetables and rice. The prompt decision to discontinue its marketing, as well as effective distribution of information, minimized damages.

TABLE 17

Excretion of Radioactivity from Mice Orally Administered with C^{14}-PCBacid[a]

	% Recovery of radioactivity[b]		
	Hours after administration		
	0-24	24-72	Total (0-72)
Urine[c]	92.7	0.07	92.8
Feces[c]	0.71	0.27	0.98
Total	93.4	0.34	93.8

[a]Sp. act. = 2.33 μc/mg. 1 mg/male mouse (approx 20 g).

[b]Average of duplicates.

[c]Urine was diluted with ethanol and feces were extracted with ethanol.

References

Dewey, O. R., R. V. Lyndsay, and G. S. Hartley: Nature 195, 1232 (1962).

Ishida, M., H. Sumi, and H. Oku: Residue Rev. 25, 139 (1969).

Kirkland, J. J.: Anal. Chem. 33, 1520 (1961).

MacRae, I. C., and M. Alexander: J. Agr. Food Chem. 13, 72 (1965).

Sakurai, Y., M. Kawamoto, C. Furuta, Y. Sekiguchi, T. Yoshizawa, K. Shibano, K. Toriyama, and K. Sekizawa: Studies on phytotoxicity caused by chlorinated hydrocarbon anti-rice blast agents. National Chugoku Agr. Exp. Sta. (1970).
Sumi, H., Y. Takahi, K. Nakagami, and Y. Kondo: Ann. Phytopathol. Soc. Japan 30, 305 (1965).
Sumi, H., Y. Takahi, K. Nakagami, and Y. Kondo: Ann. Phytopathol. Soc. Japan 34, 114 (1968).
Thimann, K. V. Ann. Rev. Plant Physiol. 14, 1 (1963).
Yasumura, S.: Sanshi-Kenkyu (Acta Sericologica) 68, 116 (1968).

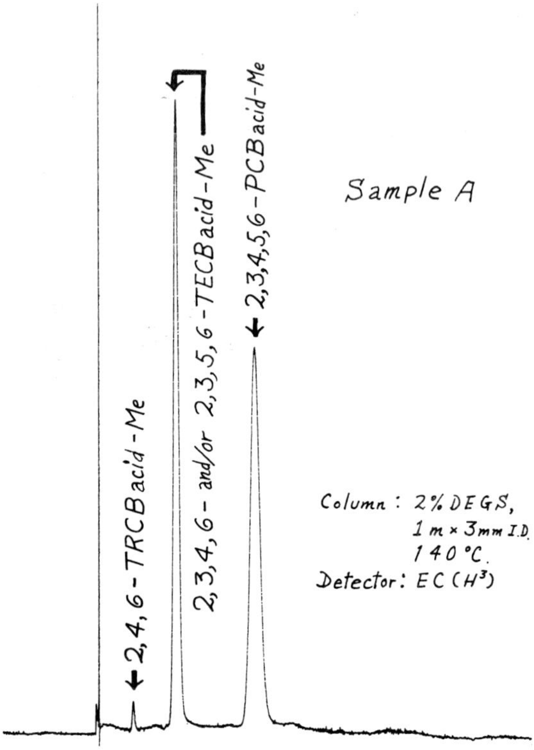

Fig. 1A. Gas-liquid chromatogram of a methylated acidic fraction of phytotoxic compost from Sanjo, Niigata, Sample A.

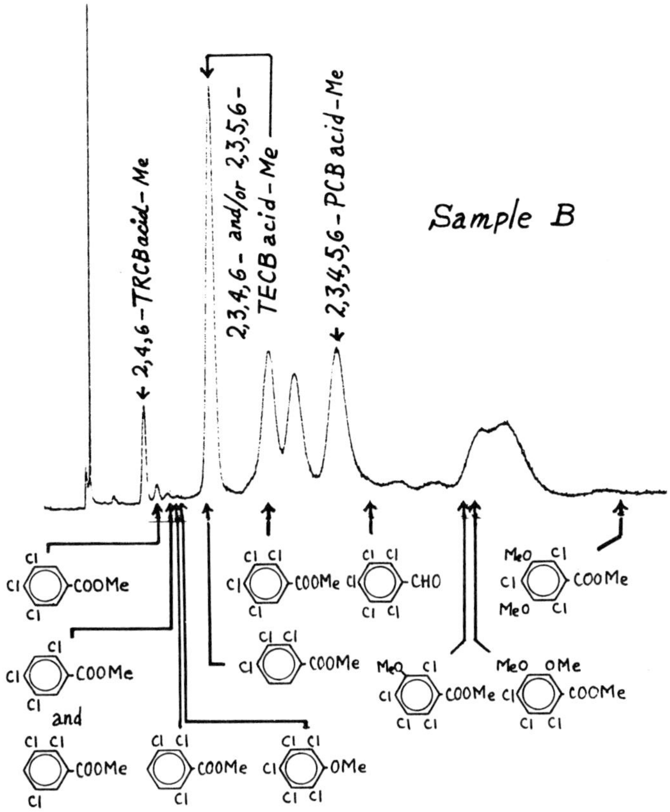

Fig. 1B. Gas-liquid chromatogram of a methylated acidic fraction of phytotoxic compost from Sanjo, Niigata, Sample B.

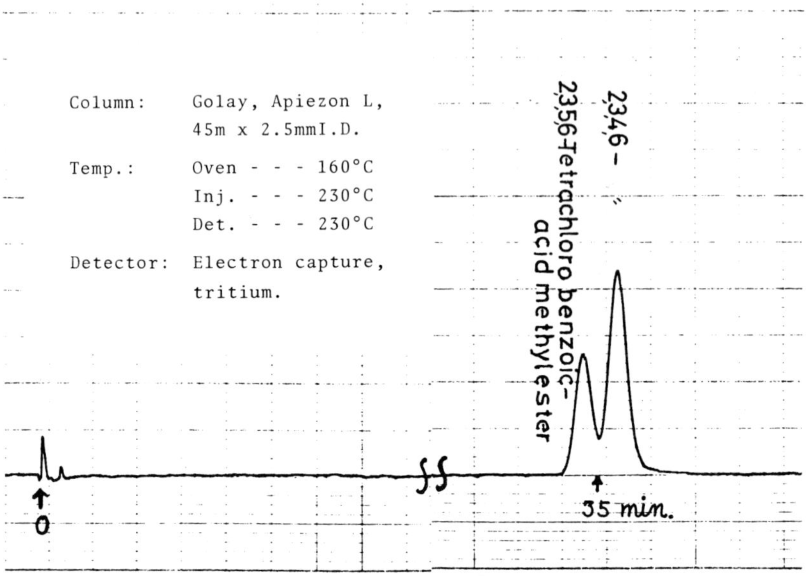

Fig. 2. Gas-liquid chromatogram of methyl esters of 2, 3, 5, 6- and 2, 3, 4, 6-tetrachlorobenzoic acid.

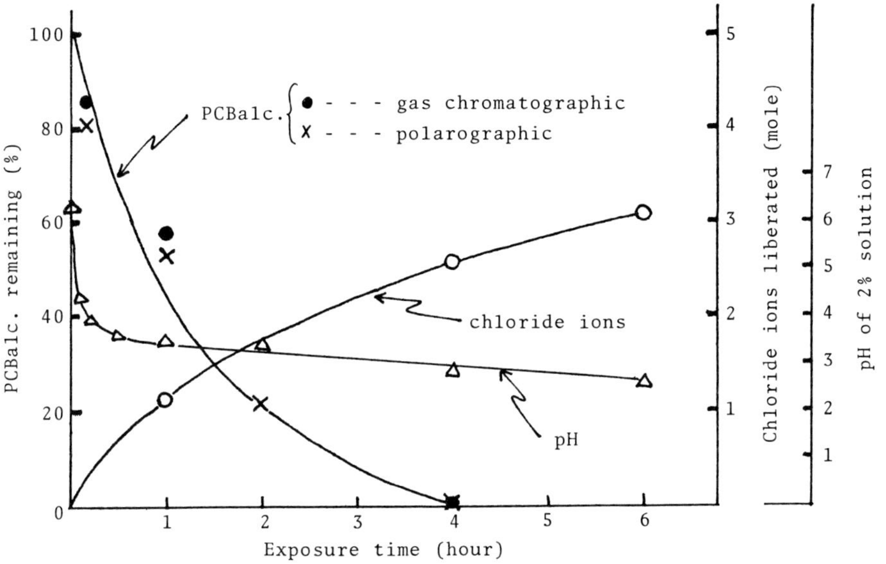

Fig. 3. Photodecomposition of PCBalc. in methanol. One gram of PCBalc. in 500 ml of methanol containing 1% of water was placed in a quartz vessel and irradiated by a low pressure mercury lamp (253.7mμ) at 15-17°C without oxygen supply.

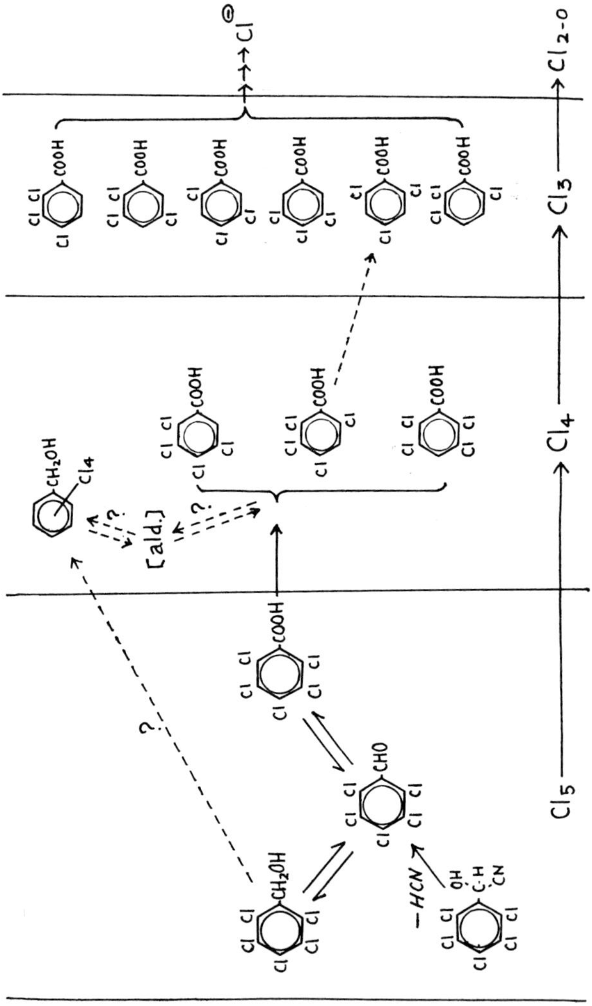

Fig. 4. Possible metabolic pathways of pentachlorobenzyl alcohol.

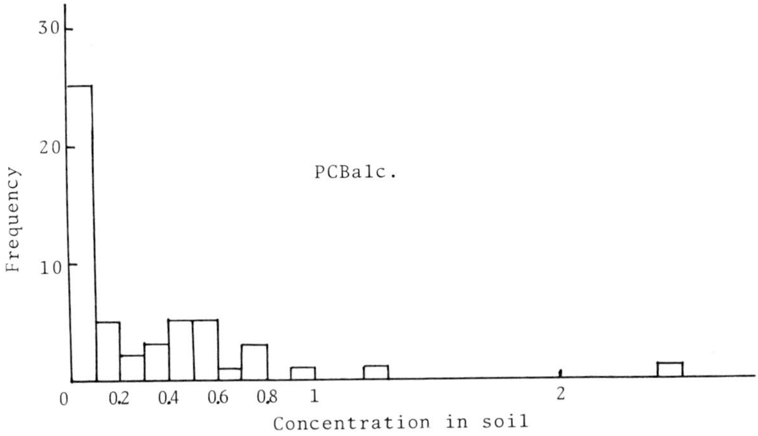

Fig. 5. Contamination of paddy field soil with PCBalc.

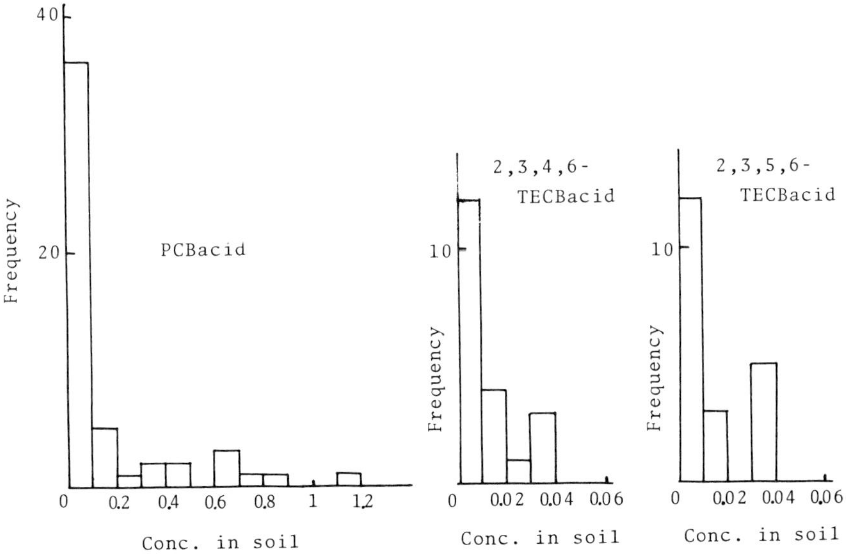

Fig. 6. Contamination of paddy field soil with PCBacid and TECBacid.

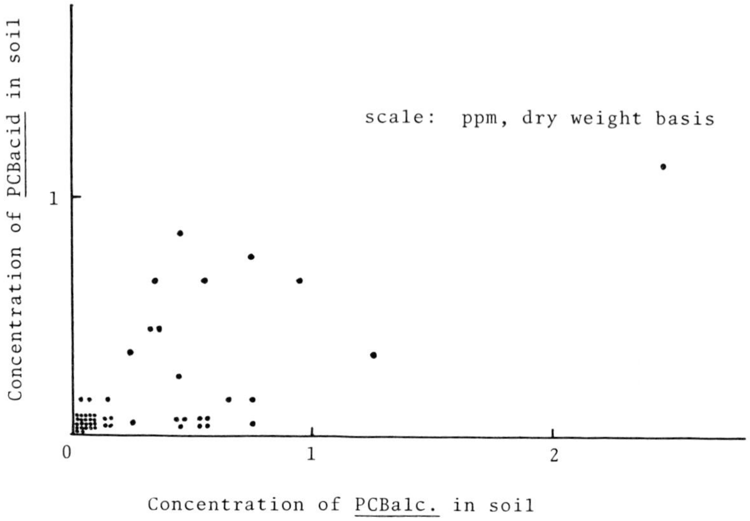

Fig. 7. Relationship between PCBalc. and PCBacid contents in paddy field soil.

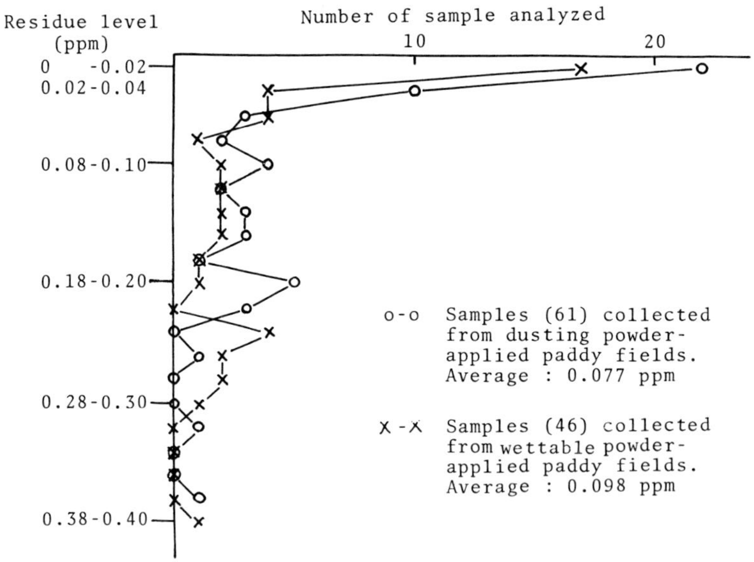

Fig. 8. Distribution of PCBalc. residues in unpolished rice grains.

METABOLISM OF ORGANOPHOSPHORUS INSECTICIDES IN PLANTS

Junshi Miyamoto

Research Department, Pesticides Division
Sumitomo Chemical Co., Ltd.
Osaka, Japan

Introduction

With increased evidence indicating the harmful environmental impact of persistent chemicals, it is apparent that organochlorine insecticides are being replaced by organophosphorus compounds. Accordingly, emphasis in recent times has been placed upon metabolic studies with the latter compounds, in plants, in an attempt to clarify the extent of biodegradation, duration of toxicity, and possible production of toxicologically significant metabolites. In foliar applications, plants generally develop conversion products by ultraviolet light and atmospheric oxygen, whereas in soil treatments, some metabolites are inevitably absorbed into the plants. Nevertheless, these factors do not devaluate the importance of metabolic studies. Rather, toxic residues in plants should be considered inclusive of these factors.

Oxidation of Phosphorothioates and Dithioates

The organophosphorus insecticides are often activated by oxidation, and detoxified by various types of hydrolytic reactions. Relatively inert phosphorothioates and phosphorodithioates are oxidatively converted to the more toxic phosphorates and phosphorothiolates, respectively (O'Brien, 1967), and although these toxic oxygen analogs are often unable to be demonstrated in plant tissues, probably due in part to the minute quantities present, and also to the intrinsic instability of the products, this activation reaction is apparently universal in plants and is similar to that occurring in mammals and insects.

David and Aldridge (1957) were among the first to demonstrate the presence of the oxygen analog of parathion, paraoxon, in the guttation fluid from plants grown in contact with parathion. Coffin (1966) detected a minute amount of paraoxon in lettuce samples shortly after spraying with parathion. He also demonstrated two other anticarboxyesterase components which he presumed to be S-ethyl and S-phenyl isomers. Eldefrawi and Hopkins (1966) observed that parathion was absorbed into bean plants from the nutrient solution, but that it was only slightly translocated into leaves, whereas up to 5 ppm of paraoxon formed in the root was translocated into the upper tissues. Methylparathion was found to be oxidized also to methylparaoxon in rice plants (Miyamoto et al., 1963). In an experiment using ^{32}P-fenitrothion, Miyamoto and Sato (1965) demonstrated the presence of fenitrooxon in rice plants, and noted that it decreased more rapidly than the parent compound. Leuck and Bowman (1968) analyzed Coastal bermuda grass and corn for fenitrooxon, as well as fenitrothion, by gas chromatograph equipped with a phosphorus sensitive detector and found that the content of fenitrooxon reached its maximum 1 day after spraying and that it decreased rapidly thereafter and was hardly detectable after 14 days (Fig. 1).

Several research groups have attempted to demonstrate diazoxon from diazinon in various plant species. Some were successful in detecting its presence (Coffin and McKinley, 1964; Ralls et al., 1966; Kansouth and Hopkins, 1968; Hirano and Yushima, 1969; Masuda and Fukuda, 1970; Nelson and Hamilton, 1970), but the quantity of the oxygen analog was usually less than 0.01 ppm. Pardue et al. (1970) identified gas chromatographically and by infrared and mass spectra, hydroxydiazinon which had been reported to be a photodecomposition product of diazinon (Fig. 2).

The formation of such biologically active oxygen analogs has been known to take place with other phosphorothioates and dithioates such as azinphosmethyl (Tietz et al., 1952), Bromophos (Stiasni et al., 1969), cidial (Hirose et al., 1971), dimethoate (Santi and Pietri-Tonelli, 1959; Santi and Giacomelli, 1962; Okui and Akiyama, 1965; Lucier and Menzer, 1968), malathion (Coffin, 1966), and Supracide (Cassidy et al., 1969); the content was by far lower than other detoxified metabolites, as will be discussed below. These oxygen analogs were also present on the plant surface after foliar applications implying oxidation by ultraviolet light and/or atmospheric oxygen, although enzymatic oxidation undoubtedly takes place.

Oxidation of Thioethers in Plants

The introduction of a thioether group into phosphorothioate or dithioate increases the possibility of metabolic attack due to the oxidation of the thioether group, leading to the formation of the respective sulfoxide and sulfone.

Between two isomers of demeton, the thiol analog more easily penetrated into the tissues of bean, cotton, and lemon, while the thiono isomer was only slowly translocated (Metcalf et al., 1954). As Metcalf and his colleagues demonstrated (Metcalf et al., 1954; Metcalf et al., 1955; Metcalf et al., 1956; Fukuto et al., 1957), these isomers were largely oxidized to their sulfoxides with an accompanying more potent anticholinesterase activity (TABLE 1). They were further oxidized, at either P=S bond or sulfoxide group, to form respective sulfones. The thiono isomer was less persistent than the thiol counterpart because there are two possible metabolic pathways available (namely oxidation of thiono S to compound VI, TABLE 1) and oxidation of sulfoxide to the sulfone (Fig. 3). For example, the half-life of thiol metabolites

TABLE 1

Oxidation Products of Demeton[a]

Product	Structure	Fly-head ChE inhibition (PI_{50})
I. Thiono-isomer	$(C_2H_5O)_2P(S)OC_2H_4SC_2H_5$	3.7
II. Thiolo-isomer	$(C_2H_5O)_2P(O)SC_2H_4SC_2H_5$	5.5
III. Phosphate	$(C_2H_5O)_2P(O)OC_2H_4SC_2H_5$	7.6
IV. Thionosulfoxide	$(C_2H_5O)_2P(S)OC_2H_4S(O)C_2H_5$	5.5
V. Thiolosulfoxide	$(C_2H_5O)_2P(O)SC_2H_4(O)C_2H_5$	5.8
VI. Sulfoxide phosphate	$(C_2H_5O)_2P(O)OC_2H_4S(O)C_2H_5$	6.0
VII. Sulfone phosphate	$(C_2H_5O)_2P(O)OC_2H_4S(O)(O)C_2H_5$	6.9
VIII. Thiolosulfone	$(C_2H_5O)_2P(O)SC_2H_4S(O)(O)C_2H_5$	6.2
IX. Thionosulfone	$(C_2H_5O)_2P(S)OC_2H_4S(O)(O)C_2H_5$	6.1

[a] Fukuto et al., 1955.

in cotton was 18 days, while that of the thiono analog was 8 days. These toxic metabolites disappeared by hydrolytic decomposition.

Thioethers of phosphorodithioates such as disulfoton and phorate are oxidized similarly. Bull (1965) observed that in cotton leaves disulfoton was converted almost quantitatively, and extremely rapidly to its sulfoxide and the thiono sulfone. Also, the thiol sulfoxide and sulfone increased with the lapse of time. After 16 days the major metabolite in a chloroform extract was thiol sulfone, amounting to 20% of the total ^{32}P. This finding was essentially identical with the results reported by Metcalf et al. (1959) in which they demonstrated that the disulfoton sulfoxide formed was further oxidized approximately 1.9 times faster for each 10°C rise in temperature. In addition, they found that tomato leaves metabolized disulfoton sulfoxide two to three times faster than cotton leaves, ultimately to innocuous hydrolysis products, and that the rate of metabolism in ten other plants appeared to be intermediate between those of tomato and cotton. The notable exception was lemon, in which intact disulfoton was dissolved in oil glands of the tissue, thus being protected from decomposition (see also Gunther, 1969).

Similar results were obtained by other workers with lemon, bean, alfalfa (Metcalf et al., 1957), stone leek (Okui and Akiyama, 1965), and spinach (Menzer and Ditman, 1968), which indicate the rapid formation and persistence of sulfoxide and sulfone.

Bowman and Casida (1957), Metcalf et al. (1957), Bowman et al. (1969), and Menzer and Ditman (1968) showed that the metabolism of phorate in various plants such as bean, beet, cabbage, carrot, lettuce, cotton, corn, and spinach followed essentially the same pathways as for disulfoton. That is, phorate was oxidized rapidly to its sulfoxide, and phoratoxon, the oxygen analog, was rarely detected (Fig. 4). Therefore, the major parts of sulfoxide and sulfone of oxy-disulfoton and phoratoxon were presumably derived from thiono sulfoxide, once formed, rather than by thioether oxidation of the oxygen analog (Fig. 3).

Fenthion is another thioether phosphorothioate of which oxidation in plants was unequivocally demonstrated. Fukuda et al. (1962) found that the major toxic metabolites of fenthion in rice plants and cabbage were its sulfoxide and sulfone, and Niessen et al. (1962 /1963) showed that following an application to young bean plants fenthion gave rise to four oxidation products, namely the sulfoxide and sulfone of the phosphorothioate and of the oxygen analog, analogous to the degradation of disulfoton and phorate already mentioned. Leuck and Bowman (1968) analyzed corn and grass forage for fenthion and five of its metabolites. They noted (Fig. 5) that fenthion decreased rapidly, from which its sulfoxide and sulfone and those of the oxygen analog were formed. Also, they detected only a trace amount of the oxygen analog. Most of these toxic metabolites disappeared more gradually than the parent compound. According to Niessen et al. (1962 / 1963) and Metcalf et al. (1963), oxidation of fenthion to its sulfoxide and sulfone was largely due to light and air, which may corroborate the rapid conversion of the aforementioned thioether insecticides to their sulfoxides and sulfones.

These thioether oxidations were confirmed or presumed also in the case of other compounds such as Abate (Blinn, 1968), Carbophenothion (Coffin, 1964), Dasanit (Katague and Anderson, 1967), Dyfonate (McBain et al., 1970), and GC-6506 (Bull and Stokes, 1970; Wendel and Bull, 1970), although the formation rate and the duration of the sulfoxide and sulfone were different depending upon the compound and/or plant species.

Ester Cleavages and Dealkylation

The organophosphorus compounds discussed so far are decomposed by cleavage of phosphate ester bonds, thus producing dealkyl phosphorothioates, dithioates and dialkylphosphorothioic and/or dialkylphosphorodithioic acid. Indeed there are many examples indicating the presence of these degradation products in plants.

Therefore, we are listing only some of them here: Abate (Blinn, 1968), bromophos (Stenersen, 1969; Stiasni et al., 1969), Cidial (Hirose et al., 1971), demeton (Metcalf et al., 1955, 1956; Thomas, 1956; Fukuto et al., 1957), diazinon (Kansouth and Hopkins, 1968; Masuda and Fukuda, 1970; Pardue et al., 1970), dimethoate (Dauterman et al., 1960; Santi and Giacomelli, 1962; Hacskaylo and Bull, 1963; Morikawa and Saito, 1966; Lucier and Menzer, 1968), disulfoton (Metcalf et al., 1957, 1959; Bull, 1965), Dursban (Smith et al., 1967a, b), Dyfonate (McBain et al., 1970), fenitrothion (Miyamoto and Sato, 1965), fenthion (Tomizawa, 1962), Imidan (Menn and McBain, 1964), malathion (Tomizawa et al., 1960), methylparathion (Tomizawa et al., 1960; Miyamoto et al., 1963), parathion (Coffin, 1966), phorate (Metcalf et al., 1957), Supracide (Bull, 1968; Cassidy et al., 1969), and GC-6506 (Bull and Stokes, 1970; Wendel and Bull, 1970). Although the oxygen analogs of these organophosphorus compounds are presumed to be hydrolyzed in plants, there have been lesser numbers of reports, as compared with mammals and insects, referring to the hydrolysis products of such oxygen analogs. This may reflect the slower or lesser formation of the oxygen analog in plants.

In the case of phosphorates, such as mevinphos and Bidrin, hydrolytic cleavage was also observed to form dealkylated phosphate esters and dialkylphosphoric acid. These metabolites were detected and/or identified, for example, with chlorofenvinphos (Beynon and Wright, 1967, 1968), mevinphos (Casida et al., 1956; Spencer and Robinson, 1960), phosphamidon (Anliker et al., 1961; Menzer and Ditman, 1963; Bull et al., 1967), Bidrin (Bull and Lindquist, 1964), Azodrin (Lindquist and Bull, 1967), and also with trichlorfon (Hassan et al., 1966; Bull and Ridgway, 1969).

Some organophosphorus compounds have an additional group in the molecule which is likely to be easily metabolized, namely the carboxyester or carboxyamide group. Dimethoate is an example. The degradation of dimethoate in plants, including corn, potato, bean, cotton, and cabbage, has been intensively studied by Dauterman et al. (1960), Santi and Giacomelli (1962), Hacskaylo and Bull (1963), Morikawa and Saito (1966), and Lucier and Menzer (1968). Dauterman et al. (1960) showed that by foliar application the oxycarboxy analog was predominant on the leaf surface, while inside the plant dimethoate was converted largely to demethyldimethoate. Hacskaylo and Bull (1963) separated ten ^{32}P metabolites of dimethoate from excised cotton leaves, among which dimethoate acid derived from carboxyamide hydrolysis accounted for 50% of the radioactivity. Furthermore, Lucier and Menzer (1968) detected in bean plants N-demethyldimethoate and presumed the presence of the N-hydroxymethyl analog of dimethoate and of the oxygen analog. Thus, metabolism of the carboxyamide group in dimethoate might proceed both hydrolytically and oxidatively (Fig. 6).

Phosphamidon, a vinyl phosphate with a carboxyamide group, was also metabolized through N-deethylation. Anliker et al. (1961) isolated from bean plants N-deethyl phosphamidon, which decreased more rapidly than the original compound (Menzer and Ditman, 1963). Bull and Lindquist (1967) demonstrated that in cotton leaves cis-phosphamidon was converted probably via a N-hydroxy analog to N-deethyl phosphamidon more rapidly than the trans isomer (TABLE 2). Besides these metabolites, demethylphosphamidon, dimethylphosphoric acid, α-chloroacetoacetic acid diethylamide, and α-chloroacetoacetic acid ethylamide were identified as hydrolysis products (Fig. 7).

Bidrin and Azodrin were similarly decomposed in bean and cotton plants by (oxidative) elimination of the N-methyl group and by hydrolytic deamidation and O-demethylation and phosphate ester cleavage (Bull and Lindquist, 1964; Menzer and Casida, 1965; Lindquist and Bull, 1967). As shown in TABLE 3, the oxidation products are more or less biologically active. However, they are of short persistency. Menzer and Casida (1965) showed that half-lives of ^{32}P from Bidrin and Azodrin in bean plant were 9 and 14 days, respectively.

TABLE 2

Relative Concentration of Phosphamidon-^{32}P and its Metabolites in Cotton Leaves[a]

Product	% of Applied dose at indicated days after treatment							
	Cis-isomer				Trans-isomer			
	1	2	4	8	1	2	4	8
H_3PO_4	3.2	2.9	2.3	2.2	5.5	4.7	4.9	3.4
Methylphosphate	0.2	0.2	0.0	0.0	0.4	0.4	0.0	0.0
Dimethylphosphate	17.2	23.0	27.8	24.0	7.4	10.8	14.3	16.3
Unknown-D	1.2	1.3	.5	.7	.7	.9	.0	.7
O-Demethyl phosphamidon	26.4	11.8	4.9	2.9	9.2	4.5	4.2	2.3
Hydroxyethyl phosphamidon	.6	2.4	2.6	1.7	3.9	3.3	3.1	2.4
N-Deethyl phosphamidon	18.4	8.6	5.8	2.5	.0	1.3	3.0	3.2
Phosphamidon	15.6	7.2	.7	3.2	38.6	13.9	3.8	4.0
Residue	6.6	7.8	8.1	7.2	14.1	15.1	14.0	11.1
Lost	10.6	34.8	47.3	55.6	20.2	45.1	52.7	56.6

[a]Bull et al., 1967.

TABLE 3

Toxicity of Bidrin and its Metabolites[a]

			PI_{50} Fly-head ChE (mol. conc.)	LD_{50} Housefly (mg/kg)	LD_{50} Mouse (mg/kg)
		$(CH_3O)_2P(=O)OC(CH_3)=CHCON(CH_3)_2$	7.2	38	14
II		$-N(CH_2OH)(CH_3)$	7.0	14	18
III	(Azodrin)	$-NHCH_3$	6.8	6.4	8
IV		$-NHCH_2OH$	6.9	30	12
V		$-NH_2$	6.5	1.0	3

[a]Menzer and Casida, 1965.

Mevinphos is degraded at three sites in the molecule, namely at the vinyl phosphate, 0-methyl, and carboxy-ester bonds. Casida et al. (1956) found that decomposition of mevinphos was very rapid, 90% being lost in 1.7 days. The more toxic cis isomer was less stable than the trans isomer (Casida et al., 1956; Spencer and Robinson, 1960). Although Casida et al. (1956) presumed that the first metabolic attack was at the carboxyester bond, followed by hydrolysis of vinyl phosphate, Spencer and Robinson (1960) confirmed that cis-mevinphos was metabolized in pea plant via two routes, the major one being the direct hydrolytic formation of dimethylphosphoric acid, and the minor pathway, mevinphos acid formation.

Hirose et al. (1971) showed that $^{14}C=O$ cidial was decomposed in cabbage, apple, and strawberry into several metabolites including cidial acid. From the finding that more cidial acid was present in the tissues shortly after application, they presumed that carboxyester cleavage was the initial step of decomposition in plants.

Butonate, n-butyric acid ester of trichlorfon, was found by Dedek (1968) to recover trichlorfon in plants. The quantity of trichlorfon relative to the applied butonate was 50% in plum, 17∿25% in cherry, 65% in immature wheat, and 5% in bean 24 hours after treatment. Although there has been no conclusive evidence presented, Bourke et al. (1968) presumed that ^{14}C-malation was rapidly degraded in bean seedlings into five metabolites, seemingly identical with those of mammals, including malation acid.

Based upon the above findings, metabolic pathways of organophosphorus compounds in plants have been essentially the same as those in mammals, although the importance of each pathway and the amount of intermediary metabolites accumulated might be different with plant species. The only exception has been that plants and animals conjugate differently some phenols derived from organophosphorus compounds. In plants they are conjugated usually as glycosides, whereas in animals they are bound with glucuronic acid and/or sulfate. If compared with organochlorine insecticides, organophosphorus compounds are degraded more rapidly, and ultimately to innocuous metabolites. These characteristic properties might ensure the safer use of organophosphorus insecticides for plant protection, causing no serious residue problems.

References

Anliker, R., E. Beriger, M. Geiger, and K. Schmidt: Helv. Chim. Acta 44, 1622 (1961).

Beynon, K. I., and A. N. Wright: J. Sci. Food Agr. 18, 143 (1967).

Beynon, K. I., and A. N. Wright: J. Sci. Food Agr. 19, 146 (1968).

Blinn, R. C.: J. Agr. Food Chem. 16, 441 (1968).

Bourke, J. B., E. J. Broderick, L. R. Hackler, and P. C. Lippold: J. Agr. Food Chem. 16, 585 (1968).

Bowman, J. B., and J. E. Casida: J. Agr. Food Chem. 5, 192 (1957).

Bowman, M. C., M. Beroza, and J. A. Harding: J. Agr. Food Chem. 17, 138 (1969).

Bull, D. L.: J. Econ. Entomol. 58, 249 (1965).

Bull, D. L.: J. Agr. Food Chem. 16, 610 (1968).

Bull, D. L., and D. A. Lindquist: J. Agr. Food Chem. 12, 310 (1964).

Bull, D. L., and R. L. Ridgway: J. Agr. Food Chem. 17, 837 (1969).

Bull, D. L., and R. A. Stokes: J. Agr. Food Chem. 18, 1134 (1970).

Casida, J. E., P. E. Gatterdam, L. W. Getzin, and R. K. Chapman: J. Agr. Food Chem. 4, 236 (1956).

Cassidy, J. E., D. P. Pyskiewich, and R. T. Murphy: J. Agr. Food Chem. 17, 558 (1969).

Coffin, D. E.: J. AOAC 47, 662 (1964).

Coffin, D. E.: J. AOAC 49, 1018 (1966).

Coffin, D. E., and W. P. McKinley: J. AOAC 47, 632 (1964).

Dauterman, W. C., G. B. Viado, J. E. Casida, and R. D. O'Brien: J. Agr. Food Chem. 8, 115 (1960).

David, W. A. L., and W. N. Aldridge: Ann. Appl. Entomol. 45, 332 (1957).

Dedek, W.: Z. Naturforsch 23b, 504 (1968).

Eldefrawi, A., and T. L. Hopkins: J. Agr. Food Chem. 14, 588 (1966).

Fukuda, H., T. Masuda, and Y. Miyahara: Jap. J. Appl. Entomol. Zool. 6, 230 (1962).

Fukuto, T. R., R. L. Metcalf, R. B. March, and M. G. Maxon: J. Econ. Entomol. 48, 347 (1955).

Fukuto, T. R., J. P. Wolf, R. L. Metcalf, and R. B. March: J. Econ. Entomol. 50, 399 (1957).

Gunther, F. A.: Residue Rev. 28, 1 (1969).

Hacskaylo, J., and D. L. Bull: J. Agr. Food Chem. 11, 464 (1963).

Hassan, A., S. M. A. D. Zayed, and I. Y. Mostafa: Z. Naturforsch. 21b, 498 (1966).

Hirano, C., and T. Yushima: Jap. J. Appl. Entomol. Zool. 13, 174 (1969).

Hirose, M., T. Miyata, and T. Saito: Botyu-Kagaku 36, 43 (1971).

Kansouth, A. S. H., and T. L. Hopkins: J. Agr. Food Chem. 16, 446 (1968).

Katague, D. B., and C. A. Anderson: Bull. Environ. Contam. Toxicol. 2, 228 (1967).

Leuck, D. B., and M. C. Bowman: J. Econ. Entomol. 61, 1594 (1968).

Lindquist, D. A., and D. L. Bull: J. Agr. Food Chem. 15, 267 (1967).

Lucier, G. W., and R. E. Menzer: J. Agr. Food Chem. 16, 936 (1968).

Masuda, T., and H. Fukuda: Botyu-Kagaku 35, 134 (1970).

McBain, J. B., L. J. Hoffman, and J. J. Menn: J. Agr. Food Chem. 18, 1139 (1970).

Menn, J. J., and J. B. McBain: J. Agr. Food Chem. 12, 162 (1964).

Menzer, R. E., and L. P. Ditman: J. Agr. Food Chem. 11, 170 (1963).
Menzer, R. E., and J. E. Casida: J. Agr. Food Chem. 13, 102 (1965).
Menzer, R. E., and L. P. Ditman: J. Econ. Entomol. 61, 225 (1968).
Metcalf, R. L., R. B. March, T. R. Fukuto, and M. Maxon: J. Econ. Entomol. 47, 1045 (1954).
Metcalf, R. L., R. B. March, T. R. Fukuto, and M. Maxon: J. Econ. Entomol. 48, 364 (1955).
Metcalf, R. L., T. R. Fukuto, R. B. March, and E. M. Stafford: J. Econ. Entomol. 49, 738 (1956).
Metcalf, R. L., T. R. Fukuto, and R. B. March: J. Econ. Entomol. 50, 338 (1957).
Metcalf, R. L., H. R. Reynolds, M. Winton, and T. R. Fukuto: J. Econ. Entomol. 52, 435 (1959).
Metcalf, R. L., T. R. Fukuto, and M. Winton: Bull. World Health Organ. 29, 219 (1963).
Miyamoto, J., and Y. Sato: Botyu-Kagaku 30, 45 (1965).
Miyamoto, J., Y. Sato, T. Kadota, A. Fujinami, and M. Endo: Agr. Biol. Chem. 27, 381 (1963).
Morikawa, O., and T. Saito: Botyu-Kagaku 31, 130 (1966).
Nelson, L. L., and E. W. Hamilton: J. Econ. Entomol. 63, 874 (1970).
Niessen, H., H. Teitz, and H. Frese: Pflanzenschutz Nachrichten XV, 125 (1962/1963).
O'Brien, R. D.: Insecticides. Academic Press, New York (1967).
Okui, S., and T. Akiyama: Shokuhin Eiseigaku Zasshi 6, 364 (1965).
Pardue, J. R., E. A. Hansen, R. P. Barron, and J-Y. T. Chen: J. Agr. Food Chem. 18, 405 (1970).
Ralls, J. W., D. R. Gilmore, and A. Cortes: J. Agr. Food Chem. 14, 387 (1966).
Santi, R., and P. de Pietri-Tonelli: Nature 183, 398 (1959).
Santi, R., and R. Giacomelli: J. Agr. Food Chem. 10, 257 (1962).

Smith, G. N., B. S. Watson, and F. S. Fischer: J. Agr. Food Chem. 15, 127 (1967a).

Smith, G. N., B. S. Watson, and F. S. Fischer: J. Agr. Food Chem. 15, 870 (1967b).

Spencer, E. Y., and J. R. Robinson: J. Agr. Food Chem. 8, 293 (1960).

Stenersen, J.: Bull. Environ. Contam. Toxicol. 4, 104 (1969).

Stiasni, M., W. Deckers, K. Schmidt, and H. Simon: J. Agr. Food Chem. 17, 1017 (1969).

Thomas, W. D. E.: J. Sci. Food Agr. 7, 20 (1956).

Tietz, H., R. L. Metcalf, and T. R. Fukuto: Hochen-Briefe 10, 279 (1952).

Tomizawa, C.: Japan. J. Appl. Entomol. Zool. 6, 237 (1962).

Tomizawa, C., T. Sato, H. Yamashina, and H. Kubo: Botyu-Kagaku 25, 99 (1960).

Wendel, L. E., and D. L. Bull: J. Agr. Food Chem. 18, 420 (1970).

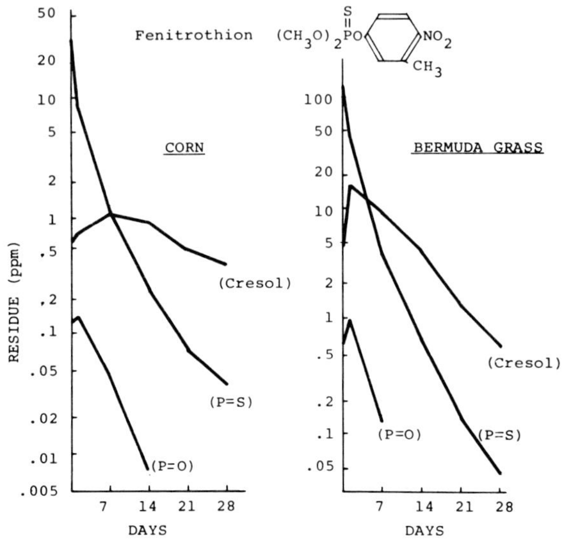

Fig. 1. Fenitrothion residues after spray application of 2 lb/acre. (Leuck and Bowman, 1969)

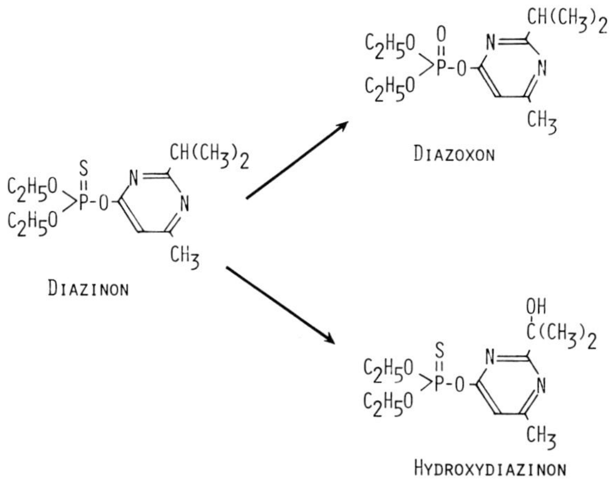

Fig. 2. Conversion of diazinon in plants.

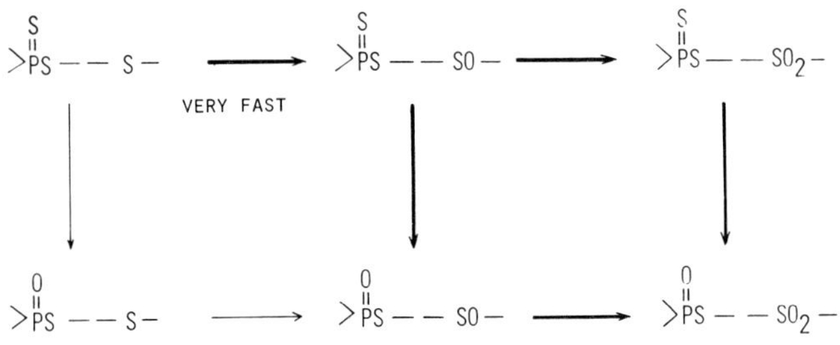

Fig. 3. Oxidation of thioether phosphoro(di)thioates.

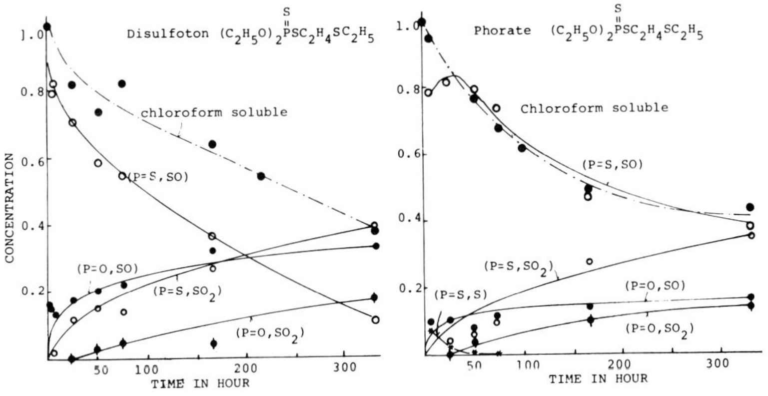

Fig. 4. Relative amount of Disulfoton and Phorate metabolites in chloroform extract of cotton leaves. (Metcalf et al., 1957)

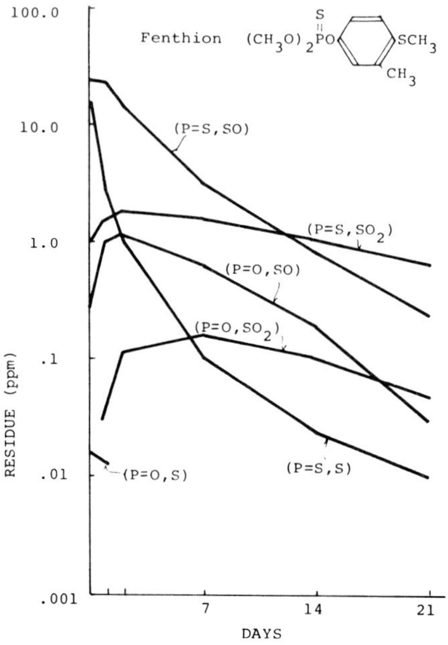

Fig. 5. Fenthion residues in Coastal Bermuda Grass from a spray application at 1 lb/acre. (Leuck and Bowman, 1968)

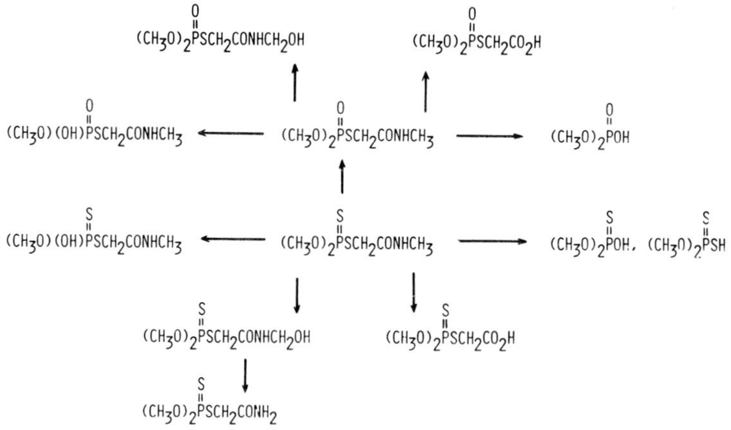

Fig. 6. Metabolic pathway of dimethoate in plants.

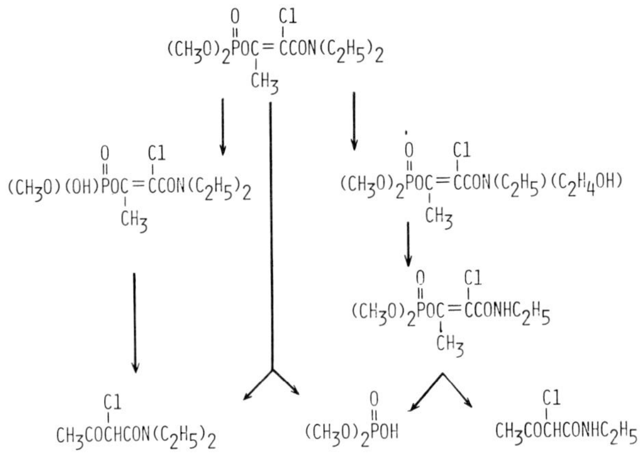

Fig. 7. Metabolic pathways of Phosphamidon in plants.

DEGRADATION OF ORGANOPHOSPHORUS FUNGICIDES

Yasuhiko Uesugi, Chojiro Tomizawa, and Toshinobu Murai

National Institute of Agricultural Sciences
Nishigahara, Kita-ku, Tokyo

Introduction

It is interesting to note that some organophosphorus compounds are effective in controlling rice blast, a serious disease to which many organomercuric compounds had been previously widely used. Various organophosphorus compounds have been tested for fungicidal activity on the rice blast fungus, *Pyricularia oryzae*, and three groups of compounds have been reported to be effective. The first group consists of phosphorothiolates and phosphonothiolates which constitute the earliest anti-blast organophosphorus fungicides (Kado and Yoshinaga, 1969).

The second group is made up of phosphates having a halogenophenyl or nitrophenyl radical in their structure. Oxons of some organophosphorus insecticides, such as methyl paraoxon, have fungicidal activity (Kuwatsuka et al., 1970).

The compounds in the third group are called phosphoramidates. They have recently been found to have fungicidal activity, especially to blast fungi resistant to phosphorothiolate fungicides. However, they are less effective on susceptible fungi. Phosphoramidates have synergistic action with the thiolate fungicides against ordinary susceptible fungi (Katagiri and Uesugi, 1971).

Among the three groups, only the first is in practical use, whereas the other two groups are under investigation. The phosphorothiolate fungicides are under intensive study regarding their behavior in the field, metabolism in organisms, and degradation under various physical conditions. Also, they have been studied from the standpoint of appropriate usage and avoiding any undesirable influences upon the environment. Kitazin P® (S-benzyl diisopropyl phosphorothiolate), Hinosan® (ethyl S,S-diphenyl phosphorodithiolate), and Inezin® (S-benzyl ethyl phenylphosphonothiolate) have been especially investigated.

Method of Investigation

Identification and determination of degradation products were made by various chromatographic methods in comparison with authentic samples. Gas-liquid chromatography, mass spectrometry, or other combined isolation-spectrometry methods are excellent when the degradation products have unknown structures. In the study of degradation products of organophosphorus fungicides, the combined isolation-spectrometry method was not used.

Identification by comparison with authentic samples using chromatographic methods coupled with radiotracer techniques was satisfactory. The fungicides were usually labeled with P^{32}, S^{35}, or C^{14}. Degradation products from labeled fungicides were co-chromatographed with authentic samples utilizing ion exchange chromatography, thin-layer chromatography, and paper chromatography.

Double labeling was sometimes very useful for two reasons. First, two portions of the fungicide molecule could be traced at the same time. Secondly, the ratio of two atoms in a molecule of a degradation product assured the identification of the product.

Gas-liquid chromatography (GLC) with a specifically sensitive detector was also used and could detect minute amounts of degradation products without interference. Degradation products containing either a phosphorus or sulfur atom were identified with authentic samples using GLC with a flame thermionic detector or flame photometric detector (Murai et al., 1971). Hydrolysates or hydroxylated products generally underwent methylation with diazomethane before GLC. Authentic sample for GLC, therefore, may be either known degradation products (i.e., hydrolysates, hydroxy derivatives, etc.) or their derivatives (i.e., methyl ester, methyl ether, and other derivatives). Methyl or other derivatives are often more easily prepared than the expected products themselves. On the other hand, there were a few cases where methyl or other derivatives of an expected degradation product were unstable. Methylation of degradation products having a PSOH group yielded two types of methyl esters, generally a larger amount of S-methyl thiolate and a lesser amount of O-methyl thionate. But on the whole, GLC with a specifically sensitive detector was convenient and could be used successfully in degradation studies though it was inferior to radiotracer techniques in tracing (i.e., GLC cannot distinguish between a product from a given organophosphorus fungicide and that from other sources, and the degradation products with larger molecular weight or polar groups might not be detected at all by GLC).

Rate of Degradation

Rates of hydrolysis of organophosphorus fungicides have not been fully examined. In our experiments, degradation in phosphate buffer of pH 7 at 27°C after 48 hours was 43% for Hinosan and 1.6% for Kitazin P. This means that Hinosan has a rate constant of 2×10^{-4} (min^{-1}) and Kitazin P has that of 6×10^{-6} (min^{-1}). Thus, the half-life for Hinosan is 60 hr and that for Kitazin P is 80 days, in an aqueous solution of pH 7 at 27°C. Inezin seems to have about the same order of rate constant as Kitazin P.

Degradation was accelerated by addition of organisms or irradiation with UV light. For instance, the half-life of Kitazin P in neutral aqueous solution is 80 days, but is shortened to only 2 hr by the addition of 50 mg fresh weight of mycelia of blast fungus per 1 μg Kitazin P.

Even when the fungicides were in hexane solutions or as dry film on the wall of a quartz tube, they degraded to water-soluble products far more rapidly by UV light, emitted from a fluorescent black-light lamp producing highest energy at 360 mμ, than in an aqueous solution without UV light (Murai et al., 1970a; Murai et al., 1970b).

Cleavage of Ester Linkage

In most cases, degradation products of these organophosphorus fungicides were mainly hydrolysates of the parent fungicides. Pathways of degradation through partial ester is illustrated in Fig. 1, 2, and 3.

Hinosan was degraded mainly by cleavage of the P-S linkage to produce ethyl S-phenyl hydrogen phosphorothiolate (H-I). This acidic ester was obtained as a main metabolite by utilizing rats, cockroaches, rice plants, P. oryzae, and Escherichia coli (Fukami et al., 1969; Tan et al., 1969; Tomizawa et al., 1970a; Ueyama et al., 1970; Uesugi and Tomizawa, 1971a). This was also the main degradation product by hydrolysis in aqueous solution of pH 7 at 27°C. Irradiation of UV light around 360 mμ under various conditions (i.e., in aqueous solution, in hexane solution, and as a dry film on the wall of a quartz tube, and incubation with soils) also yielded this acidic ester (Murai et al., 1970a; Tomizawa et al., 1970b). Further cleavage of the P-S linkage yielded ethyl dihydrogen phosphate (H-II), which occurred subsequently. Cleavage of the S-C linkage of Hinosan has not yielded O-ethyl S-phenyl hydrogen phosphorodithioate or O-ethyl dihydrogen phosphorodithioate so far.

On the other hand, cleavage of the S-C linkage occurred in metabolism and degradation studies of Kitazin P by rats, cockroaches, rice plants, and P. oryzae, in soils, and by UV light, yielding O,O-diisopropyl hydrogen phosphorothioate (K-I) (Murai et al., 1970a; Tomizawa et al., 1970b). Cleavage of S-C linkage also occurred in degradation of Inezin by rice plants, P. oryzae, and UV light as well as in neutral and acidic solutions yielding

0-ethyl hydrogen phenylphosphonothioate (I-I) (Endo et al., 1970; Murai et al., 1970b; Murai et al., 1971; Uesugi and Tomizawa, 1971b). The acidic esters produced from *S*-benzyl thiolate fungicides by cleavage of the S-C linkage, K-I and I-I, seemed to be stable as compared with other acidic phosphorus esters. Thus, they were usually found in large quantities in most cases, but their further degradation pathway has not been fully clarified.

Partial esters produced by cleavage of the P-S linkage of *S*-benzyl thiolate fungicides was also routinely found. Diisopropyl hydrogen phosphate (K-II) was found as one of the degradation products of Kitazin P by rice plants (Yamamoto et al., 1970) and UV light (Murai et al., 1970a), but it was apparently a rather minor metabolite in rats, cockroaches, and *P. oryzae* (Tomizawa et al., 1970a; Tomizawa and Uesugi, 1971). Ethyl hydrogen phenylphosphonate (I-II) was also found as one of the main degradation products by *P. oryzae* (Uesugi and Tomizawa, 1971b), by UV light (Murai et al., 1970b), and in neutral and acidic solutions.

Cleavage of the P-0-alkyl linkage to produce *S,S*-diphenyl hydrogen phosphorodithiolate (H-III) from Hinosan, *S*-benzyl isopropyl hydrogen phosphorothiolate (K-III) from Kitazin P, or *S*-benzyl hydrogen phenylphosphonothiolate (I-III) from Inezin occurred but was a minor pathway. However, it did occur to some extent. Rice plants seemed to produce H-III (Tan et al., 1969) and I-III (Endo et al., 1970) more easily than other organisms or non-biological forces. While H-III or K-III was found more or less under various degradation conditions, I-III was not detected in degradation utilizing *P. oryzae* (Uesugi and Tomizawa, 1971b) and UV light (Murai et al., 1970b) and in neutral or acidic solution (Murai et al., 1971).

Besides the above-mentioned degradation products, the following hydrolysates were found. As one of the degradation products of Hinosan, *S*-phenyl dihydrogen phosphorothiolate (H-IV) was often found. Also, it was one of the main degradation products by UV light (Murai et al., 1970a) and by rice plants (Ueyama et al., 1970). From Kitazin P, isopropyl dihydrogen phosphate K-IV) was found by degradation studies with the rice plant (Yamamoto et al., 1970) and by UV light (Murai et al., 1970a), which seemed to be derived from K-II and K-III. A degradation product having a sulfur atom was found at the initial stage of degradation of Kitazin P in rice plants and soils, which was eluted by the gradient of pH 2 to pH 1 from an ion exchange column of Dowex 1-X8. However, it has not yet been identified (Tomizawa et al., 1970b). Phenylphosphonic acid (I-IV) and phenylphosphonothioic acid (I-V) were found as degradation products of Inezin. I-IV is presumed to be a further degraded compound from unstable I-II, but it did not increase with a decrease of I-II, so that I-IV was suggested to be degraded to a further product. I-V was reported as a metabolite by rice plant studies (Endo et al., 1970), but it was not found in the degradation products by *P. oryzae* (Uesugi and Tomizawa, 1971b) and by UV light (Murai et al., 1970b), but I-V was found as a degradation product in acidic solutions (Murai et al., 1971).

The phosphorus partial esters which were found as degradation products of organophosphorus fungicides were almost non-toxic to rice blast fungus. Cleavage of the ester linkage, therefore, means detoxication to the fungus.

Fate of Removed Phenylthio, Benzylthio, and Benzyl Radicals

As mentioned above, the initial step in the degradation of organophosphorus fungicides generally involves the removal of the phenylthio, benzylthio, or benzyl radical. In fact, this degraded radical is often found in the form of corresponding thiols or alcohol to some extent. However, larger amounts of the oxidation products of these radicals were also found in most cases.

Diphenyl disulfide, benzenesulfonic acid, and sulfuric acid were found as degradation products of Hinosan by rice plant metabolism (Ueyama et al., 1970), UV light action

(Murai et al., 1970a), and soils (Tomizawa et al., 1970b). The process of oxidation is shown in Fig. 4.

From Inezin, toluene-α-sulfonic acid and benzoic acid were also found through metabolism by P. oryzae, and the oxidation process shown in Fig. 4 is suggested from the fact that toluene-α-sulfonic acid increased at an earlier stage but decreased afterward, whereas benzoic acid increased during the entire course of the experiment (Uesugi and Tomizawa, 1971b).

Sulfuric acid was found in the degradation of Inezin and Kitazin P (Murai et al., 1970a; Murai et al., 1970b) and the metabolism of Kitazin P by rice plants (Yamamoto et al., 1970). This acid may be produced in the process from phosphoro- or phosphonothioate to phosphate or phosphonate, such as from K-I to K-II, from I-I to I-II, and from I-V to I-IV, as well as from toluene-α-sulfonic acid to benzoic acid.

Hydroxylation

Hydroxy derivatives have been found as fat-soluble metabolites among metabolites of organophosphorus fungicides. However, the pathway resulting in these metabolites is generally minor.

Thus, we can see that a part of Hinosan was metabolized through its p-hydroxy derivative, ethyl S-p-hydroxyphenyl S-phenyl phosphorodithiolate (H-V) by P. oryzae (Uesugi and Tomizawa, 1971a). Rat and cockroach metabolic studies also produce a metabolite which may be identical with H-V (Fukami et al., 1969; Tomizawa et al., 1970). It was also produced by microsomal fraction of rat livers, and in cockroaches in the presence of $NADPH_2$ (Fukami et al., 1969).

It seems strange that apparently hydroxylation of S-benzyl thiolate fungicides (Kitazin P and Inezin) occurs on the m-position of the benzyl radical by P. oryzae action yielding S-m-hydroxybenzyl diisopropyl phosphorothiolate (K-V) and ethyl S-m-hydroxybenzyl phenylphosphonothiolate (I-VI), respectively, and o- or p-hydroxylation does not occur (Tomizawa and Uesugi, 1971; Uesugi and Tomizawa, 1971b).

These experimental results are strikingly similar to studies showing that hydroxylation of benzoic acid occurs on the p-position, and that of phenylacetic acid occurs on the m-position by the action of the fungus, Rhizoctonia solani (Kohmoto and Nishimura, 1970). Both the phosphorus esters and the aromatic carboxilic acids are hydroxylated on the m-position of the benzyl radical when linked to acid groups, and on the p-position of the phenyl radical when linked to acid groups.

Fungicidal activity of hydroxylated metabolites of organophosphorus compounds to the rice blast fungus was tested. With Kitazin P, decrease in the activity by hydroxylation was observed. Therefore, hydroxylation of Kitazin P resulted in the detoxication of this compound to the fungus.

Isomerization

Isomerization of Kitazin P to its isomer, benzyl diisopropyl phosphorothionate (K-VI), by UV light of 254 mμ was reported (Shinohara and Akazaki, 1970). This type of isomerization was also observed under UV irradiation of 254 mμ with Inezin, and yielded benzyl ethyl phenylphosphonothionate (I-VII). However, this was not observed in the case of Hinosan (Murai, unpublished) (Fig. 5).

This isomerization seems to be strictly related to the wavelength of the UV light as UV light of 360 mμ scarcely causes the isomerization.

Transesterification

Besides the above-mentioned degradation reactions, there may be transesterification in some cases. By degradation of Hinosan-S^{35} in soils and by rice plants, two substances seemingly identical to triphenyl phosphorotrithiolate (H-VI) and diethyl S-phenyl phosphorothiolate (H-VII) were detected (Fig. 6). They have not been found in metabolism through *P. oryzae* action.

This type of transesterification has not been found in the case of *S*-benzyl thiolate fungicides.

References

Endo, K., Y. Mori, K. Kakiki, and T. Misato: *J. Agr. Chem. Soc. Japan* **44**, 356 (1970).

Fukami, J., T. Shishido, and K. Fukunaga: *Abstr. Ann. Meet. Agr. Chem. Soc. Japan*, 149 (1969).

Kado, M., and E. Yoshinaga: *Residue Rev.* **25**, 133 (1969).

Katagiri, M., and Y. Uesugi: *Ann. Phytopathol. Soc. Japan* **37**, 188 (1971).

Kohmoto, K., and S. Nishimura: *Ann. Phytopathol. Soc. Japan* **36**, 360 (1970).

Kuwatsuka, S., Y. Uesugi, M. Katagiri, and K. Fukunaga: *Abstr. Ann. Meet. Agr. Chem. Soc. Japan*, 112 (1970).

Murai, T., Y. Uesugi, C. Tomizawa, and T. Tanaka: *Abstr. Ann. Meet. Agr. Chem. Soc. Japan*, 113 (1970a).

Murai, T., Y. Uesugi, C. Tomizawa, and T. Tanaka: *Noyaku Seisan Gijutsu* **22** (Supplementary), 35 (1970b).

Murai, T., Y. Uesugi, C. Tomizawa, and T. Tanaka: *Noyaku Seisan Gijutsu* **24**, 23 (1971).

Shinohara, R., and K. Akazaki: *Noyaku Seisan Gijutsu* **22** (Supplementary), 33 (1970).

Tan, K., I. Takase, K. Ishizuka, and S. Mitsui: *Abstr. Ann. Meet. Agr. Chem. Soc. Japan*, 149 (1969).

Tomizawa, C., and Y. Uesugi: *Agr. Biol. Chem.*, submitted for publication (1971).

Tomizawa, C., Y. Uesugi, S. Miyamoto, H. Dambara, I. Ueyama, and H. Yamamoto: *Abstr. Ann. Meet. Agr. Chem. Soc. Japan*, 115 (1970a).

Tomizawa, C., Y. Uesugi, H. Yamamoto, H. Sugiyama, and I. Ueyama: *Noyaku Seisan Gijutsu* **22** (Supplementary), 37 (1970b).

Uesugi, Y., and C. Tomizawa: Agr. Biol. Chem. 35, 941 (1971a).

Uesugi, Y., and C. Tomizawa: Agr. Biol. Chem., submitted for publication (1971b).

Ueyama, I., Y. Uesugi, C. Tomizawa, and I. Takase: Abstr. Ann. Meet. Agr. Chem. Soc. Japan, 114 (1970).

Yamamoto, H., C. Tomizawa, Y. Uesugi, H. Sugiyama, E. Yoshinaga, and K. Fukunaga: Abstr. Ann. Meet. Agr. Chem. Soc. Japan, 114 (1970).

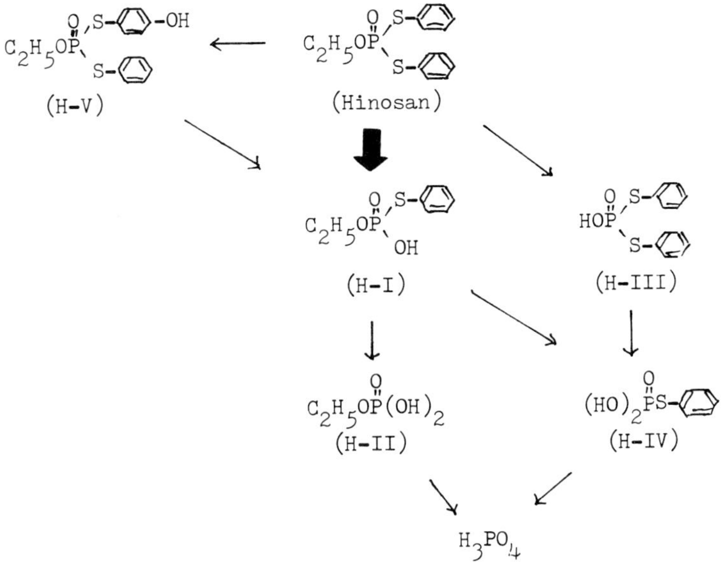

Fig. 1. Degradation pathway of Hinosan.

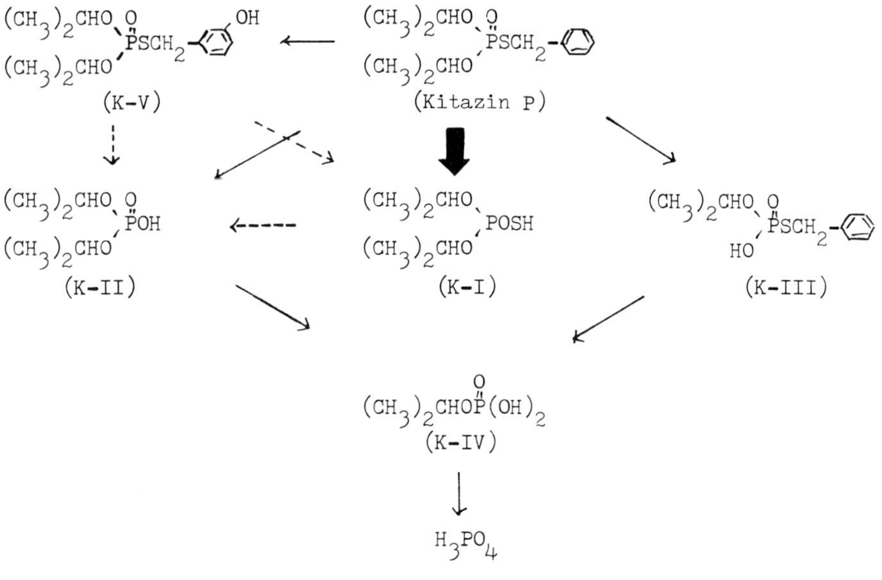

Fig. 2. Degradation pathway of Kitazin P.

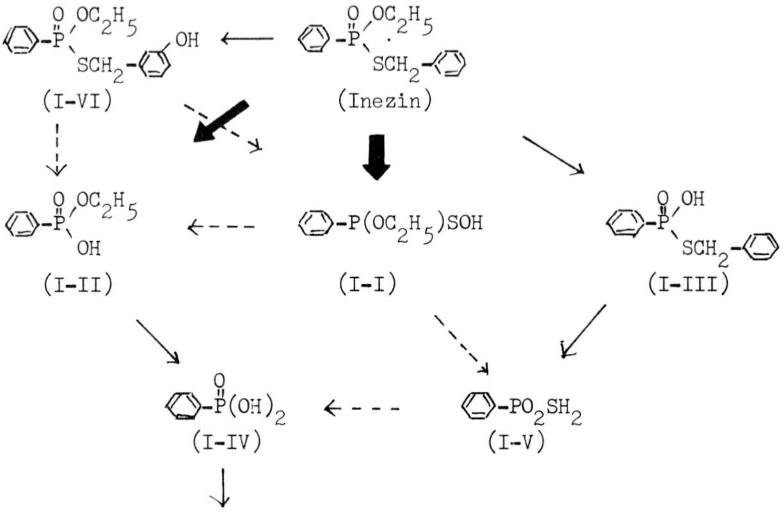

Fig. 3. Degradation pathway of Inezin.

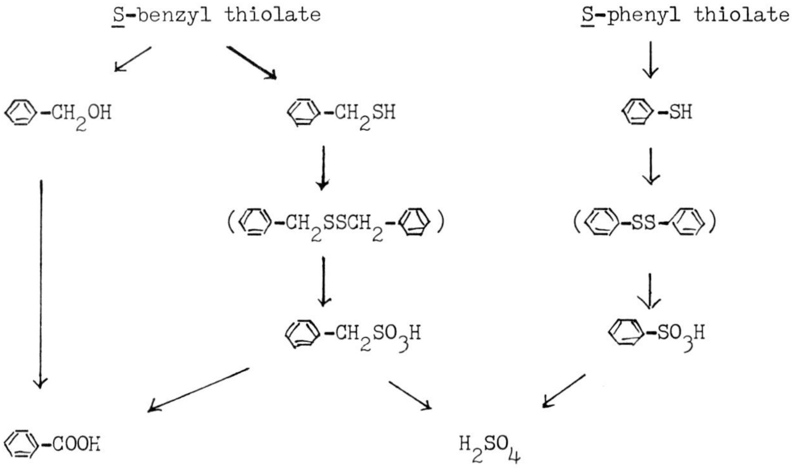

Fig. 4. Oxidation of phenylthio, benzylthio, and benzyl groups.

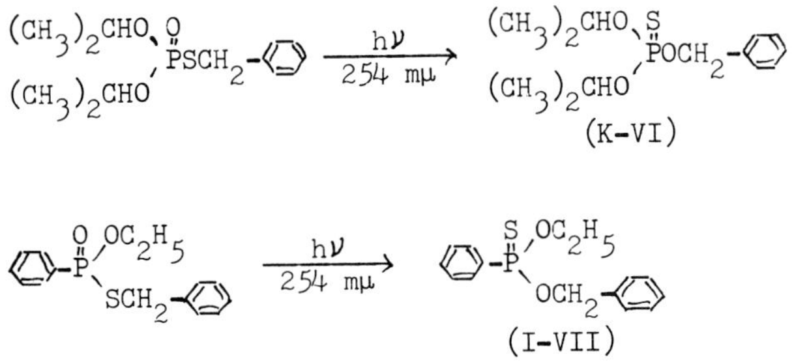

Fig. 5. Isomerization of S-benzyl thiolate fungicides.

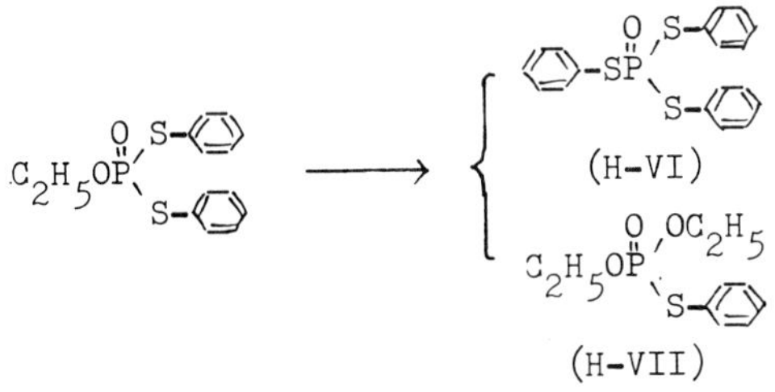

Fig. 6. Transesterification of Hinosan.

METABOLISM OF PESTICIDES IN HIGHER PLANTS

Shooichi Matsunaka

National Institute of Agricultural Sciences
Konosu, Saitama, Japan

Introduction

Microorganisms in soil and water can metabolize or degrade pesticides. Also, photolysis by solar radiation is an important factor in transformation. Recently, purely chemical reactions, such as the free radical reaction, were found to be participating in such transformation (Kearney and Helling, 1969).

Pesticides are usually applied to crop plants or weeds. Some of the material may, after application, be washed off by rain-fall, volatilize spontaneously, or be degraded by solar radiation on the leaf surfaces. However, some pesticides are systemic and can penetrate into plant tissue and thus may be metabolized to other compounds, or remain in the plant. In general, rates of metabolism by higher plants are slower than those of microorganisms or animals. However, we should be aware that although their intensity is relatively low their mass is large. Thus, the total amount of pesticide metabolized by higher plants cannot be neglected.

Moreover, sometimes such metabolism occurs only in higher plants. For instance, the photoreduction (Homer et al., 1960) of paraquat takes place with the aid of photosynthesis of plants. This means that unique metabolites may be found only in higher plants, making such studies especially meaningful.

Several articles (Hilton et al., 1963; Casida and

Lykken, 1969; Kearney and Kaufman, 1969; Matsunaka, 1969b; Frear and Shimabukuro, 1970) deal with pesticide metabolism in higher plants. We will attempt here to classify the metabolic patterns performed. There are many types of metabolism of pesticides in higher plants. In order to recognize them, it is useful to arrange and classify the metabolic patterns. In higher plants, almost all the reactions are mediated by enzymes. Therefore, the metabolic patterns can be classified on the type of enzyme action. For instance, one has been devised by the International Union of Biochemistry (Report of the Enzyme Commission, 1964). However, some of the reactions in higher plants are simple chemical reactions mediated by a special plant component or photochemical factor. I will use in this paper the types of reactions rather than the name of the enzyme as the standard of classification. Because of space limitations, the metabolism of herbicides is stressed. In the case of reactions not found in herbicide metabolism, insecticides or fungicides are used as examples.

The following reactions have been selected to describe the metabolic pattern of pesticides in higher plants: (a) hydrolysis, (b) oxidation (hydroxylation, beta-oxidation, epoxidation, etc.), (c) reduction, (d) decarboxylation, (e) N-dealkylation, and (f) conjugation. Although O-dealkylation, dehalogenation, dehydrohalogenation, or cleavage of the aromatic ring may also be included in pesticide metabolism in higher plants, they are omitted because of a paucity of practical examples.

Hydrolysis

Hydrolysis of amides - The best known example of hydrolysis of an amide pesticide is that of propanil by rice plants. Two pathways for the metabolism of propanil in rice plants (Yih et al., 1968a) are assumed (Fig. 1). In both cases the aromatic product is the same compound, 3,4-dichloroaniline (DCA), which is converted into glucosylamine (Still, 1968) or lignin conjugates (Yih et al.,

1968b) in rice plants. The author (Matsunaka, 1969a) has recently provided direct evidence to show that the propanil-hydrolyzing enzyme system in rice plants is genetically controlled. This was accomplished by the selection of a rice mutant which was susceptible to propanil and lacked the enzyme responsible for the hydrolysis.

The substrate specificity of the aryl acylamidase was reported by Frear and Still (1968) and Akatsuka et al. (1969). As to the ring substitution, 2,3- or 2,4-dichloro-substituted propionanilides were better substrates than propanil itself, a 3,4-dichloro compound. In the case of crabgrass, *Digitaria sanguinalis* (Akatsuka and Hattori, 1969) or barnyard grass, *Echinochloa crusgalli* (Akatsuka, 1971), almost no activity to propanil was found, but 2,4- or 2,3-dichloropropionanilide could be hydrolyzed. Here, the substrate specificity of the aryl acylamidase may reflect their selective toxicity to propanil.

The rice aryl enzyme, acylamidase, was inhibited by organophosphates (Matsunaka, 1968; Akatsuka et al., 1970) or by carbamate insecticides (Akatsuka et al., 1969) having inhibitory activity on acetylcholinesterase.

A new herbicide, CMPT, can be used as a highly selective one with wheat. This herbicide is a typical photosynthesis inhibitor, but in tolerant plants the inhibition will disappear 10 days after application (Matsunaka and Nakamura, 1972). The presumed hydrolyzed compound, an aminothiazole, is shown in Fig. 2. It showed no inhibitory activity either in the Hill reaction or as a herbicide. Therefore, hydrolysis of CMPT in wheat plants, as in the case of propanil in rice plants, was assumed, but the identification of the product has not been examined.

Hydrolysis of esters - The hydrolysis of the 2,4-D esters is a typical example of this type. Morré and Rogers (1960) found that the octyl ester was hydrolyzed by extracts from cucumber cotyledons, pumpkin seeds, and corn plants, or by acetone powder from spinach. Their enzyme system was inhibited by the addition of 2 x 10^{-4} M copper sulfate which concentration was known to inhibit the esterase activity. They also measured the development of color from the hydrolysis of the colorless p-nitrophenol ester of 2,4-D by several plant enzyme preparations.

Jooste and Moreland (1963) reported that the rate of ester hydrolysis decreased with chlorination of the phenoxyacetic acid derivatives when purified enzymes from a variety of plants were used. Others reported the hydrolysis of 2,4-D esters in barley (Crafts, 1960) and in lemons (Erickson et al., 1963).

As shown above, the ability to hydrolyze 2,4-D esters seems to be widespread among higher plants. This ability has contributed to the high utility of the ester form of a phenoxy type herbicide in practical use.

In the case of insecticides, malathion, for example (Bourke et al., 1968), its hydrolysis is carried out by carboxyesterase and other enzymes.

Many N-alkyl or N,N-dialkyl carbamates are used as pesticides. They are not rapidly hydrolyzed in plants (e.g., carbaryl or CIPC). Recently the author (Matsunaka, in press) tried to examine the susceptibility of the above-described propanil-susceptible rice mutant to swep (methyl 3,4-dichlorocarbanilate). Half inhibition of plant growth by swep was noted at a dosage of 28 kg/ha in the case of normal rice plants, while it required 3 kg/ha in the propanil-susceptible mutant. Residues of swep in the top portions of soil-treated rice plants was increased to 22 ppm in the mutant; but in the normal plants a maximum value of 7 ppm was obtained 2 days after soil application. Swep seems to be hydrolyzed to some extent in the intact rice plant, and the degradation mechanism is assumed to be the same as in the case of propanil, and may contribute to the tolerance of the rice plant to this herbicide.

Hydrolysis of s-triazines - The chlorinated s-triazine herbicides, simazine, atrazine, propazine, or ipazine, can be degraded by corn plants. As shown in Fig. 3, one of the detoxification processes was found to be a non-enzymatic reaction by a special component in corn plants,

2,4-dihydroxy-7-methoxy-1,4-benzoxazin-3-one or its 2-glucoside (Roth and Knüsli, 1961; Tipton et al., 1971). This process is a kind of hydrolysis. Also, by this process simazine is converted to hydroxysimazine, which is inactive as a herbicide. As shown in TABLE 1, Hamilton (1964) showed that the amount of hydroxysimazine produced by excised roots of several grasses had a high correlation with benzoxazinone content.

TABLE 1

The Ability of Excised Roots of Several Species of *Graminea* to Produce Hydroxysimazine, and the Content of Benzoxazinone Derivative in their Roots or Shoots[a]

Species	Hydroxysimazine produced	Benzoxazinone content[b]	
		Root	Shoot
Corn (Pa-812)	43.7	3.47	9.21
Redcoat wheat	9.5	0.56	0.83
Rye (Tetra Petkus)	25.9	1.12	2.90
Coix lacrymaojobi	39.8	3.84	13.00
Sorghum (Kafer-60)	0	0	0
Oats (Clinton)	0	0	0
Barley (Hudson)	0	0	0

[a] From Hamilton (1964).

[b] μ moles/g fresh weight.

Oxidation

Hydroxylation - The hydroxylation process occupies an important place in the detoxification or degradation of toxic or foreign compounds in animals, microorganisms, and in higher plants. In Fig. 4, the ring hydroxylation of 2,4-D in beans (Thomas et al., 1964b) is shown. In this case, the para-position is hydroxylated, the chlorine atom is shifted from the 4-position to the 5- or 3-position. Hydroxylated 2,4-D will then form glucoside or amino acid conjugates, as will be described later.

N-methyl carbamate insecticides are also hydroxylated in higher plants. We can classify the hydroxylation into three groups, ring hydroxylation, N-methyl hydroxylation, and benzylic carbon hydroxylation. Kuhr (1970) made a table classifying the carbamate insecticide metabolisms in higher plants (TABLE 2). In Fig. 5, the reaction

TABLE 2

Hydroxylation of Carbamate Insecticides in Plants[a]

Type	Insecticides
Ring hydroxylation	carbanolate, carbaryl, 3-keto-carbofuran, propoxur
N-methylhydroxylation	carbanolate, carbaryl, carbofuran, landrin, propoxur, UC-10854
Benzylic carbon hydroxylation	carbanolate, carbofuran, landrin, UC-10854

[a]From Kuhr (1970).

on the left side is the N-methylhydroxylation of carbaryl and on the right side is ring hydroxylation in higher plants.

A herbicide, diphenamid, was found by Schultz and Tweedy (1971) to form glucose conjugates, especially in tomato shoots. In their experiment they speculated N-methylhydroxylation was an intermediate reaction (Fig. 6).

Beta-oxidation - Beta-oxidation of fatty acids is a well known biological oxidation process (Wain, 1955). Figure 7 shows the beta-oxidation of MCPB into MCPA, an active auxin-type herbicide, by weeds susceptible to the original MCPB. Weeds having high beta-oxidation activity will be killed, but crop plants with low or no beta-oxidation activity will survive. TABLE 3 shows the classification of tolerant or susceptible plants to phenoxybutyric acid herbicides. Omega-oxidation in flax was also purposed by Wain's group (Fawcett et al., 1954).

TABLE 3

Susceptibility of Several Plants to Phenoxybutyric Acids

Plants tolerant to		Susceptible plants
2,4-DB & MCPB	2,4,5-TB	
Alfalfa	Flax	Annual nettle (*Urtica urens*)
Carrot	Rape	Canada thistle (*Cirsium arvense*)
Celery	Soybeans	Charlock (*Sinapis arvensis*)
Clover	Sweet clover	Fat hen (*Chenopodium album*)
Flax		Fumitory (*Fumaria* spp.)
Parsnips		Pigweed (*Amaranthus* spp.)

Epoxidation - The epoxidation of aldrin to form dieldrin is a famous example of the epoxidation process (Fig. 8). Oloffs and Lichtenstein (1969) showed the distribution of the epoxidative activity in higher plants. Recently the reaction system was checked in cell-free preparations from plants by Yu et al. (1971), and found to be characteristic of the mixed function oxidase system (TABLE 4). This is in general agreement with animal systems and soil microorganisms (Matsumura and Boush, 1967; Matsumura et al., 1968).

TABLE 4

Inhibition of Aldrin Metabolism in Bean Root Homogenates[a]

Additions	Concentration in M.	Formation of	
		Dieldrin	Aldrindiol
None	---	100%	100%
$CuSO_4$	10^{-4}	17	25
KCN	10^{-3}	73	50
Sesamex	10^{-3}	61	60
Safrole	10^{-3}	14	0
SKF-525-A	5×10^{-4}	18	18
N_2-atmosphere	---	21	19

[a]From Yu et al. (1971).

Sulfur oxidation - Blinn (1968) showed the oxidation of an insecticide, Abate (0, 0, 0', 0' - tetramethyl 0, 0'-thiodi-p-phenylene phosphorothioate), in which the reaction P=S was converted to P=O, and the sulfur atom of the diphenylthioether was oxidized to $=SO_2$ or =SO in bean leaves. Bowman and others (1969) also surveyed the fate of an insecticide, phorate, and found five metabolites through reactions containing the oxidation of sulfur atom(s) (Fig. 9).

Reduction

Herbicides, paraquat or diquat, may represent a special case of pesticide reduction in higher plants. As shown in Fig. 10, the electron excited by solar radiation through photochemical systems in chloroplasts will reduce paraquat to its free radical, which can then be oxidized by molecular oxygen (Homer et al., 1960). During the course of oxidation, produced peroxides give herbicidal activity. Following this, the reduction of paraquat is inhibited by photosynthesis inhibitors; for example, monuron (Mees,

1960) or propanil (Matsunaka, 1969b). Thus the inhibitors stop the formation of electrons at the first step of photosynthesis.

Some of the naphthoquinone pesticides also receive electrons produced by photosynthesis, and are reduced, as shown by Zweig (1969).

Nitro radicals on the benzene ring of some pesticides are occasionally reduced by higher plants. Diphenylether compounds having para-nitrosubstitutions are now being utilized as herbicides. A well known example, nitrofen (2, 4-dichlorophenyl p-nitrophenylether), was reported to be reduced in the rumen of dairy cows (Gutenmann and Lisk, 1967), and flooded paddy field soils (Kuwatsuka, 1971). Rogers (1971) recently found two reduced compounds from a diphenylether herbicide, C-6989. One of them was noted in soybean plants, as shown in Fig. 11.

Colby (1966) also reported the nitro radical reduction of a herbicide, dinoben, to another one, amiben, which was conjugated with glucose as described later (Fig. 12).

Golab et al. (1967) showed N-dealkylation of trifluralin was followed by reduction of the nitro radical to an amino radical in carrot roots (Fig. 13). Biswass and Hamilton (1969) also reported that one of the nitro radicals on trifluralin was reduced by a crude enzyme preparation from sweet potato.

As shown above, the reduction of pesticides in higher plants seems to be limited to the reduction of the nitro radical on benzene rings. Other reductive processes are hardly found in metabolism in higher plants. Perhaps such inability to reduce metabolism may come from aerobic conditions in higher plants.

Decarboxylation

The decarboxylation process was studied with 2, 4-D having a ^{14}C-labeled side chain and measurement of ^{14}C-CO_2 caught by an alkaline trap. Sometimes the decarboxylation process seems to explain the difference of susceptibility between cultivarieties of apples, strawberries, or weeds (Luckwill and Lloyd-Jones, 1960a, 1960b).

N-Dealkylation

N-dealkylation is usually the detoxification process of herbicides. In plants, this process is found in the metabolism of phenylurea and s-triazine herbicides. As shown in Fig. 14, in cotton plants (Swanson and Swanson, 1968), diuron is N-demethylated by two steps of demethylation and cleavage of the C-N bond to form 3, 4-dichloroaniline. These processes were also found in other plants (Onley et al., 1968) or other phenylurea herbicides, monuron (Smith and Sheets, 1967), fluometuron (Rogers and Funderburk, 1968), chloroxuron (Geissbühler et al., 1963), linuron (Kuratle et al., 1969), and chlorbromuron (Nashed et al., 1970). The latter two compounds are degraded also by the N-demethoxylation process.

These processes also usually relate to the tolerance mechanisms to herbicides. Tolerant plants have high activities of N-dealkylation, while susceptible ones have low activity. The degree of metabolism is correlated with the recovery rates of photosynthetic inhibition. A microsomal mixed-function oxidase which catalyzes the initial N-demethylation steps in the metabolism of phenylurea herbicides has been isolated and partially characterized from tolerant plant tissues (Frear et al., 1969). It has been noted (Russell et al., 1968; Knapp et al., 1969) that plant microsomal oxidative systems are similar in many respects to animal oxidase systems.

In the N-demethylation of diuron or monuron, several N-methyl carbamate insecticides were found to be potent inhibitors of the mixed function oxidase system (the N-demethylation step), while SKF-525, a noted inhibitor of mixed function oxidase systems in animal tissue, had almost no inhibitory activity (Frear et al., 1969).

A s-triazine herbicide, atrazine, is also N-demethylated. This process is dominant in peas (Shimabukuro, 1967) and cotton (Shimabukuro and Swanson, 1970) and contributes to the moderate tolerance of these plants. Other s-triazine herbicides may be N-demethylated in the same way as atrazine. Figures 6 and 13 show the N-dealkylation process of diphenamid and trifluralin, respectively.

Conjugation

Higher plants, as compared with animals, have lower activities in the excretion of used or unfavorable metabolites. Here, conjugation with other metabolites or polymerization *per se*, followed by precipitation at inactive sites in the cell, are important. In some cases, the conjugation process contributes to their transfer through tissue.

Conjugation of pesticides in plant tissues can be classified in four groups, depending upon the conjugation partners: (a) sugars, especially glucose, (b) amino acids, (c) peptides, and (d) high molecular compounds. When glucose is the partner it may be divided into two conjugated forms--glucoside and the glucose ester. The former form of conjugation can take place through two different routes: i.e., conjugation with amino radical or the hydroxyl radical of pesticides.

A good example of the first case is the conjugation between the amino radical of amiben, a herbicide used in soybean culture. This is catalyzed by arylamine N-glucosyl-transferase (Frear, 1968) which is specific for uridine diphosphate-5'-glucose (UDPG). This enzyme is present in a number of plants, associated with root tissues, and has a broad specificity for substituted anilines and heterocyclic compounds with amino radical. Another herbicide, dinoben (Colby, 1966), will conjugate with glucose after being reduced to amiben in soybean plants, as shown in Fig. 12.

The heterocyclic amine herbicide, pyrazone (5-amino-4-chloro-2-phenyl-3(2H)-pyridazinone) also forms N-glucosyl-pyrazon in red beets (Ries et al., 1968). In rice plants, 3,4-dichloroaniline, which was produced by the hydrolysis of propanil by a special enzyme aryl acylamidase, conjugates with glucose to form N-(3,4-dichlorophenyl)-glucose amine (Still, 1968).

Some pesticides, after being hydroxylated by the above-described system, will conjugate with glucose. Sometimes 2, 4-D is ring-hydroxylated to form 2, 5-dichloro-4-hydroxyphenoxy acetic acid or 2, 3-dichloro-4-hydroxyphenoxy acetic acid (Fig. 4), then the hydroxyl radical will conjugate with glucose (Feung et al., 1971). Beta-D-glucoside of 4-hydroxy-2, 5-dichlorophenoxyacetic acid was detected in bean plants.

CIPC (chloropropham) (James and Prendeville, 1969) and diphenamid (Schultz and Tweedy, 1971) may form glucoside compounds in corn or sorghum and in tomato plants. N-methyl carbamate insecticides may also form glucoside conjugates, not glucuronides, after being hydroxylated in plants (Kuhr and Casida, 1967).

The compound 2, 4-D, itself, can form an ester conjugate with glucose in higher plants (Thomas et al., 1964a).

The conjugate of 2, 4-D with glutamic acid was found in soybean callus (Feung et al., 1971), and that with aspartic acid, in wheat or pea. Amitrole (3-amino-s-triazole) has been known to conjugate with glycine and serine in plants.

In the case of peptides, conjugation of atrazine with glutathione is best known. The enzyme responsible for catalyzing this reaction, a glutathion S-transferase, is located primarily in the shoots of tolerant plants (Shimabukuro et al., 1970). Corn and sorghum plants have high activities of the glutathion-S-transferase in their shoots but have low activities in their roots. This enzyme was partially purified and its properties surveyed (Frear and Swanson, 1970).

Very recently Lamoureux et al. (1971) reported that a 2-chloroacetoanilide herbicide, propachlor, (2-chloro-N-isopropylacetoanilide), conjugated with glutathione or γ-glutamylcysteine in corn, sorghum, sugar cane, and barley (Fig. 15). The terminal chlorine atom in propachlor may be active in this conjugation as in the case with the chlorine atom at the 2-position of atrazine molecule.

In the case of high molecular components in plants,

conjugations of swep with lignin, and of 2, 4-D with protein are known to occur.

Conclusion

In conclusion, in the metabolism of pesticides in higher plants, three reactions may occur: hydrolysis, oxidation hydroxylation, and conjugation processes.

The above-described microsomal mixed function oxidase system is very important but also very labile in the cell-free condition (Frear et al., 1969), making the biochemical research on this system difficult. However, since most of the oxidative metabolism of pesticides, hydroxylation, epoxidation, and also N-dealkylation depends upon the microsomal oxidase system, extensive researches on this subject are needed. Considering the differences in excretion mechanisms between animals and plants, the conjugation of pesticides or of their metabolites has special meaning.

Knowledge of pesticide metabolism seems to be of dual importance; first, the residue problems involved, and secondly, the development of new pesticides. We should consider at least two items in discussing the residue problem: (a) the toxicity of metabolites, and (b) the nature of the conjugated products. It is always possible that a metabolite from a pesticide may be more toxic to animals than the parent pesticide, and also that the conjugated products may remain, as the parent chemical structure, in the tissues of crop plants. The reactions in pesticide metabolism have a close relationship with crop injury or their selectivity. The enzymes for degradation of herbicides by crop plants can be assumed to be the detoxification mechanism.

On the other hand, long residual pesticides are not favorable today. Therefore, rapid metabolism or degradation by crop plants to CO_2 will be welcome. Fortunately, higher plants seem to handle the oxidative reactions easily, which means degradation to CO_2 in the case of complete oxidation. Now we should consider the metabolism of pesticides by microorganisms, animals, and higher plants, along with photoreactions and purely chemical reaction, to understand the complex behavior of pesticides in the environment.

References

Akatsuka, T.: Symposium on Plant Enzymes, Osaka, Japan, October (1971).

Akatsuka, T., and M. Hattori: Sci. Report Faculty Agr. Ibaraki Univ. 17, 49 (1969).

Akatsuka, T., M. Ozeki, and S. Kuwatsuka: Sci. Report Faculty Agr. Ibaraki Univ. 18, 23 (1970).

Akatsuka, T., K. Suzuki, and S. Kuwatsuka: Sci. Report Faculty Agr. Ibaraki Univ. 17, 45 (1969).

Biswass, P. K., and W. Hamilton, Jr.: Weed Sci. 17, 206 (1969).

Blinn, R. C.: J. Agr. Food Chem. 16, 441 (1968).

Bourke, J. B., E. J. Broderick, L. R. Hackler, and P. C. Lippold: J. Agr. Food Chem. 16, 585 (1968).

Bowman, M. C., M. Beroza, and J. A. Harding: J. Agr. Food Chem. 17, 138 (1969).

Buffs, J. S., and S. C. Fang: A Conference on Radioactive Isotopes in Agriculture, Atomic Energy Comm., Rep. TID-7512 (1956).

Casida, J. E., and L. Lykken: Ann. Rev. Plant Physiol. 20, 607 (1969).

Chin, W. T., R. P. Stanovick, T. E. Cullen, and G. C. Holsing: Weeds 12, 201 (1964).

Colby, S. R.: Weeds 14, 197 (1966).

Crafts, A. S.: Weeds 8, 19 (1960).

Erickson, E. A., B. L. Brannaman, and C. W. Coggins: J. Agr. Food Chem. 11, 437 (1963).

Fawcett, C. H., J. M. A. Ingram, and R. L. Wain: Proc. Royal Soc. (London) B142, 60 (1954).

Feung, C. S., R. H. Hamilton, and F. H. Witham: J. Agr. Food Chem. 19, 475 (1971).

Frear, D. S.: Phytochemistry 7, 381 (1968).

Frear, D. S., and R. H. Shimabukuro: A background paper, 1st FAO Int. Conf. Weed Control, Davis, California, June (1970).

Frear, D. S., and G. G. Still: Phytochemistry 7, 913 (1968).

Frear, D. S., and H. R. Swanson: Phytochemistry 9, 2123 (1970).

Geissbühler, H., G. Haselbach, H. Aebi, and L. Ebner: Weed Res. 3, 277 (1963).

Golab, T., R. J. Herberg, S. J. Parka, and J. B. Tepe: J. Agr. Food Chem. 15, 638 (1967).

Gutenmann, W. H., and D. J. Lisk: J. Dairy Sci. 50, 1516 (1967).

Hamilton, R. H.: J. Agr. Food Chem. 12, 14 (1964).

Hilton, J. L., L. L. Jansen, and H. M. Hull: Ann. Rev. Plant Physiol. 14, 353 (1963).

Homer, R. F., G. C. Mees, and T. E. Tomlinson: J. Sci. Food Agr. 6, 309 (1960).

James, C. S., and G. N. Prendeville: J. Agr. Food Chem. 17, 1257 (1969).

Jooste, J., and D. E. Moreland: Phytochemistry 2, 263 (1963).

Kearney, P. C., and C. S. Helling: Residue Rev. 25, 25 (1969).

Kearney, P. C., and D. D. Kaufman (Eds.): Degradation of Herbicides. Marcel Dekker, Inc., New York (1969).

Knapp, F. F., R. T. Aexel, and H. J. Nicholas: Plant Physiol. 44, 442 (1969).

Kuhr, R. J.: J. Agr. Food Chem. 18, 1023 (1970).

Kuhr, R. J., and J. E. Casida: J. Agr. Food Chem. 15, 814 (1967).

Kuratle, H., E. M. Rahn, and C. W. Woodmansee: Weed Sci. 17, 216 (1969).

Kuwatsuka, S.: Environmental Toxicology of Pesticides, U.S.-Japan Seminar, Oiso, Japan, October (1971).

Lamoureux, G. L., L. E. Stafford, and F. S. Tanaka: J. Agr. Food Chem. 19, 346 (1971).

Luckwill, L. C., and C. P. Lloyd-Jonesi: Ann. Appl. Biol. 48, 613 (1960a).
Luckwill, L. C., and C. P. Lloyd-Jonesi: Ann. Appl. Biol. 48, 626 (1960b).
Matsumura, F., and G. M. Boush: Science 156, 959 (1967).
Matsumura, F., G. M. Boush, and A. Tai: Nature 219, 965 (1968).
Matsunaka, S.: Science 160, 1360 (1968).
Matsunaka, S.: Biochemical Approaches to Weed Control, U.S.-Japan Seminar, Raleigh, N. C., March (1969a).
Matsunaka, S.: Residue Rev. 25, 45 (1969b).
Matsunaka, S.: Pure Appl. Chem., special issue (in press).
Matsunaka, S., and H. Nakamura: Weed Res. (Tokyo), 13, 29 (1972).
Mees, G. C.: Ann. Appl. Biol. 48, 601 (1960).
Morré, D. J., and B. J. Rogers: Weeds 8, 436 (1960).
Nashed, R. B., S. E. Katz, and R. D. Ilnicki: Weed Sci. 18, 122 (1970).
Oloffs, P. C., and E. P. Lichtenstein: J. Agr. Food Chem. 17, 143 (1969).
Onley, J. H., G. Yip, and M. H. Aldridge: J. Agr. Food Chem. 16, 426 (1968).
Ries, S. K., M. S. Zabik, G. R. Stephenson, and T. M. Chen: Weed Sci. 16, 40 (1968).
Rogers, R. L.: J. Agr. Food Chem. 19, 32 (1971).
Rogers, R. L., and H. H. Funderburk, Jr.: J. Agr. Food Chem. 16, 434 (1968).
Roth, W., and E. Knüsli: Experientia 17, 312 (1961).
Russell, D. W., E. E. Conn, A. Sutter, and H. Grisebach: Biochim. Biophys. Acta 170, 210 (1968).
Schultz, D. P., and B. G. Tweedy: J. Agr. Food Chem. 19, 36 (1971).
Shimabukuro, R. H.: J. Agr. Food Chem. 15, 557 (1967).
Shimabukuro, R. H., and H. R. Swanson: Weed Sci. 18, 231 (1970).
Shimabukuro, R. H., H. R. Swanson, and W. C. Walsh: Plant Physiol. 46, 103 (1970).

Smith, J. W., and T. J. Sheets: J. Agr. Food Chem. 15, 577 (1967).
Still, G. G.: Science 159, 992 (1968).
Swanson, C. R., and H. R. Swanson: Weed Sci. 16, 137
Thomas, E. W., B. C. Loughman, and R. G. Powell: Nature 204, 286 (1964a).
Thomas, E. W., D. C. Loughman, and R. G. Powell: Nature 204, 884 (1964b).
Tipton, C. C., R. R. Husted, and F. H-C. Tsao: J. Agr. Food Chem. 19, 484 (1971).
Wain, R. L.: Ann. Appl. Biol. 42, 151 (1955).
Yih, R. Y., D. H. McRae, and H. F. Wilson: Plant Physiol. 43, 1291 (1968a).
Yih, R. Y., D. H. McRae, and H. F. Wilson: Science 161, 376 (1968b).
Yu, S. J., U. Kiigenamagi, and L. C. Terriere: J. Agr. Food Chem. 19, 5 (1971).
Zweg, G.: Residue Rev. 25, 69 (1969).

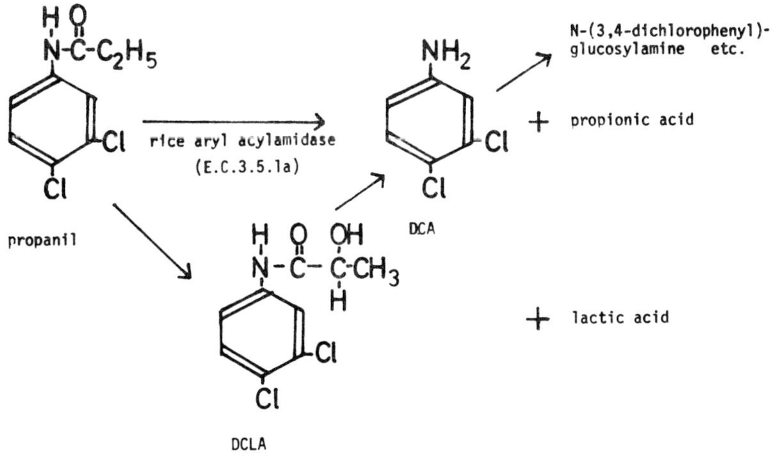

Fig. 1. Metabolism of propanil in rice plants.

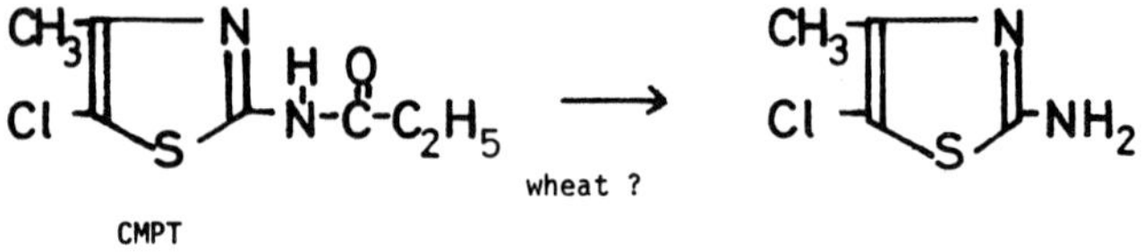

Fig. 2. Speculative mechanism of the detoxification of CMPT by wheat.

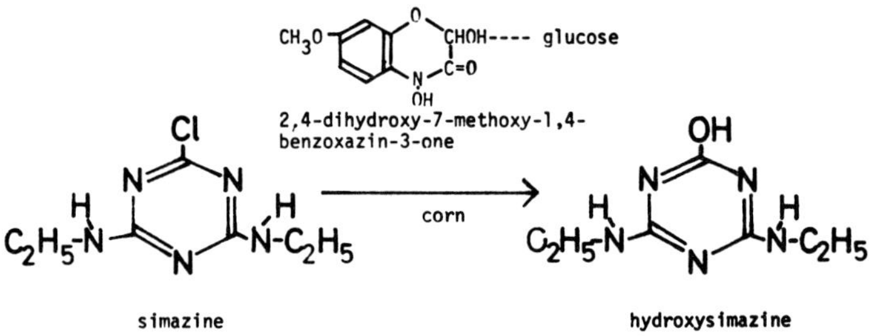

Fig. 3. Hydrolysis of simazine by 2,4-dihydroxy-7-methoxy-1,4-benzoxazin-3-one or its 2-glucoside.

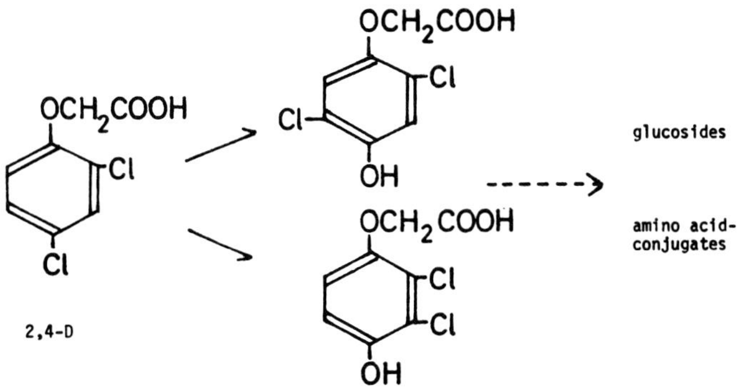

Fig. 4. Ring hydroxylation of 2,4-D with chlorine shift to the 5- or 3- position.

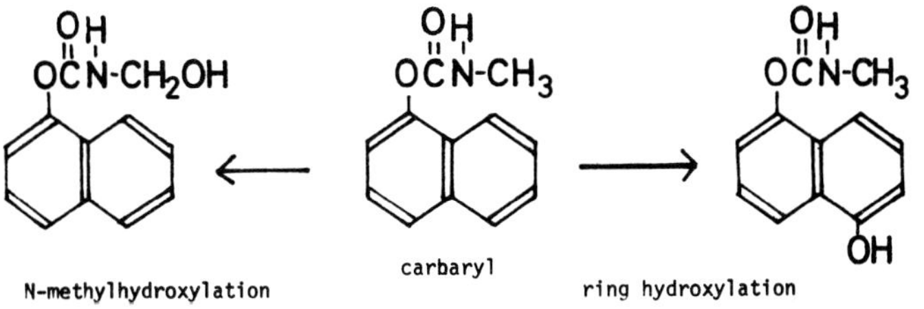

Fig. 5. Hydroxylation of carbaryl in plants.

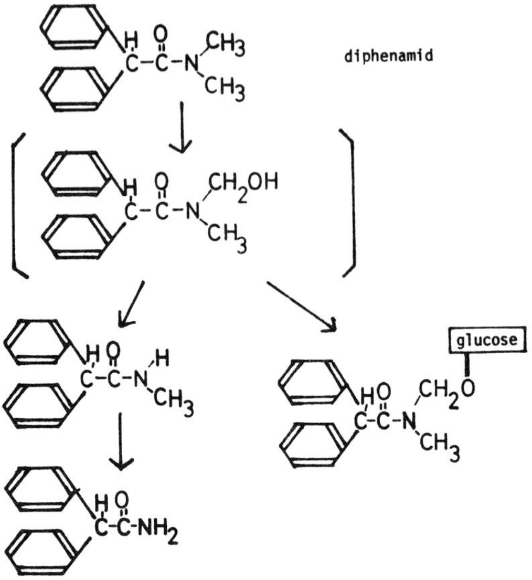

Fig. 6. Diphenamid metabolism in wheat and tomato plants.

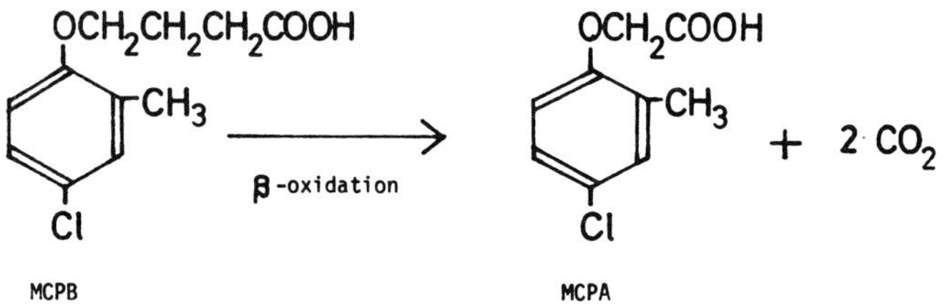

Fig. 7. Beta-oxidation of MCPB to MCPA.

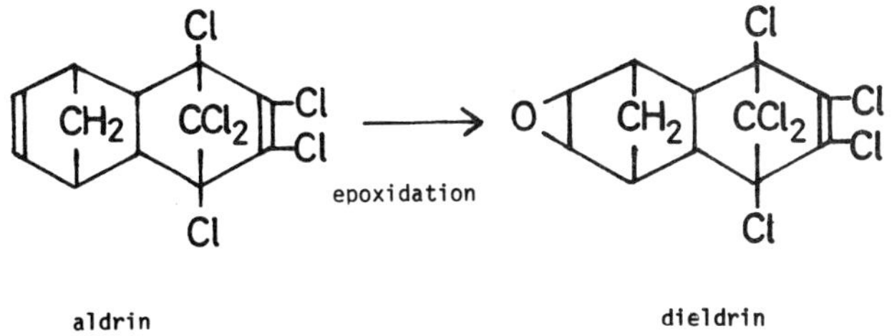

Fig. 8. Epoxidation of aldrin to dieldrin.

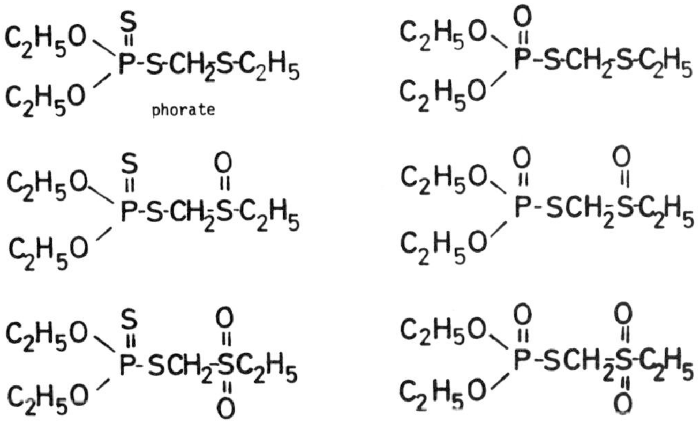

Fig. 9. Phorate and five of its metabolites in corn plants.

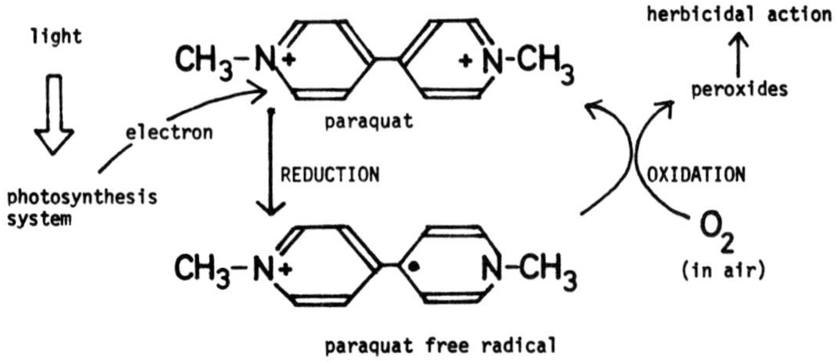

Fig. 10. Photoactivation mechanism of paraquat in higher plants.

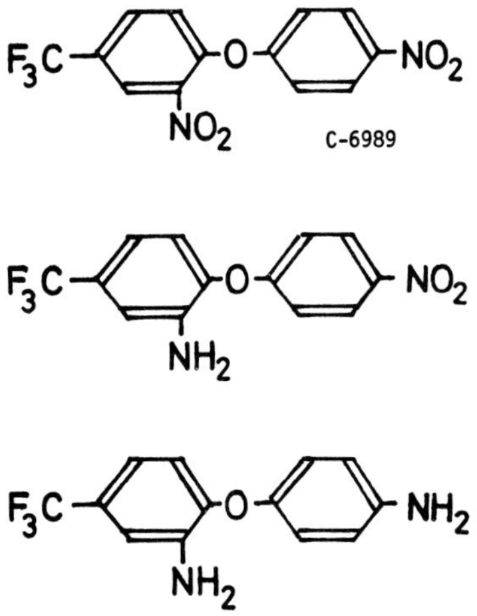

Fig. 11. Reduction of nitro radical on a diphenylether herbicide, C-6989.

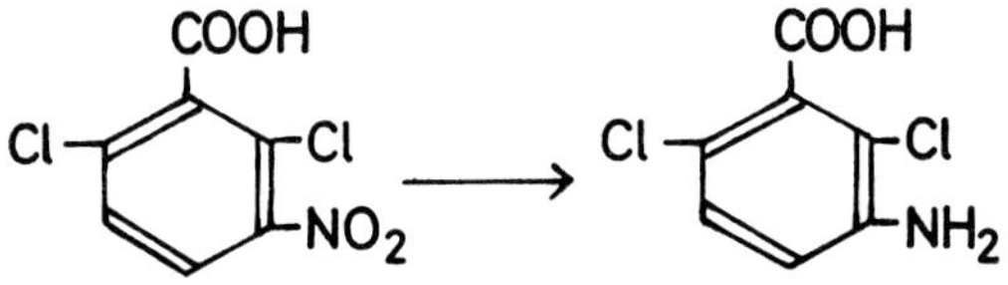

dinoben amiben

Fig. 12. Reduction of dinoben to amiben.

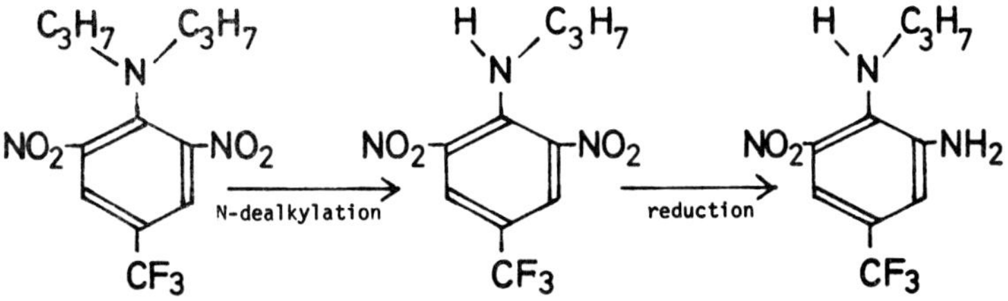

trifluralin

Fig. 13. N-dealkylation and reduction of trifluralin.

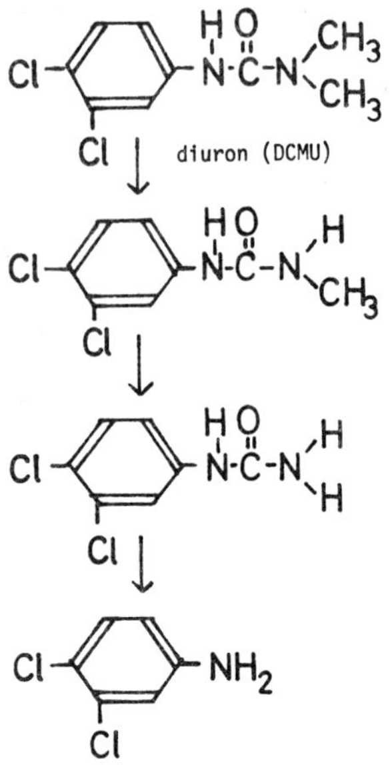

Fig. 14. N-demethylation of diuron by higher plants.

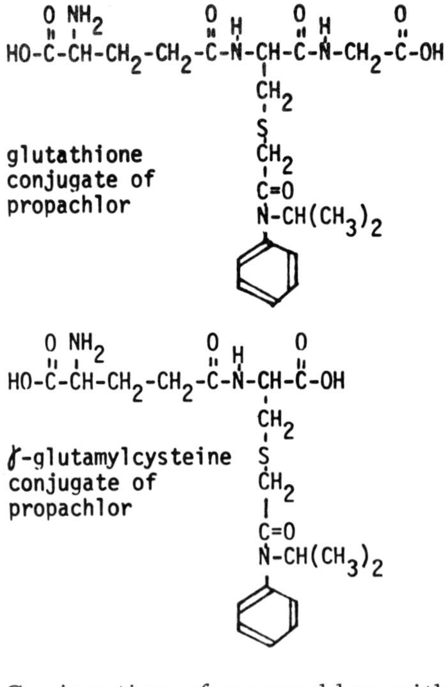

Fig. 15. Conjugation of propachlor with peptides.

PART V

MICROBIAL INFLUENCE ON PESTICIDE DEGRADATION

MICROBIAL DEGRADATION OF PESTICIDES

M. Alexander

Department of Agronomy, Cornell University
Ithaca, New York

Introduction

Microorganisms are important to any discussion of the behavior of pesticides in natural environments, both because of what they do and what they are unable to do. Microorganisms are frequently the major and sometimes the only means by which pesticides are eliminated from a variety of ecosystems, and they are therefore important in governing persistence. During the degradation, a number of products are formed, some of which may be persistent and several of which may be quite toxic. If an intermediate is both persistent and toxic, a major environmental hazard may appear. The mere fact that a pesticide is persistent suggests that it is intrinsically unsuitable or is unsuitable in that particular region as a substrate for microorganisms.

Environments supporting large and heterogeneous communities of microorganisms have always been deemed to be ideal biological incinerators, provided oxygen is available. The residents of these habitats have been viewed as agents for the destruction of any type of organic compound which might be introduced. And, indeed, because of the catabolic versatility of the indigenous species, biologically synthesized organic molecules do not long persist in aerated localities. The indigenous populations have thus acquired the reputation of being infallible in regard to their degradative abilities. Owing to the remarkable range of carbonaceous compounds that are decomposed, the species present in soil, water, and sewage are unquestionably major agents of environmental decontamination inasmuch as they prevent the accumulation of countless compounds which otherwise might appear at toxic levels.

The public concern with the persistence of chlorinated hydrocarbon insecticides and certain herbicides arises because of the invalidity of the early view that soil and water are perfect waste disposal media. It has now become evident that man cannot continue to pour a diversity of organic compounds into these microbial habitats and assume that the substances will be degraded. Furthermore, evidence of the biochemical fallibility of microorganisms is found not only in the area of pesticide behavior in nature. Plastics, packaging materials, and industrial wastes also frequently endure for surprisingly long periods in areas where one might have assumed that the bacteria or fungi would cause destruction at reasonably rapid rates. Synthetic materials are also not the only ones which are refractory since many biological products are not destroyed at reasonable rates, and some may persist for years, decades, and sometimes even centuries or millenia. Thus, the organic fraction of soil or humus, the portion of pollen known as exine, peat deposits, materials residing within organic soils, petroleum hydrocarbons, bacterial endospores, and fungal chlamydospores and sclerotia last for periods of from several to many thousands of years. Their longevity in environments where the temperature, moisture, aeration, and pH are conducive to microbial development demonstrates that natural products are likewise sometimes refractory (Alexander,

1965).

The doctrine of microbial infallibility probably arose because of the frequent successes in obtaining isolates capable of destroying innumerable natural and several synthetic organic products. To find such an isolate, an enrichment culture is established usually with the compound of interest as the sole source of carbon and energy or, on occasion, as the source of another essential nutrient. After periods ranging from a few days to several weeks, growth frequently appears in these enrichments and an organism can be obtained with little difficulty. Sometimes, growth and degradation require one or more growth factors, but these needs can readily be met. It is not clear whether a genetic modification or a physiological adaptation takes place so that a species present in the soil or water sample gains the enzymatic capacity to carry out the decomposition of synthetic materials. These many observations that biochemicals as well as a number of herbicides sustained the replication of suitable bacteria and fungi supported the view of microbial infallibility. Therefore, when microbiologists failed in their attempts to obtain isolates which would grow at the expense of a given pesticide, their failures were usually considered to result from the use of inappropriate techniques or making enrichments from samples of the wrong environment. Rarely has credence been given to the view that microorganisms may be unable to degrade the pesticide in question.

An extensive literature exists on the persistence of a wide range of insecticides and herbicides. Unfortunately, less is known about the longevity in soil of fungicides. These studies demonstrate that DDT, BHC, aldrin and its epoxide dieldrin, chlordane, and heptachlor and its epoxide are extremely durable. Occasionally the insecticide is still detectable in soils more than a decade after the time of application. As a rule, herbicides do not show the longevity of the chlorinated hydrocarbon insecticides, but compounds like 2, 4, 5-T, triazine herbicides, picloram, and monuron and related substituted ureas often remain in sufficiently high concentrations in soil to be detectable and occasionally to cause crop damage the year after application. In each instance, the very fact of persistence is a priori evidence that the compounds are either intrinsically poor substrates or they are not acted upon by indigenous microorganisms of that particular area. Indeed, with most such pesticides, isolates able to use the compounds as sources of carbon, nitrogen, or energy have yet to be obtained. Thus, field evidence suggests that there are refractory molecules which fail to be metabolized under natural conditions at rates sufficient to preclude concern about environmental pollution.

Two general causes of such microbial failings can be postulated. First, the environment, or possibly sometimes the microenvironment, may not be conducive to microbial life or to the carrying out of a given degradative reaction; that is, an organism able to carry out a particular transformation is excluded or a reaction which potentially can be catalyzed by an indigenous species does not take place. Second, certain chemicals may be resistant to biodegradation under all conditions. For these molecules, which have been termed recalcitrant (Alexander, 1965), microorganisms do not serve as incinerating agents, and inherent features of the molecule prevent rapid decomposition.

Environmental factors undoubtedly contribute to or govern molecular recalcitrance. In almost all instances, the evidence that these factors may be important is derived from investigations of compounds other than pesticides. However, it is likely that such environmental variables do frequently regulate the rate of destruction of many of the pest-control agents also. Adsorption of many organic compounds to colloids increases their resistance to attack; for example, proteins bound to clays are frequently far more slowly depolymerized by proteolytic enzymes than are the same proteins free in solution (Pinck et al., 1954). Complexing reactions likewise may make a normally metabolizable substance into one which is refractory; thus, proteins or polysaccharides are less available for attack by hydrolytic enzymes excreted by microorganisms when the polymers are bound to lignin or melanin than when they are free of the polyaromatic complexing agents. This type of complexing, at least in part, seems to explain why the polysaccharide-containing walls of melanized fungi are so resistant and hence allow the fungus to retain its viability for prolonged periods (Bloomfield and Alexander, 1967; Kuo and Alexander, 1967). The deposition of a chemical in a microenvironment too small for even small bacteria to penetrate might lead to a protection of a usually available substrate; because it is located in an essentially sterile microenvironment, the compound is not destroyed. Indirect evidence suggests that carbonaceous materials in soil may be protected because they are deposited in such inaccessible microenvironments. Poor water solubility may contribute to the persistence of certain compounds since natural microbial communities probably utilize more readily soluble than many comparable insoluble substrates. In addition, for reasons largely unknown, organic accumulations tend to be associated with anaerobic sites, and persistence of certain substances in areas where there is little or no free oxygen, as below the oxidized layer of flooded soils, may be attributable to the unknown factors preventing the microflora from utilizing the otherwise suitable substrates.

Chemical factors also determine or contribute to molecular recalcitrance. Thus, with fatty acids and simple aromatics, the type, number, and position of substituents regulate whether members of a class of substances which contain biodegradable members will be susceptible to biological attack. Studies in our laboratory have been concerned with a number of substituted aliphatics, substituted benzenes, benzoic acid compounds, phenoxy herbicides, and phenolic fungicides (Alexander and Aleem, 1961; Alexander and Lustigman, 1966; Dias and Alexander, 1971; Hammond and Alexander, unpublished observations).

Investigators in industrial, governmental, and university laboratories have established many correlations between chemical structure and selective toxicity to insects, plants, or pathogens affecting crops. The justification for the latter studies is to permit a rational delineation of which compounds may be useful for the control of economically important pests. It is also essential to determine what structural characteristics are correlated with or account for resistance to microbial degradation. With this kind of information, it may be possible to design pesticides which are not only selectively toxic but also which would persist for periods most suited for individual agricultural or environmental purposes.

The phenoxy herbicides and related phenolic fungicides illustrate how modest changes in chemical structure modify the suitability of organic compounds for decomposition. It is well known that 2,4-D (2,4-dichlorophenoxyacetic acid) is far less persistent in soil than the structurally related 2,4,5-T (2,4,5-trichlorophenoxyacetic acid). Similarly, 4-(2,4-DB) (4-(2,4-dichlorophenoxy)butyric acid) is more quickly inactivated by natural communities of soil organisms than is the related 4-(2,4,5-trichlorophenoxy)butyric acid (Whiteside and Alexander, 1960). These observations in the field, in the greenhouse, or with soil suspensions in the laboratory suggested that either the number or the position of the chlorines on the benzene ring affected biodegradability. The compounds which are destroyed quickly are detoxified by microbial means, and hence the longevity of the structurally related pesticides is attributable to the microbiological failing.

To determine whether it was the position or the number of halogens, a survey was made of the persistence of several related chemicals. These studies revealed that omega-substituted phenoxyalkanoic acids which contained a ring chlorine meta to the ether-oxygen were long-lived whereas analogous omega-substituted compounds with no meta halogen were rapidly detoxified and their rings cleaved. The alpha-substituted phenoxyalkanoic acids, by contrast, were reasonably persistent, regardless of the position of the chlorines on the ring (TABLE 1). Because the omega-substituted 4-chloro- and 2,4-dichloroalkanoic

TABLE 1

Effect of Chemical Structure on Decomposition of Phenoxy Compounds in Soil Suspensions

Fatty acid portion	Ring portion of molecule			
	4-chloro-phenoxy	2,4-dichloro-phenoxy	3,4-dichloro-phenoxy	2,4,5-tri-chlorophenoxy
Acetic	11[a]	26	> 205	> 205
ω-Propionic	11	4	> 81	> 81
ω-Butyric	53	11	> 205	> 205
α-Propionic	> 205	> 205	> 205	> 205
α-Valeric	> 81	> 81	> 81	> 81

[a]Days for ring cleavage.

acids were degraded whereas the corresponding 3, 4-dichloro and 2, 4, 5-trichlorophenoxyalkanoic acids were long-lived, it was suggested that the meta-chlorine on the ring contributed to or determined resistance to attack. These findings were supported by tests with a variety of mono-, di-, tri-, and tetrachloro- or bromophenols. As shown in TABLE 2, phenols with no meta-halogen are destroyed at reasonably rapid rates, while those containing a halogen on the 3 or 5 position, that is meta to the phenolic hydroxyl, are cleaved slowly (Alexander and Aleem, 1961; Burger et al., 1962; Alexander and Lustigman, 1966).

TABLE 2

Effect of Chemical Structure on Decomposition of Halogenated Phenols by Soil Microorganisms

Time for ring cleavage	
2 weeks	10 weeks
Phenol	3-Chlorophenol
2-Chlorophenol	3-Bromophenol
4-Bromophenol	2, 3-Dichlorophenol
4-Chlorophenol	3, 4-Dichlorophenol
4-Bromophenol	2, 5-Dichlorophenol
2, 4-Dichlorophenol	3, 5-Dichlorophenol
2, 6-Dichlorophenol	2, 4, 5-Trichlorophenol
2, 4, 6-Trichlorophenol	2, 3, 4, 6-Tetrachlorophenol

Thus, one can demonstrate structural correlations with microbial failings. Such observations need to be extended as the need grows more pressing to replace long-lived compounds with biodegradable pesticides.

The resistance of the alpha- as compared with the omega-substituted phenoxy compounds was not surprising, considering the single widespread mechanism for the metabolism of fatty acids. Aliphatic acids are usually decomposed by beta-oxidation, a sequence which requires two protons on both the alpha- and the beta-carbons. If either one of these carbons has a substituent other than a hydrogen, either the offending substituent must be removed or the organism will find difficulty in carrying out beta-oxidation. A substituted acetic acid might be produced, but these are not known. The alpha-substituted phenoxy compounds bear a large substituent on the alpha-carbon of the fatty acid, namely the phenoxy moiety of the molecule. This probably is the reason why such compounds are not decomposed rapidly. Any chemical feature which prevents the functioning of a widespread, dominant, or sole mechanism of biodegradation would be a characteristic probably changing a susceptible molecule into a resistant one.

Recent studies of the decomposition of simple substituted aliphatic acids and alcohols by the microbial communities of soil and water support the hypothesis that substituents on the alpha-carbon of the acids or on a carbon near the hydroxyl group of alcohols make these chemicals more resistant to degradation than the corresponding unsubstituted compounds. However, if a disubstituted carbon is somewhat removed from the carboxyl end of the aliphatic acid, degradation does in fact proceed, but the reaction sequence yields an end-product which contains a disubstituted carbon adjacent to the carboxyl group, as would be expected for such a hindrance to beta oxidation (Dias and Alexander, 1971; Hammond and Alexander, unpublished observations).

Environmental monitoring has shown that various pesticides are modified in soil apparently as a result of biological action, but the compounds which are modified to a modest extent fail to support the growth of any organisms when they are provided in culture media as sources of carbon or energy. The inability to obtain an organism may be because the microbiologist is inept. On the other hand, it is possible that another one of our traditional concepts is at fault. Microbiologists have commonly assumed that a substrate which is metabolized by a bacterium, actinomycete or fungus must serve as a nutrient source, and it would be converted to simple products which allow the organism to multiply, products like organic acids, ammonia, phosphate, or alcohols. Yet, plant and animal biochemists do not believe that each substrate of plant or animal enzymes is used as a nutrient source by these species. Some of our observations on cometabolism seem to aid in resolving the dilemma. Cometabolism refers to the capacity of a microorganism to metabolize a compound which it cannot use for growth. In one instance that has been investigated, a bacterium able to grow on 2,4-D could metabolize other phenoxyacetates and convert them to the corresponding phenols, but the phenols formed would not support growth; by contrast, the dichlorophenol formed from 2,4-D was further degraded and apparently was used as a growth substrate (Loos et al., 1967a). A likely explanation of many instances of cometabolism is differing substrate specificities of enzymes involved in a multistep reaction sequence. If the catalyst for the first step in the sequence has a reasonably wide substrate specificity but a later enzyme has a narrower specificity, then although the growth substrate would be converted to simple products, a structurally related compound might be degraded only partially, yielding intermediates which would not be further metabolized by the latter enzymes having narrower specificities.

An attempt was made to isolate a DDT-degrading bacterium based upon the view that cometabolism might be operating. An enrichment culture was provided with diphenylmethane, a compound structurally related to DDT, as the carbon source, and from a sewage inoculum a diphenylmethane-utilizing bacterium was obtained. When the bacterium was provided with a number of substituted diphenylmethanes, a distinct pattern of growth and cometabolism became evident. Compounds which had no or one chlorine were used for growth, and increasing the number of chlorines or changing the substituent on the bridge carbon made the molecule, initially a growth substrate, first into one which would not support replication but which was cometabolized and later into one which was completely resistant to aerobic metabolism by the *Hydrogenomonas* strain (Focht and Alexander, 1970). These findings demonstrate that it is not difficult to determine the influence of chemical structure on growth and cometabolism with individual cultures, but the studies are limited to showing the idiosyncrasies of an individual isolate rather than the capacities or failings of natural communities.

The initial studies on DDT metabolism by *Hydrogenomonas* sp. were conducted using aerobic incubation conditions. However, if the organism or enzymes derived from the cells were incubated in the absence of oxygen, DDT was metabolized, and several products could be detected including *p,p'*-dichlorodiphenylmethane and *p,p'*-dichlorobenzophenone. These observations demonstrate that organisms which are unable to cometabolize substrates under aerobic conditions may be able to do so in the absence of oxygen. However, ring cleavage reactions typically require oxygen, but yet the reaction sequence catalyzed by *Hydrogenomonas* sp. yields chlorinated diphenylmethane and benzophenone under anaerobic circumstances. Inasmuch as the bacterium might be able to cleave one of the rings of DDT aerobically, provided it were first allowed to cometabolize the substrate anaerobically, the bacterial enzymes were incubated with DDT in the absence of air, then fresh cells were added and the mixture was exposed to oxygen to allow for the functioning of the ring-cleavage enzymes. In this way, it was found that a single organism is able to convert DDT to a product, *p*-chlorophenylacetic acid, which results from the cleavage of one of the rings of the insecticide. In turn, *p*-chlorophenylacetic acid is an excellent substrate for a variety of microorganisms, which cleave the ring of this intermediate produced from DDT. Thus, it has been demonstrated for the first time that DDT is essentially completely decomposed by microorganisms (Pfaender and Alexander, unpublished observations).

Since DDT is potentially completely biodegradable, why is it so persistent in nature? This probably reflects cometabolism. When a pesticide is used for growth, the cells multiply as they utilize the compound so that a large population appears from the initially low cell numbers, the reaction becoming increasingly more rapid as the cells become more numerous in the environment. On the other hand, a compound subject to cometabolism does not support the proliferation of the organisms and, should there initially be a small number of cells, the reaction would proceed at a very slow rate; moreover, the rate would not increase with time because the presence of the pesticide does not provide the metabolizing species with a selective advantage. They get no energy and derive no benefit from the conversions they bring about.

Although several pesticides are decomposed by microorganisms growing on those chemicals, in only a few instances are the pathways of degradation established. In these reaction sequences, a compound with low potency may be converted to a substance of greater toxicity. For example, 4-(2, 4-DB) is transformed microbiologically in soil to 2, 4-D, the former being nonherbicidal while the latter is the active toxicant (Gutenmann et al., 1964). Sometimes a toxic compound may be converted to an inactive product, and this type of detoxication has been demonstrated with 2, 4-D as well, for this compound is cleaved in culture (Loos et al., 1967b) and in soil (Sharpee and Alexander, unpublished observations) to yield 2, 4-dichlorophenol; however, the product is not itself without toxicity, for it has been reported to have an effect on microorganisms, so that an intermediate may be generated which has a different action spectrum on living organisms than the original substance. Finally, microorganisms may be able to defuse a potentially active compound. Defusing refers to a reaction destroying the potential toxicity of a compound which requires activation before it serves as a toxicant. Defusing has been demonstrated in bacterial cells incubated with 4-(2, 4-DB) (Macrae et al., 1963).

Studies of metabolic processes effected by microorganisms are inherently interesting, and demonstrations of pathways of degradation of exotic molecules provide for an understanding of physiological processes. However, these investigations have great practical importance as well. Thus, new toxins may be formed as a result of the decomposition in soil, fresh water, sediments, the open sea, food, or the rumen of the herbivore. The new toxins may inhibit different species than the parent pesticide. The intermediates formed may have a longer persistence than the initial compound and hence may constitute a more serious environmental hazard than the substance from which they were derived. In addition, the analytical techniques usually employed may not disclose the presence of the intermediates and hence analytical problems may be created because of these conversions.

Demonstration of a pathway for the microbial metabolism of pesticides has a number of unique problems, however: (a) the chemicals often have esoteric structures, (b) the intermediates are formed in extremely small amounts because the pesticides are applied at low concentrations, (c) biochemical investigations have yet to be conducted with such classes of chemicals, (d) the microbial community is frequently extremely heterogeneous, and (e) the environment commonly contains a variety of inanimate and colloidal materials capable of reacting with both the parent and intermediary compounds. For these reasons, pathways are frequently studied with the purpose of initially establishing appropriate *in vitro* precedents, biochemical models which may serve as a guide to what happens under natural conditions.

To illustrate the use of *in vitro* prototypes, the pathway of degradation of phenoxyalkanoic acids will be considered. A bacterium was isolated which was capable of growing on 4-(2,4-DB). This bacterium, a strain of *Flavobacterium*, oxidized the herbicide and released chloride in the process. The ring was cleaved as indicated by the disappearance of ultraviolet absorbancy. Sequential induction studies suggested that 2,4-D, the likely intermediate, was not formed during the decomposition, but rather 2,4-dichlorophenol and 4-chlorocatechol were probably generated. The oxidation was intriguing, moreover, because oxygen consumption by the bacterial cell suspensions was biphasic, the initial phase suggesting that the ring was first metabolized followed by oxidation of the fatty acid side-chain. On the basis of these preliminary manometric observations, it was proposed that the bacterium cleaved the ether linkage of the herbicide to yield 2, 4-dichlorophenol, 4-chlorocatechol, and the corresponding fatty acid. Chromatographic evidence supported this hypothesis (Macrae et al., 1963).

This bacterium oxidized phenoxyalkanoic herbicides with side-chains of varying length. In each instance, it appeared that the substrate was initially cleaved at the ether linkage. By means of gas chromatography, it was established that the corresponding fatty acids were in fact produced by the bacterium; hence, the organism catalyzed a previously unknown reaction involving these herbicides, one initiated by ether cleavage (Macrae and Alexander, 1964). Taylor and Wain (1962), working with other microorganisms, had already demonstrated that pesticides of this sort could be metabolized in vitro by an initial beta-oxidation of the fatty acid portion of the molecule, yielding a series of compounds each with side-chains having two less carbon atoms. Thus, two in vitro models were available prior to investigations of the metabolic sequence in nature, one showing that the class of pesticides was acted upon by cleavage of the ether linkage, the second showing that they could be transformed by beta-oxidation of the side chain before the fatty acid moiety was separated from the benzene ring (TABLE 3).

With this background information, it was simple to define the microbial processes in natural soil. Soil was treated with the herbicides, and extracts of the soil were made at regular intervals and examined by gas chromatography. As shown in TABLE 4, the data demonstrate that the beta-oxidation process was the dominant if not the sole means by which these herbicides were decomposed in the test soils (Gutenmann et al., 1964).

Nevertheless, no information was available on how 2, 4-D or the corresponding dichlorophenol was degraded. The total destruction of these compounds in soil appears

TABLE 3

Microbial Metabolism of Phenoxyalkanoic Acids[a]

Substrate	Products	
	Flavobacterium	Nocardia
PhO-propionate	Propionate	-
PhO-butyrate	Butyrate, PhOH	PhO-acetate
PhO-valerate	Valerate	PhO-propionate
PhO-hexanoate	Hexanoate	PhO-butyrate, PhO-acetate
PhO-heptanoate	Heptanoate, hexanoate	PhO-propionate
PhO-octanoate	Octanoate, hexanoate	-

[a]PhO: 2,4-dichlorophenoxy.

TABLE 4

Products of β-Oxidation in Soil[a]

Substrate	Products
PhO-butyrate	PhO-acetate
PhO-hexanoate, PhO-octanoate PhO-decanoate	PhO-butyrate, PhO-acetate
PhO-pentanoate	PhO-propionate, PhOH
PhO-heptanoate, PhO-nonanoate PhO-undecanoate	PhO-pentanoate, PhO-propionate, PhOH

[a]PhO: 2,4-dichlorophenoxy.

to be largely, if not entirely, the result of microbial actions. From soil, a 2,4-D-utilizing bacterium was obtained. Inducible enzymes of this organism, a strain of *Arthrobacter*, acted upon 4-chlorophenoxyacetic acid (4-CPA) and 2-methyl-4-chlorophenoxyacetic acid (MCPA). During growth of the organism on 2,4-D, a compound giving a reaction for phenol was found in the culture fluid. This compound was identified as 2,4-dichlorophenol by gas chromatography. Enzyme preparations derived from the *Arthrobacter* acted on 2,4-D, 4-CPA, and MCPA, and the products resulting from these conversions were identified as the corresponding phenols. Because there was no activity on the anisole by enzyme preparations converting the phenoxy compound to the phenol, anisoles were excluded as intermediates. The phenols accumulated because the enzyme preparations required a cofactor for the subsequent conversion of the phenol formed from the herbicide. The phenol oxidation required the presence of NADPH and oxygen, and it thus appears to be the result of a mixed function oxidase. The phenol-metabolizing enzyme was separated from the next enzyme in the sequence on Sephadex G-200, and by this means the products of phenol oxidation could be isolated. These products were found to be the corresponding catechols (Bollag et al., 1967; Loos et al., 1967b; Bollag et al., 1968).

Subsequent studies revealed how bacteria ultimately dispose of the ring of the phenoxy herbicides or phenolic fungicides. The ether bond is cleaved to yield what appears to be glyoxylic acid, which is then converted to α-alanine. The catechol, be it 3,5-dichloro- or 4-chlorocatechol, is acted upon by intracellular enzymes to yield the muconic acids. These are then converted, possibly partly by nonenzymatic means, to the corresponding lactones. When provided as substrates, the catechols, muconic acids, and lactones are further degraded to yield the appropriate maleylacetates; the latter intermediates are converted to succinic acid. These products were identified by gas chromatography, mass spectrometry, thin-layer chromatography, and infrared and ultraviolet spectrometry (Tiedje and Alexander, 1969; Tiedje et al., 1969; Duxbury et al., 1970). These investigations serve as another biochemical model for the process in nature. Samples from natural habitats have been taken, treated with the herbicide and then examined for the formation of intermediates. To date, clear evidence is at hand for the identification of only one product, 2,4-dichlorophenol, but other compounds are apparently formed too (Sharpee and Alexander, unpublished observations).

To illustrate how microorganisms may make hazardous compounds during their development in nature, the example of nitrosamines can be cited. If they are formed, nitrosamines may be important environmental pollutants because they are carcinogenic, mutagenic, and teratogenic. Evidence has been obtained in several laboratories that such compounds are formed in nitrite-treated fish and in the gastrointestinal tract of animals following ingestion of nitrate or nitrite together with certain amines. The nitrosamine may be generated from a tertiary amine and nitrate by a three-step sequence as shown in Fig. 1. The nitrosation may be enzymatic at neutral pH values, or it may be nonbiological in acid circumstances.

It has now been found that such reactions may occur in waters and in soil. For example, nitrosamines appear in samples of such microbial habitats when treated with nitrate or nitrite and either di- or trimethylamine. The reactions are largely or entirely the consequence of microbial metabolism. Furthermore, raw sewage kept at pH 4.0 and treated with 0.1% thiram (tetramethylthiuram disulfide) yielded, after incubation for 6 days, 13.3 μg of dimethylnitrosamine/ml. The product was identified by thin-layer chromatography, gas chromatography, and photochemical cleavage techniques. Although the last stage in nitrosamine formation from this fungicide may be nonenzymatic, the evidence suggests that part of the conversion results from microbial processes (Ayanaba, Verstraete, and Alexander, unpublished observations).

Microorganisms are therefore important because of what they can and cannot do to pesticides in nature. Their inactivity is reflected in the persistence of the pest-control agent. Microbial cometabolism may be a reason for the slow disappearance of some of these compounds, with the accumulation of closely related structures. Finally, intermediates may be produced as the microflora grows at the expense of such environmentally important substances. Unfortunately, however, our knowledge of the metabolic sequences involved and the intermediates formed in natural habitats still are largely unknown.

References

Alexander, M.: Adv. Appl. Microbiol. 7, 35 (1965).

Alexander, M., and M. I. H. Aleem: J. Agr. Food Chem. 9, 44 (1961).

Alexander, M., and B. K. Lustigman: J. Agr. Food Chem. 14, 410 (1966).

Bloomfield, B. J., and M. Alexander: J. Bacteriol. 93, 1276 (1967).

Bollag, J.-M., C. S. Helling, and M. Alexander: Appl. Microbiol. 15, 1393 (1967).

Bollag, J.-M., C. S. Helling, and M. Alexander: J. Agr. Food Chem. 16, 826 (1968).

Burger, K., I. C. Macrae, and M. Alexander: Soil Sci. Soc. Amer. Proc. 26, 243 (1962).

Dias, F. F., and M. Alexander: Appl. Microbiol. (in press, 1971).

Duxbury, J. M., J. M. Tiedje, M. Alexander, and J. E. Dawson: J. Agr. Food Chem. 18, 199 (1970).

Gutenmann, W. H., M. A. Loos, M. Alexander, and D. J. Lisk: Soil Sci. Soc. Amer. Proc. 28, 205 (1964).

Kuo, M.-J., and M. Alexander: J. Bacteriol. 94, 624 (1967).

Loos, M. A., R. N. Roberts, and M. Alexander: Can. J. Microbiol. 13, 679 (1967a).

Loos, M. A., R. N. Roberts, and M. Alexander: Can. J. Microbiol. 13, 691 (1967b).

Macrae, I. C., and M. Alexander: Agron. J. 56, 91 (1964).

Macrae, I. C., M. Alexander, and A. D. Rovira: J. Gen. Microbiol. 32, 69 (1963).
Pinck, L. A., R. S. Dyal, and F. E. Allison: Soil Sci. 78, 109 (1954).
Taylor, H. F., and R. L. Wain: Proc. Royal Soc. (London) B156, 172 (1962).
Tiedje, J. M., and M. Alexander: J. Agr. Food Chem. 17, 1080 (1969).
Tiedje, J. M., J. M. Duxbury, M. Alexander, and J. E. Dawson: J. Agr. Food Chem. 17, 1021 (1969).

(1) $(CH_3)_3N \longrightarrow (CH_3)_2NH$

(2) $NO_3^- \longrightarrow NO_2^-$

(3) $(CH_3)_2NH + NO_2^- \longrightarrow (CH_3)_2N\text{-}N{=}O$

Fig. 1. Pathway for microbial formation of dimethylnitrosamine.

DEGRADATION OF SEVERAL HERBICIDES IN SOILS UNDER DIFFERENT CONDITIONS

Shozo Kuwatsuka

Department of Agricultural Chemistry, Nagoya University
Nagoya, Aichi, Japan

Introduction

The behavior and fate of herbicides in soils have been fairly well studied during the last decade. The degradation of herbicides in soils under upland conditions have been studied especially in the United States of America (Kearney and Kaufman, 1969; Anonymous, 1970). In Japan, about 40% of all herbicides (about 60% of all agricultural use, excluding forests) are consumed in paddy fields, and most of these herbicides are different in chemical structure from those used for upland purposes. Remarkable differences also exist between flooded and upland soils in their physical and chemical properties and microflora. The latter, especially, may greatly affect the behavior and fate of herbicides. Therefore, it is not unusual that the degradation of herbicides should be different on uplands versus paddy fields. Studies on their behavior and fate in paddy fields is restricted to the last few years.

This paper deals with the fate of several important herbicides used in rice fields: PCP (pentachlorophenol), swep (MCC, SWEP®) of the carbamate type; propanil (DCPA) of the acid amide type; benthiocarb (Saturn®) of the thiolcarbamate type; and CNP (MO®) and nitrophen (NIP) of the diphenyl ether group. The chemical structures of these herbicides are shown in Fig. 1. All studies were performed in the laboratory under various conditions, and employed 11 soil types possessing different properties (TABLE 1). We especially directed our efforts in comparing the degradation patterns of these chemicals between flooded and upland (nonflooded) soil conditions. These studies are still in progress.

Wet soils, or air-dried soils, were placed in flasks, adjusted to flooded or upland water condition, and preincubated at 30°C for 3 weeks for wet soils and for 5 weeks for air-dried soils. Then, herbicide chemicals were introduced into the soils. They were then further incubated. At various intervals, the chemicals were extracted with organic solvents, and analyzed by gas chromatography or colorimetry.

Pentachlorophenol (PCP)

PCP made up a major part of the herbicides used in paddy fields until a few years ago, and still is widely used. Photodecomposition is thought to be the major degradation process in paddy fields. Munakata and Kuwahara (1969) studied the photochemical process in detail. Although a considerable portion of the PCP applied to the soil surface in flooded paddy fields infiltrates into the soil with percolating water, PCP degradation was not studied for a long period of time. PCP adsorption on soil particles was studied in detail by Nose et al. (1963, 1964). Hayashi and Watanabe (1970) studied PCP degradation in soils, and isolated seven species of bacteria which degraded PCP. Suzuki and Nose (1971) found PCP methyl ether and tetrachlorohydroquinone dimethyl ether as metabolic products in the culture media of these bacteria. Niki et al. (1971) detected three tetrachlorophenols as methyl ether on gas chromatograms in extracts of PCP-treated soil.

Kuwatsuka and Igarashi (1971a and 1971b) studied the degradation of PCP in several soils under various conditions, and detected three tetrachlorophenols and four trichlorophenols from PCP-treated soils. These products were measured during the PCP-incubation period. The degradation of PCP in various soils is shown in Fig. 2.

TABLE 1

Properties of Soils Used

Field obtained (abbreviation)			Clay Mineral	Texture	Clay Content (%)	pH (H_2O)	Total Carbon (%)	C.E.C. (me/100 g)
Tochigi	paddy	(T)	Allo.	SiC	5.4	5.8	7.87	40.5
	upland	(T')	Allo.	SiC	4.5	6.0	7.52	46.4
Nagano	paddy	(N)	Mont.	CL	20.5	4.6	1.81	21.3
	upland	(N')	Mont.	CL	21.0	5.2	1.45	24.3
Shizuoka	paddy	(S)	Kaol.	CL	22.7	5.9	3.64	19.9
	upland	(S')	Kaol.	CL	22.8	4.8	1.91	13.2
Anjo	paddy	(A)	Kaol.	SCL	23.1	5.8	1.93	13.6
	upland	(A')	Kaol.	CL	27.5	6.4	1.23	15.2
Kikugawa	paddy	(K)	Kaol.	CL	22.4	6.1	1.21	14.7
	upland	(K')	Kaol.	CL	15.4	4.7	1.04	13.0
Higashiyama	(B-C)	(H)	Kaol.	CL	19.1	4.6	0.04	5.0

PCP disappeared more rapidly from soils at higher temperatures under flooded conditions than under nonflooded upland conditions. Under flooded conditions degradation was faster in soils obtained from paddy fields than soils from upland fields of the same area. The reverse was true for upland conditions, although differences in the rates of degradation between soils from the two types were not very significant. The half-life of PCP at 100 ppm (initial concentration in soils) was 10 to 40 days at 30°C under flooded conditions of arable soils. In the case of Higashiyama soil, which has a B-C horizon of red-yellow soil, and is used as a soil sample representing an extremely low organic matter content, PCP remained almost 100% even after a 2-month incubation under similar conditions. Figure 3 shows the relation between the degradation rate and the organic matter content of various soils under different incubation periods. The degradation rate was largely related to the organic content, except for two montmorillonitic Nagano soils in which PCP degraded slowly at first, then rapidly during the latter stages of incubation. Cation exchange capacity and soil pH were less related to chemical degradation. Soil texture, clay content, degree of base saturation and free iron oxides were not closely related to the rate of PCP degradation.

PCP degradation was assumed to proceed by both chemical and microbial means based on the results of experiments on soil sterilization, soil temperature, and the degradation products of PCP. Chemical degradation is, however, presumed to be caused and promoted by microbial action as no degradation occurred in Higashiyama soil containing almost no organic matter.

The products of PCP degradation detected by gas chromatography were as follows: 2, 3, 4, 5-, 2, 3, 4, 6-, and 2, 3, 5, 6-tetrachlorophenols; and 2, 3, 6-, 2, 4, 6-, 2, 3, 5-, and 2, 3, 4-; and/or 2, 4, 5-trichlorophenols. Figure 4 shows the changes in amounts of remaining PCP and the compounds produced in Anjo soil under flooded and upland conditions. The major products were 2, 3, 4, 5-tetrachlorophenol and 2, 3, 6- and 2, 4, 6-trichlorophenols. Other compounds were present only in small or trace amounts. The production curve of 2, 3, 4, 5-tetrachlorophenol showed two or three peaks during the course of the incubation period. Further studies concerned with degradation are in progress.

Swep and Propanil

Swep (MCC, methyl N-3, 4-dichlorophenylcarbamate) and propanil (DCPA, 3, 4-dichlorophenylpropionanilide) are a carbamate and an acid amide, respectively, of 3, 4-dichloroaniline. Swep is used mainly for soil surface treatments; propanil is used mainly as a foliar treatment because of rapid degradation in soil. Propanil degradation in upland soils has been largely studied by Bartha and Pramer (1967), Bartha et al. (1968), Chisaka and Kearney (1970), and Plimmer et al. (1970). They showed that propanil was converted into 3, 4-dichloroaniline and then coupled to 3, 4, 3', 4'-tetrachloroazobenzene and 1, 3-bis (3, 4-dichlorophenyl) triazene in soils.

The degradation of carbamate herbicides such as CIPC in upland soils has been studied by Kaufman and Kearney (1965), and Kaufman (1967), but few reports have been published on swep degradation, especially in flooded soils.

In the present investigation (Kuwatsuka and Hayakawa, 1971), 3, 4-dichloroaniline and coupled azobenzene were detected as degradation products of swep in flooded and nonflooded soils, as well as by propanil degradation. As shown in Fig. 5, swep was degraded more slowly than propanil, and the amount of 3, 4-dichloroaniline produced from swep was much less than the latter. At 30°C, swep degradation under flooded soil conditions was slower than under upland conditions. However, the moisture level in nonflooded soils varied from 40 to 80% of maximum water holding capacity and had little effect on degradation. Propanil degradation was more rapid under flooded conditions than under upland circumstances at 30°C. On the contrary, at lower temperatures (12°C and 4°C) swep degradation in nonflooded soils was slower than under flooded soil conditions, although degradation at lower temperatures was much slower under both conditions. Changes in pH between pH 5 and 8 did not affect the degradation rate.

Benthiocarb

Benthiocarb (Saturn®, p-chlorobenzyl N,N-diethylthiolcarbamate) has been commercially used this past year, and is now one of the herbicides in greatest use in paddy fields. The metabolism of thiolcarbamates such as EPTC in organisms, and their degradation in upland soils have been studied, but their degradation in flooded soils has not been investigated. Metabolism and degradation of benthiocarb have been studied very little. Yamada and Tamura (1971) determined the residual amount of this compound in rice grain and in Tochigi soil. Ishikawa and Kuwatsuka (unpublished results) studied the degradation of benthiocarb in soils as well as *in vivo* and *in vitro* metabolism studies in plants and animals. Figure 6 shows the degradation of benthiocarb in various soils under both flooded and upland conditions. The incubation periods (days) are shown on log scale. In general, this chemical is fairly stable in soils. The half-life, from an initial concentration of 20 ppm, was more than 6 days at 30°C in flooded soils. However, the period varied widely with the kind of soil. With certain soils, it took more than 40 days for 50% of this chemical to disappear. Also, generally degradation was faster under flooded soil conditions than under upland conditions. For the study on the degradation process, ^{14}C-benthiocarb, labeled as benzylmethylene, was used. Even after long periods of incubation, benzene or acetone extracts from the acidified soils contained little other than benthiocarb as the radioactive material. On the other hand, at least nine metabolites were found in mouse liver homogenate, four in mouse urine, and three in plants. Some of them were N-monoethyl derivatives; bis(p-chlorobenzyl)thio- or dithioether, p-chlorobenzoic acid and p-chlorohippuric acid. These facts suggest that the degradation products of benthiocarb are readily degraded to smaller or very polar materials unextractable with organic solvents, although benthiocarb itself is stable in the soil. In addition, benthiocarb is not absorbed to any extent by rice plants through the roots, from soil or aqueous solutions. Also, it hardly translocates within rice plants, although it is readily absorbed and translocated by barnyard grass or dicotylendon plants (Ishikawa, Nakamura, and Kuwatsuka, unpublished results). These facts suggest that the possibility of translocation and residual pollution of benthiocarb in rice grains may be low, in spite of its comparatively long persistence in soils.

Diphenyl Ethers

CNP (MO®, 2,4,6-trichlorophenyl 4'-nitrophenyl ether) and nitrophen (NIP, 2,4-dichlorophenyl 4'-nitrophenyl ether), especially CNP, are largely used in paddy fields. Ichihashi et al. (1971) compared the residual effects in soil of CNP, nitrophen, and X-52 (2,4-dichloro 3'-methoxy-4'-nitrophenyl ether) by bioassay with barnyard grass and other weeds. They found that X-52 had a long residual effect and that in enriched soil solution, the reduced amino derivatives formed from all these diphenyl ethers.

In the author's laboratory, CNP, nitrophen, X-52, and MO-500 (2,4-dichloro-6-fluorophenyl 4'-nitrophenyl ether) were compared as to their degradation in soils under flooded and upland conditions. Amino compounds of the diphenyl ethers were detected only in flooded soils treated with diphenyl ethers. Figure 7 shows the results of the degradation of diphenyl ethers, and the production of amino compounds under flooded and upland conditions. Degradation was much slower under upland conditions than under flooded ones, and no amino derivatives were detected in soils under upland conditions. Under flooded conditions, nitrophen and MO-500 were degraded much faster than CNP and X-52. Gas chromatograms and other analytical data showed that amino derivatives were the major metabolites in the earlier stages of incubation, whereas later, amino compounds were also metabolized to other compounds.

General Discussion

The comparison of degradation rates in soils between upland and flooded conditions and the major products detected in flooded soils are summarized in TABLE 2. Two ester compounds, among the investigated chemicals--swep of the carbamate type and benthiocarb of the thiolcarbamate type--showed different degradation patterns than did the other chemicals. Swep was degraded faster under upland conditions than under flooded ones, and no definite comparison was possible on the degradation of benthiocarb in soils under the two conditions. The other chemicals were degraded much faster under flooded conditions. DDT and BHC had been found to be degraded much faster under flooded conditions (Guenzi and Beard, 1968; Tatsukawa et al., 1969, 1970; Sethunathan et al., 1969; Yoshida and Castro, 1970), probably reflecting the ability of microbes to degrade them under such conditions. Also, it is known that more bacteria and fewer fungi and actinomycetes live in flooded soils than in upland soils (Ishizawa and Toyoda, 1964). On the other hand, hydroxylation, fission of the benzene ring, dechlorination, and dehydrochlorination are caused chiefly by bacteria, and esterase activity is shown by both fungi and bacteria. These facts might explain the reason why pesticides with ester linkages are degraded faster in nonflooded soils than in flooded soils, and chemicals with no ester linkages are degraded much faster under the reverse situation. The reduction of the nitro groupings of diphenyl ethers may be a reflection of the redox state of soils as flooded soils are in a strongly reduced state. Examples of studies are few, and investigations are still incomplete. Therefore, this suggestion should be confirmed by further study.

TABLE 2

Comparison of Degradation Rate of Pesticides in Soils under Flooded and Upland Conditions and the Main Products

Pesticides	Degradation rate			Main products
PCP	flooded	$>$	upland	dechlorinated comps.
DCPA	flooded	$>$	upland	DCA and TCAB
Swep	upland	$>$	flooded	DCA and TCAB
Benthiocarb	upland	$\geq$ or $\leq$	flooded	(p-chlorobenzoic acid)
Diphenyl ethers	flooded	$\gg$	upland	amino derivatives
BHC[a]	flooded	$\gg$	upland	PCCH
DDT[a]	flooded	$\gg$	upland	DDD

[a]Guenzi and Beard, 1968; Sethunathan et al., 1969; Tatsukawa et al., 1969, 1970; Yoshida and Castro, 1970.

The degradation of the herbicides investigated was more rapid in those soils containing high amounts of organic matter, and was more related to this aspect than to their other soil properties. Our results suggest that degradation in soils is mainly a function of soil microbial activity. However, we cannot say whether degradation in the soil is caused by microorganisms directly or if the chemicals are degraded as a result of favorable conditions indirectly created by the microbes. Kaufman et al. (1968) has shown, for example, that amitrole degradation is mainly through chemical means, although degradation stops by soil sterilization. We do not yet know which of the chemical and biological reactions contributes more to the degradation of the herbicides investigated.

Studies of herbicide degradation in flooded soils have been recently started.

References

Anonymous: Pesticides in the soil, ecology, degradation and movement, International Symposium of Pesticides in the Soil, Michigan State University (1970).

Bartha, R., and D. Pramer: Science 156, 1671 (1967).

Bartha, R., H. A. B. Linke, and D. Pramer: Science 161, 582 (1968).

Chisaka, H., and P. C. Kearney: J. Agr. Food Chem. 18, 854 (1970).

Guenzi, W. D., and W. E. Beard: Soil Sci. Soc. Amer. Proc. 32, 522 (1968).

Hayashi, H., and I. Watanabe: Abstr. Ann. Meet. Soc. Soil Manure, Japan, 16, 44 (1970).

Ichihashi, M., R. Takahashi, F. Kimura, and Y. Yokomichi: Abstr. 10th Meet. Weed Soc., Japan, 35 (1971).

Ishizawa, S., and K. Toyoda: Bull. Nat. Inst. Agr. Sci. (Japan), N14, 203 (1964).

Kaufman, D. D.: J. Agr. Food Chem. 15, 582 (1967).

Kaufman, D. D., and P. C. Kearney: Appl. Microbiol. 13, 443 (1965).

Kaufman, D. D., J. R. Plimmer, P. C. Kearney, J. Blake, and F. S. Guardia: Weed Sci. 16(2), 226 (1968).

Kearney, P. C., and D. D. Kaufman: Degradation of herbicides. Mercel Dekker, Inc., New York (1969).

Kuwatsuka, S., and M. Hayakawa: Abstr. Ann. Meet. Soc. Soil Manure, Japan, 17, 23 (1971).

Kuwatsuka, S., and M. Igarashi: Meet. Soc. Soil Manure, Kanazawa, Japan, Chubu Branch (1971a).

Kuwatsuka, S., and M. Igarashi: Abstr. Ann. Meet. Soc. Soil Manure, Japan, 17, 24 (1971b).

Munakata, K., and M. Kuwahara: Residue Reviews, 25, 13 (1969).

Niki, Y., I. Watanabe, A. Ide, and H. Watanabe: Abstr. Ann. Meet. Agr. Chem. Soc. Japan, 387 (1971).

Nose, K., T. Suzuki, and K. Fukunaga: J. Soc. Soil Manure, Japan, 34, 234, 291, 367 (1963).

Nose, K., T. Suzuki, and K. Fukunaga: J. Soc. Soil Manure, Japan, 35, 306 (1964).

Plimmer, J. R., P. C. Kearney, H. Chisaka, J. B. Yount, and U. I. Klingebiel: J. Agr. Food Chem. 18, 859 (1970).

Sethunathan, N., E. M. Bautista, and T. Yoshida: Can. J. Microbiol. 15, 1349 (1969).

Suzuki, T., and K. Nose: Abstr. 10th Meet. Weed Soc., Japan, 114 (1971).

Tatsukawa, R., M. Kubo, and T. Ogawa: Abstr. Ann. Meet. Soc. Soil Manure, Japan, 16 (1969).

Tatsukawa, R., T. Wakimoto, and T. Ogawa: J. Food Hyg. Soc., Japan, 11, 1 (1970).

Yamada, T., and H. Tamura: Abstr. 10th Meet. Weed Soc., Japan, 108 (1971).

Yoshida, T., and T. F. Castro: Soil Sci. Soc. Amer. Proc. 34, 440 (1970).

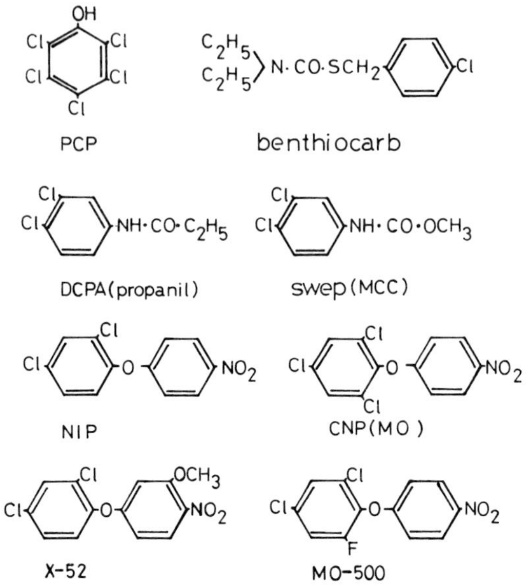

Fig. 1. Chemical structures of herbicides investigated.

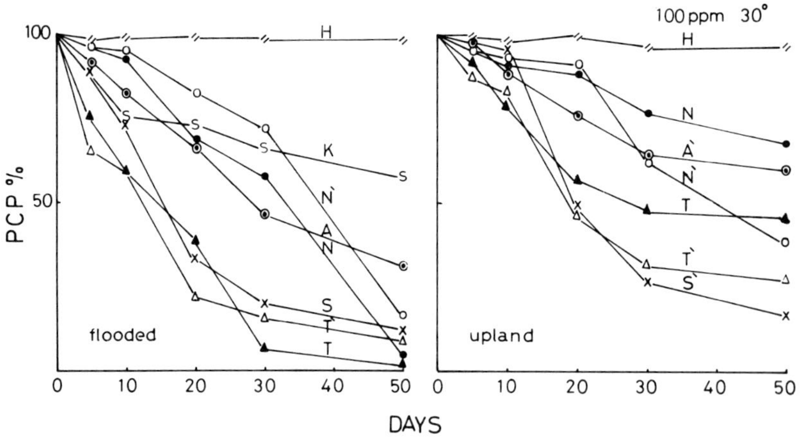

Fig. 2. PCP degradation in various soils under flooded and upland conditions.

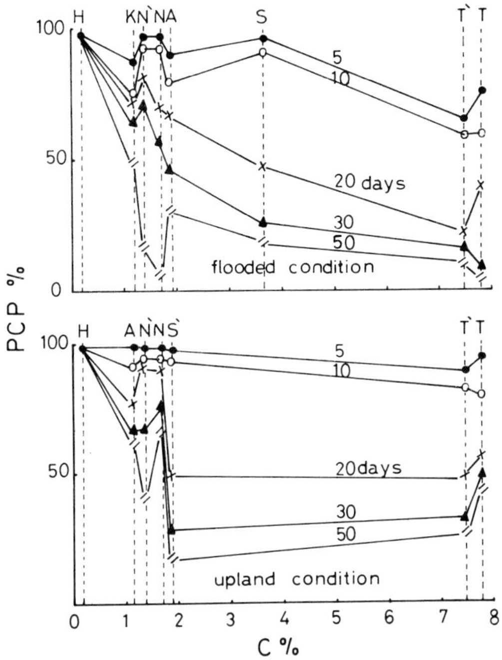

Fig. 3. Relation between degradation rate of PCP and organic matter content in various soils for different incubation periods (days).

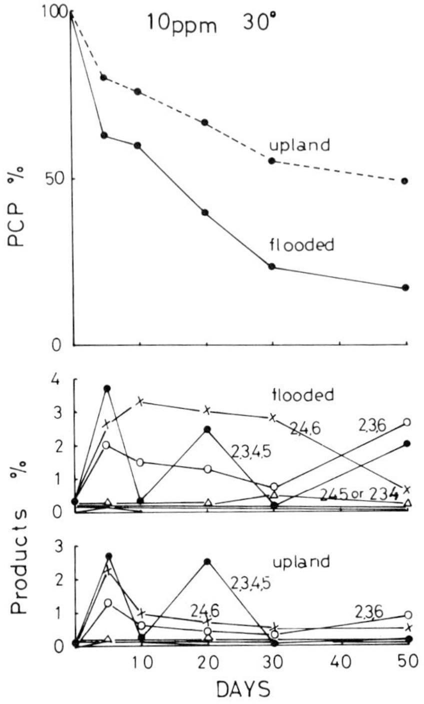

Fig. 4. Changes of degraded PCP and produced polychlorophenols during the incubation under flooded and upland conditions.

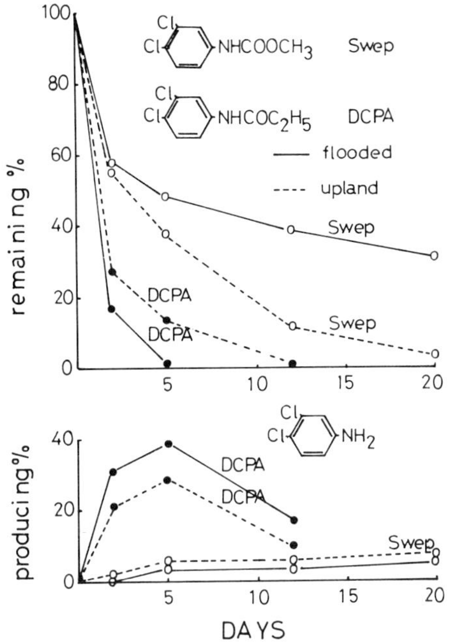

Fig. 5. Degradation of swep and DCPA and production of 3, 4-dichloroaniline in Anjo soil under flooded and upland conditions.
50 ppm at initial concentration, 30°C.

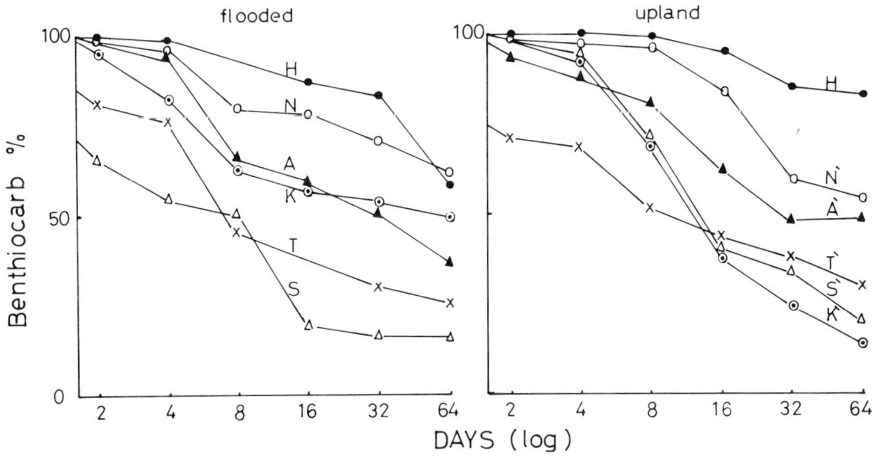

Fig. 6. Benthiocarb degradation in various soils under flooded and upland conditions.
20 ppm at initial concentration, 30°C.

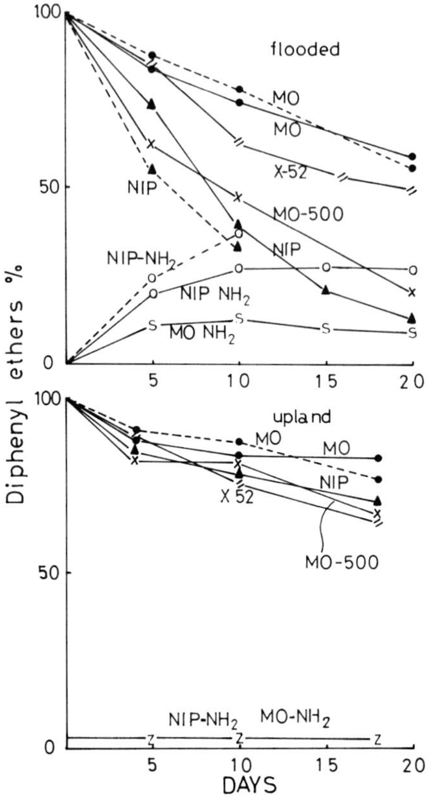

Fig. 7. Degradation of diphenyl ethers in Anjo soil (solid line) and Tochigi soil (dashed line) under flooded and upland conditions.

10 ppm at initial concentration, 30°C.

ECOLOGICAL ASPECTS OF PESTICIDE-MICROBIAL RELATIONSHIPS

G. Mallory Boush and John C. Batterton

Department of Entomology, University of Wisconsin
Madison, Wisconsin 53706

Introduction

Probably the first concerted interests and efforts in determining the interactions between synthetic organic pesticides and soil microorganisms in the United States occurred in the late 1940's and early 1950's. This interest was not principally because of altruistic concern over possible harmful effects of the toxicants on the microflora, but because of erratic herbicidal results in the field, and the supposition that microbial degradation or tie-up might be involved. Indeed, microorganisms were found generally to be quite important--probably the major factor in the breakdown of many herbicides in nature (Audus, 1949, 1951; Rogoff and Reid, 1955; Kearney and Kaufman, 1965). Consequently, considerable effort over the past 25 years has been directed toward better understanding of herbicidal-microbial ecology, especially in soil, with the result that this area is somewhat advanced over similar insecticidal studies. However, during this period investigations on the possible effects of chlorinated hydrocarbons (especially BHC and DDT) on soil bacteria and fungi were conducted. Because of their importance or relative occurrence, emphasis was placed on the nitrifiers and on commonly isolated forms. Although disturbances were noted, especially in laboratory studies with high dosages, most workers felt that toxicant levels normally used in field insect control did not appreciably alter the microflora--at least to the extent that there was a significant reduction in soil fertility (Fletcher and Bollen, 1953; Martin et al., 1959). A high degree of broad-spectrum insecticidal effectiveness, persistence, and thus good long-term insect control, together with little or no information at the time concerning undesirable environmental effects of these chlorinated hydrocarbons, created a climate conducive to experimental stagnation. However, the situation has changed markedly within the past decade. We are increasingly conscious of the indispensable contributions of the microflora.

Indispensable Role of Microorganisms

Microorganisms, especially bacteria and fungi, are the principal agents in nature involved in the continual cycling of organic matter. This essential turnover or recycling of all the principal biologically important nutritive elements through plant and animal growth, death, decay--and regrowth, is of obvious and incalculable value. Without them, it would just be a matter of time before we were buried in our own refuse, to say nothing of exhausting our nutritive base. Also, certain bacteria and algae, singly and with symbiotic associations with higher plants, are crucial in maintaining soil and water fertility through conversion of atmospheric nitrogen, and nitrogen from a vast array of organic compounds, to nitrate or ammonia. Even the continued regeneration of carbon dioxide in our biosphere, necessary for photosynthesis, and oxygen, necessary for aerobic life, is a result of the familiar microbially-associated carbon and oxygen cycles. Finally, microbes, especially aquatic algae and bacteria, form the base of most food chains.

It is no wonder then that we are concerned with not only what effects microorganisms might have on pesticide degradation, but what changes might occur in microbial make-up or physiology by the action of toxicants. Indeed, probably the major findings over the past two decades on the effects of pesticides on the environment have been that (1) many microorganisms and lower animals are capable of rapid pick-up as well as concentrating a toxicant from low levels to levels many-fold higher than ambient; (2) that these levels can be further concentrated through food chain members--eventually to the point of becoming lethal; (3) that degradation of a toxicant, either through environmental or microbial means, is not necessarily to a less toxic or less stable form; and finally, (4) the phenomenon of co-distillation as a factor in movement of certain toxicants throughout our ecosphere. All of these factors are obviously interrelated, and if our recent experiences with PCB's in the environment are pertinent, more widespread than anticipated.

There is little question but that persistent pesticides, at least as recently employed, are contributors to environmental pollution (Pimentel, 1971). This is especially unfortunate as certain of these chemicals, correctly used, are important weapons in man's fight against harmful and destructive insect pests.

We have not attempted to review the literature in this broad, multidisciplinary field, but have attempted rather to cite some of the major contributions and representative studies which relate to the over-all problem and to specific studies in our laboratory.

Direct Effect of Pesticides on Microbes

Inhibitions - Studies of direct effects of pesticides on microorganisms are often done in terms of percent growth inhibition of one or more test organisms. Many have demonstrated that various pesticides affect the growth of bacteria (Worth and McCabe, 1948; Jones, 1956; Fahralus and Johnson, 1960; Collins and Langlois, 1968; Balicka, 1969; Balicka and Krezel, 1969) and fungi (Brian, 1949; Warren et al., 1951; Richardson, 1954; Petruk, 1964; Arnold et al., 1966; Rodriguez-Kabana et al., 1967; Wilkinson and Lucas, 1969; Cowley and Lichtenstein, 1970).

Several papers also note effects of pesticides upon algal growth (Eipper, 1959; Martin and Wiggans, 1959; Tatum and Blackburn, 1962; Ukeles, 1962; Cowell, 1965; Cairns, 1968; Wurster, 1968; Christie, 1969; Vance and Drummond, 1969; Vance and Smith, 1969; Menzel et al., 1970; Moore, 1970).

From such studies it is apparent that different species vary in their sensitivity to individual pesticides (Ukeles, 1962; Wilkinson and Lucas, 1969; Cowley and Lichtenstein, 1970; Menzel et al., 1970). Thus an interesting aspect of pesticidal-microbial interactions is the subtle influence of toxicants on microbial population structure. On the one hand, certain pesticides cause some microorganisms to decrease in number, whereas other studies show that the growth of some microbes appears to be unaffected or even stimulated in the presence of pesticides (Worth and McCabe, 1948; Il'in, 1962; Petruk, 1964; Balicka, 1969; Dalton et al., 1970).

Menzel et al. (1970) has suggested that the varying resistance of marine phytoplankton to chlorinated hydrocarbons could have far-reaching effects in terms of phytoplankton population balance. Studies in this laboratory (Batterton et al., 1971) have also shown varied growth inhibition of planktonic blue-green algae by chlorinated hydrocarbons. TABLE 1 shows the effects of aldrin, dieldrin, and endrin on growth rates (generations per 24 hr) of *Anacystis nidulans* (freshwater species) and *Agmenellum quadruplicatum* (marine species). It is noticeable that generally the marine isolate is more tolerant than the freshwater isolate. This may be due to the influence of the growth medium on the insecticide. The toxicity of a pesticide in aquatic environments may vary according to the physical characteristics of water (Lichtenstein et al., 1966).

Although much variation is noted in the data, the general trend indicates both algae are tolerant to these insecticides except at concentrations higher than reported in natural waters (Weaver et al., 1965). Also notable is the sensitivity of *A. nidulans* to dieldrin, an isomer of aldrin.

TABLE 1

Growth Response of *Agmenellum quadruplicatum* and *Anacystis nidulans* to Aldrin[a], Dieldrin, and Endrin[b]

ppb	Aldrin		Dieldrin		Endrin	
	A. quadruplicatum[c]	*A. nidulans*[c]	*A. quadruplicatum*	*A. nidulans*	*A. quadruplicatum*	*A. nidulans*
950	6.2±0.7	6.4±0.4	5.8±0.9	3.2±0.8	3.5±0.9	2.2±0.7
475	7.1±0.5	6.7±0.2	6.0±0.8	3.9±1.5	4.8±1.5	3.2±1.0
95	6.6±1.2	6.8±0.5	6.0±1.0	6.9±0.6	4.9±2.2	6.3±0.3
19	6.8±0.3	7.1±0.4	6.5±0.7	7.2±0.9	5.6±1.3	6.6±0.4
0.2	6.4±1.0	7.2±0.3	5.3±1.2	6.7±1.3	4.8±0.3	7.0±0.5
Control	6.6±0.5	6.6±0.4	6.2±0.9	6.9±0.6	6.6±0.5	6.6±0.4

[a] Concentrations for aldrin 910, 455, 91, 18, and 0.2 ppb.

[b] Values reported as number of generations per 24 hours, represent mean of 3 to 5 replicate cultures.

[c] *Agmenellum quadruplicatum* (strain PR-6), *Anacystis nidulans* (strain TX20).

In preliminary experiments the growth response of these two algae was also tested against phenylmercuric acetate (PMA), an algicide and fungicide once used extensively in the wood-pulp industry. The results are summarized in TABLE 2.

TABLE 2

Susceptibility of Two Species of Blue-Green Algae against Phenylmercuric Acetate

	Phenylmercuric acetate (ppb)					
	0.10	0.25	0.50	0.75	1.00	10.00
A. nidulans	109	100	112[b]	87[c]	68[d]	0
A. quadruplicatum	109	-	100	-	112	0

[a]Expressed in % relative growth against controls as 100.

[b]Only 3 of 4 replicates grew during the experiment.

[c]Only 2 of 4 replicates grew.

[d]Only 3 of 6 replicates grew.

Thus results showed A. nidulans to be affected by as little as 0.50 ppb PMA. At this and higher concentrations growth was irregular and was preceded by lag phases. In view of mercury contamination reported in oceanic environments it was of interest to also consider the toxicity of PMA to A. quadruplicatum. Duplicate cultures in two experiments yielded the growth values as compared to controls (TABLE 2). Thus it is evident that A. quadruplicatum is more tolerant to PMA than A. nidulans; however, neither organism showed any growth at 10 ppb PMA.

Much research has shown that in addition to growth, beneficial activities of microorganisms can be affected by pesticides as well. Jones (1956) states that bacteria in soil which convert ammonia to nitrates are more sensitive to aldrin than bacteria which convert organic matter to ammonia. Several herbicides have been seen to influence the process of soil nitrification (Slepecky and Beck, 1950; Quastel and Scholefield, 1953; Jones, 1956; Hale et al., 1957; Otten et al., 1957; Teater et al., 1958; Geller and Khariton, 1961; Casely and Luckwill, 1965; Farmer et al., 1965; Balicka, 1969; Thompson and Corke, 1969; Winely and Clemente, 1970). Cellulose decomposition in soil is reported to be impaired by chlorpropham (Sobieszczanski, 1969--see Balick, 1969) and EPTC (Balicka and Sobieszczanski, 1969--see Balicka, 1969). Magee and Colmer (1955) found respiration of _Azotobacter_ sp. inhibited by herbicides at high concentrations (1500 to 2000 ppm). Vance and Drummond (1969) reported no effect on respiration of four algal species by 1 ppm dieldrin or DDT. However, photosynthesis by planktonic marine algae was reduced by less than 1 ppm DDT, dieldrin, or endrin (Wurster, 1968; Menzel et al., 1970). It has been shown that DDT reduces Hill reaction activity in barley chloroplasts (Lawler and Rogers, 1967).

Effect of degradation products - Pesticides, as they may adversely affect microorganisms, involve not only the parent compound but the intermediate and terminal residues of these compounds as well. Recent investigations (Rosen and Sutherland, 1966) have pointed out the potential of certain "terminal" residues to be as toxic as the original pesticides. Data from this laboratory also support this observation. _Anacystis nidulans_ and _Agmenellum quadruplicatum_ were grown in media containing microbial degradation products of aldrin, dieldrin, and endrin. The data in TABLE 3 show that _A. nidulans_ continues to be sensitive to photoaldrin and ketoendrin, two metabolites of aldrin and endrin, respectively. _Agmenellum quadruplicatum_ appears resistant to both compounds.

However, both organisms show continued sensitivity to metabolites of dieldrin, as shown in TABLE 4. While metabolites F and G are only formed microbially,

TABLE 3

Growth Response[a] of *Agmenellum quadruplicatum* and *Anacystis nidulans* to Metabolites of Aldrin and Endrin

ppb	Photoaldrin		Ketoendrin	
	A. quadruplicatum	*A. nidulans*	*A. quadruplicatum*	*A. nidulans*
950	6.2±1.0	5.3±0.8	7.4±0.4	4.5±1.1
475	6.4±0.4	6.3±0.7	6.8±0.2	3.5±0.1
95	6.5±0.7	6.4±0.7	7.3±0.3	6.0±0.8
19	5.9±0.7	7.0±0.6	7.1±0.8	6.6±0.2
0.2	6.0±0.6	7.0±0.6	7.3±0.6	6.3±0.1
Control	6.6±0.5	6.8±0.4	6.6±0.5	6.8±0.4

[a]Conditions as in TABLE 1.

TABLE 4

Growth Response[a] of *Agmenellum quadruplicatum* and *Anacystis nidulans* to Metabolites of Dieldrin

	Metabolite F[b]		Metabolite G[b]		Photodieldrin[b]	
ppb	*A. quadruplicatum*	*A. nidulans*	*A. quadruplicatum*	*A. nidulans*	*A. quadruplicatum*	*A. nidulans*
950	5.4±0.7	3.5±0.4	4.9±1.0	4.2±1.4	5.6±1.0	4.0±0.4
475	5.9±0.9	4.8±0.3	6.4±0.8	5.9±0.6	6.9±0.5	5.3±0.8
95	4.6±0.7	6.5±0.5	6.4±0.9	6.7±0.1	6.6±0.4	7.2±0.1
19	6.4±1.2	7.2±0.5	6.4±1.2	6.7±0.5	6.4±0.6	7.1±0.9
0.2	6.8±0.8	6.7±0.3	6.5±1.1	6.4±0.1	7.1±0.2	7.1±0.8
Control	6.2±0.9	6.9±0.6	6.2±0.9	6.9±0.6	6.2±0.9	6.9±0.6

[a]Conditions as in TABLE 1.

[b]Matsumura et al., 1968.

photodieldrin is also known to form on plant surfaces by the action of UV or sunlight. Hence it was of interest to assay the response of other algal species to photodieldrin. TABLE 5 shows that of the algae tested, Nostoc sp. and the green alga Chlorella soporensis appeared only slightly affected by photodieldrin at high concentration, while A. nidulans was most inhibited. Continued growth of A. nidulans in successive cultures in medium plus various levels of photodieldrin did not improve initial growth rates. Other attempts to improve the organisms' tolerance by varying growth conditions and dark incubation intervals were unsuccessful.

TABLE 6 incorporates data of several growth-response experiments of A. quadruplicatum and A. nidulans to DDT and five DDT-analogs. Again, although much variation occurred, the data show little growth-rate depression of either alga at concentrations below 100 ppb insecticide. Both species show continued, perhaps greater, sensitivity to the two analogs DDE and DDD, as well as DDT. However, the more polar compounds DDA, DBP, and FW152 apparently have little effect as seen in these experiments.

Microbial uptake and accumulation - Toxicants, once taken up by primary producers such as marine algae, can be passed up the food chain to higher trophic levels (Rudd, 1964; and Woodwell et al., 1967). In addition, many toxicants can, depending upon the environmental conditions and species involved, be accumulated within the cell to levels many-fold higher than ambient (Vance and Drummond, 1969). From the few studies available, accumulation appears to be primarily by inactive surface adsorption (Krauskopf, 1956; Hannerz, 1968; Södergren, 1968). However, Glooschenko (1969) found that dividing marine diatom cells in light accumulated ^{203}Hg longer than did non-dividing cells, thus indicating the possibility of some active uptake mechanism. The exotic and demanding minor-element nutritional requirements of many organisms would tend to support active uptake in some instances.

We have found that yeast cells of Rhodotorula gracilis

TABLE 5

Growth Response of Several Blue-Green Algae[a] to Photodieldrin[b, c]

Alga*	Control	0.2 ppb	19 ppb	95 ppb	475 ppb	950 ppb
I	2.6±0.1 (5)	2.7±0.1 (3)	2.8±0.2 (3)	2.7±0.2 (6)	2.9±0.3 (6)	2.7±0.2 (6)
II	3.5±0.2 (2)	3.6±0.1 (2)	3.4±0.3 (3)	3.6±0.0 (3)	3.3±0.2 (3)	2.6±0.6 (3)
III	7.0±0.7 (6)	6.7±0.1 (3)	---	7.1±0.2 (3)	4.7±0.2 (5)	3.1±0.3 (8)
IV[a]	6.2±0.6 (6)	6.9±0.3 (6)	6.5±0.5 (6)	5.0±0.3 (5)	5.2±0.2 (5)	5.2±0.4 (6)
V[d]	3.3±0.3 (4)	3.3±0.4 (4)	3.3±1.0 (5)	3.1±0.6 (4)	3.2±0.8 (5)	3.4±0.3 (6)
VI[d]	6.5±0.2 (7)	7.1±0.5 (5)	---	7.2±0.3 (3)	7.1±0.5 (4)	6.8±0.2 (7)
VII[d]	3.6±0.5 (6)	3.7±0.7 (6)	3.5±0.5 (6)	4.0±0.6 (5)	3.4±0.1 (5)	3.5±0.3 (6)

*I = Anabaena variabilis; II = Nostoc sp.; III = Anacystis nidulans; IV = Chlorella soporensis; V = Agmenellum quadruplicatum (strain BG-1); VI = Agmenellum quadruplicatum (strain PR-6); VII = Coccochloris elabans.

[a] All algae tested here are blue-green species, except C. soporensis, a green alga.

[b] Values reported as number generations per 24 hours.

[c] Numbers in parentheses indicate replicate cultures.

[d] Indicates marine isolate; others are freshwater isolates.

TABLE 6

Growth Response of *Agmenellum quadruplicatum* and *Anacystis nidulans* to DDT and its Metabolites[a, b]

	ppb	*A. quadruplicatum*	*A. nidulans*
DDT	885	5.4±1.1(13)	5.8±1.1(16)
	442	5.4±1.3(12)	5.8±0.8 (5)
	88	6.3±0.6 (4)	6.5±0.6 (4)
	18	6.5±0.4 (4)	6.2±0.2 (4)
	0.2	6.1±0.6 (4)	6.8±0.7 (4)
	0.0	6.1±0.7(16)	6.8±0.5(16)
DDE	791	5.3±1.0 (6)	4.6±0.0 (2)[c]
	395	5.4±0.9 (5)	5.4±2.0 (3)
	79	6.2±0.4 (3)	6.5±1.5 (4)
	8	6.2±0.3 (4)	6.5±0.4 (2)
	0.8	6.7±0.4 (4)	7.2±0.0 (2)
	0.0	6.3±0.6 (8)	6.8±0.5 (7)
DDD	791	3.4±0.0 (2)[c]	4.2±0.3 (1)[c]
	395	5.3±0.8 (4)	5.3±0.3 (4)
	79	6.7±0.1 (2)	6.2±0.3 (6)
	8	6.8±0.3 (2)	6.8±0.3 (2)
	0.8	7.0±0.4 (2)	6.4±0.3 (2)
	0.0	6.4±0.7 (6)	7.0±0.3 (7)
DDA	700	6.9±0.2 (2)	6.2±0.1 (2)
	350	6.7±0.0 (2)	5.9±0.2 (2)
	70	7.4±0.2 (2)	6.5±0.4 (2)
	7	7.1±0.0 (2)	6.6±0.6 (2)
	0.7	7.4±0.2 (2)	6.8±0.0 (2)
	0.0	6.9±0.2 (3)	6.0±0.3 (3)

TABLE 6 - Continued

	ppb	*A. quadruplicatum*	*A. nidulans*
DBP	720	6.5±0.3 (2)	6.4±0.1 (2)
	360	6.9±0.3 (2)	6.2±0.0 (2)
	72	6.6±0.1 (2)	6.8±0.2 (2)
	7	6.8±0.4 (2)	7.1±0.4 (2)
	0.7	6.6±0.1 (2)	6.6±0.1 (2)
	0.0	6.9±0.2 (3)	6.0±0.3 (3)
FW152	821	6.1±0.1 (4)	6.3±0.4 (2)
	410	6.8±0.3 (3)	7.2±0.2 (2)
	82	6.6±0.4 (3)	7.4±0.2 (2)
	8	6.3±0.4 (4)	7.2±0.1 (2)
	0.8	6.8±0.4 (3)	7.4±0.2 (2)
	0.0	6.6±0.3 (4)	7.0±0.3 (3)

[a]Values expressed as number of generations/24 hours.
[b]Numbers in parentheses represent numbers of replicates.
[c]Growth occurred in only one or two of several replicates.

TABLE 7

Percent Radioactivity of ^{14}C-DDT in Yeast Cells

Culture	Time (minutes)					
	2.5	7.5	12.5	17.5	32.5	Average
Control	21	42	56	38	--	39
Torulopsis utilis	92	96	91	95	--	94
Rhodotorula gracilis	97	98	97	98	96	97
Extract - *R. gracilis*	98	--	--	--	94	96
Protein producing medium - *R. gracilis*	97	--	--	--	96	97
Lipid producing medium - *R. gracilis*	95	--	--	--	98	97

rapidly accumulated 97% of the DDT in a 2-ppm aqueous solution. Likewise, another yeast, Torulopsis utilis, took up 94% of the DDT in 3 minutes.

In TABLE 7, showing the percent radioactivity of the ^{14}C-DDT in the cellular fraction, the control values showed a random distribution of DDT between the medium and the cellular fractions, ranging from 21 to 56%. These results were obtained by centrifuging, decanting, and filtering the aqueous medium, and the distribution of ^{14}C-DDT between the medium and cellular fractions was determined by liquid scintillation counting.

In contrast to the erratic control values, the cellular fractions accumulated DDT at a constant rate over the 90% level after 3 minutes.

An extract of R. gracilis was prepared by sonicating the cells before the addition of the ^{14}C-DDT. The sonicated pellet was found to accumulate an average of 96% of the DDT.

We also attempted to correlate cellular lipid content with the uptake of DDT. Rhodotorula gracilis when grown on a medium rich in carbohydrates and deficient in nitrogen and phosphorus will produce approximately 60% lipids whereas when not deprived of N and P, lipid production is reduced to approximately 40% (Enebo et al., 1946). When cultures were grown under both circumstances, no differences in DDT pick-up were noted.

It is apparent that the complexities of pesticidal-microbial interrelationships warrants continued study. It has been amply demonstrated that members of the microbial world vary widely in their response to pesticides and that several factors may influence the toxicity of pesticides (Durham, 1969). Likewise, the microbial tolerance of pesticides may be affected by growth conditions, physiological condition of the cells, and various stress factors which might exist in natural populations (e.g., temperature, limited nutrients, competition). For example, growth experiments with A. nidulans established separate tolerances of 1% NaCl (Batterton and Van Baalen, 1971) and 800 ppb DDT (Batterton, Boush and Matsumura, in preparation). However, growth of this alga is severely inhibited in medium containing both 1% NaCl and 800 ppb DDT. Figure 1 illustrates relative growth of _A. nidulans_ in various concentrations of DDT and NaCl. The resulting growth pattern indicates the combined stresses of NaCl and DDT significantly changes the tolerance of _A. nidulans_ to either substance. However, in similar experiments with test-tube cultures, growth inhibition was contra-indicated when the calcium concentration of the growth medium was increased five-fold.

It is particularly interesting to note the similarities in nearly all of the uptake-accumulation studies. First, pick-up of the toxicant is extremely rapid--varying from a matter of seconds to a few minutes. And secondly, removal of the toxicant from the medium is quite high--usually more than 90% of the total being removed by the cells (dead or alive), even when ambient levels were many-fold higher than those usually encountered in nature. However, one factor should not be ignored. Few if any studies have included competitive adsorptive substrates. Might not DDT, for example, readily adsorb to organic matter, silica, etc., if available (Pfister et al., 1969; Wershaw et al., 1969; Leshniowsky et al., 1970)? The apolarity, affinity for lipids, and low water solubility of virtually all of the persistent insecticides make studies in aqueous substrates difficult. Of even greater importance, can we extrapolate from our work, even to a limited degree, to conjecture as to what occurs in nature?

It is doubtful if we can overstress the importance of microbial accumulation. After more than 25 years of worldwide use and study, the real threat from persistent pesticides is in their unfortunate ability to concentrate within food chains. This would not occur were the toxicants not picked up from low background levels, concentrated in the cell, and finally, stable for considerable periods of time.

Effect of Microorganism on Pesticide

A consideration of pesticides and microbes should also include the aspect of microbial influence upon pesticides. Aside from uptake, as previously discussed, certain microorganisms may degrade the parent compound to various metabolites. These metabolites may be either more or less toxic or more or less stable than the initial compound. As might be expected, considerable variation exists among both microorganisms and pesticides as to the occurrence and ease of the degradative process. Also, although little data is available, it appears more likely that degradation is not often an active energy accumulation mechanism but rather requires the expenditure of energy. In other words, the pesticide is probably not often a nutritional energy or carbon source, but rather is incidentally metabolized (Alexander, 1965; Horvath and Alexander, 1970; Matsumura and Boush, 1971). Perhaps the chemical characteristics of the toxicant molecule, together with the biochemical make-up of the organism, allow penetration of the cell wall with degradation or alteration incidental to the process of excretion.

As we have recently reviewed the general status of the microbial degradation of insecticides (Matsumura and Boush, 1971), as well as the specific contributions of our laboratory (Matsumura and Boush, 1966; Matsumura and Boush, 1967; Boush and Matsumura, 1967a, 1967b; Boush and Matsumura, 1968; Matsumura and Boush, 1968; Matsumura, Boush, and Tai, 1968; Miyazaki et al., 1969; Matsumura et al., 1970; Miyazaki et al., 1970; Matsumura et al., 1971a, 1971b, 1971c; Bixby et al., 1971), we will make no attempt to go over the material again.

Summary

Undesirable persistent pesticidal residues in the environment share several characteristics. They are widespread in nature, although not uniformly in amount, being omnipresent from the ocean depths to the arctic.

Residues are characteristically in minute amounts--often in the part per billion range. Additionally, initial residues are subject to movement, concentration, and alteration of chemical structure. Thus, they may be found great distances from their origin, be greatly concentrated through microbial pick-up and food chain passage, and be in an altered perhaps more toxic form. The methylation of mercury and epoxidation of certain of the cyclodiene insecticides are examples illustrating all three phenomena.

Also, contrary to earlier reports, many pesticides as well as certain of their breakdown metabolites can adversely affect microorganisms, often by retarding growth or photosynthesis. Additionally, microorganisms can pick up, by absorption, and concentrate through largely unknown mechanisms, toxicants to levels many times that of ambient, thus forming the basis for eventual toxic levels in food chains.

References

Alexander, M.: Adv. Appl. Microbiol. 7, 35 (1965).

Arnold, W. R., P. W. Santelmann, and J. Q. Lynd: Weeds 14, 89 (1966).

Audus, L. J.: Plant Soil 2, 31 (1949).

Audus, L. J.: Plant Soil 3, 170 (1951).

Balicka, N.: The action of several herbicides on the complexes of microflora in maize crop rotation. Report derived from PL 480 Project EZ1-CR-30 (FG-Po-172), (1969).

Balicka, N., and Z. Krezel: Weed Res. 9, 37 (1969).

Batterton, J. C., and C. Van Baalen: Arch. Mikrobiol. 76, 151 (1971).

Batterton, J. C., G. M. Boush, and F. Matsumura: DDT: Inhibition of NaCl tolerance by the blue-green alga, Anacystis nidulans. In preparation (1972).

Batterton, J. C., G. M. Boush, and F. Matsumura: Bull. Environ. Contam. Toxicol. 6, 589 (1971).

Bixby, M. W., G. M. Boush, and F. Matsumura: Bull. Environ. Contam. Toxicol. 6, 491 (1971).

Boush, G. M., and F. Matsumura: Proc. VI Int. Congr. Plant Protection (Aug. 30-Sept. 6, 1967), Vienna, Austria, p. 580 (1967a).
Boush, G. M., and F. Matsumura: J. Econ. Entomol. 60, 918 (1967b).
Boush, G. M., and F. Matsumura: Proc. XIII Int. Congr. Entomol. (Aug. 2-9, 1968), Moscow, Russia (1968).
Brian, P. W.: Ann. Botany (London) 13, 59 (1949).
Cairns, J., Jr.: Mosquito News 28, 177 (1968).
Casely, J. C., and L. C. Luckwill: Rep. Agr. Horticul. Res. Sta., University of Bristol, 78 (1965).
Christie, A. E.: Water Sewage Works 116, 172 (1969).
Collins, J. A., and B. E. Langlois: Appl. Microbiol. 16, 799 (1968).
Cowell, B. C.: Trans. Amer. Fish. Soc. 94, 371 (1965).
Cowley, G. T., and E. P. Lichtenstein: J. Gen. Microbiol. 62, 27 (1970).
Dalton, S. A., M. Hodkinson, and K. A. Smith: Appl. Microbiol. 20, 662 (1970).
Durham, W. F.: Chemical Fallout. M. W. Miller and G. G. Berg, Eds. C. C. Thomas, Springfield, Illinois, p. 433 (1969).
Eipper, A. W.: N.Y. Fish Game J. 6, 46 (1959).
Enebo, L., L. G. Anderson, and H. Lundin: Arch. Biochem. 11, 383 (1946).
Fahraeus, G., and A. Johnson: Ann. Roy Agr. Coll. Sweden 26, 323 (1960).
Farmer, F. H., R. E. Benoit, and W. E. Chappell: N.E. Weed Control Conf. Proc. 19, 350 (1965).
Fletcher, D. W., and W. B. Bollen: Soc. Amer. Bacteriol. Proc., 14 (1953).
Geller, I. A., and E. G. Khariton: Microbiology 30, 494 (1961).
Glooschenko, W. A.: J. Phycol. 5, 224 (1969).
Hale, M. G., F. H. Hulcher, and W. E. Chappell: Weeds 5, 331 (1957).
Hannerz, L.: Inst. Freshwater Rec., Rep. No. 48, pp. 120-176, Drottningholm, Sweden (1968).

Horvath, R. S., and M. Alexander: Can. J. Microbiol. 16, 1131 (1970).

Il'in, A. M.: Microbiology 30, 1050 (1962).

Jones, L. W.: Utah State Agr. College Bull. 390, 1 (1956).

Kearney, P. C., and D. D. Kaufman: Science 147, 740 (1965).

Krauskopf, K. B.: Geochim. Cosmochim. Acta 9, 1 (1956).

Lawler, P. D., and L. J. Rogers: Nature 215, 1515 (1967).

Leshniowsky, W. O., P. R. Dugan, R. M. Pfister, J. I. Frea, and C. I. Randles: Science 169, 993 (1970).

Lichtenstein, E. P., K. R. Schulz, R. F. Skrentny, and Y. Tsukano: Arch. Environ. Health 12, 199 (1966).

Magee, L. A., and A. R. Colmer: Appl. Microbiol. 3, 288 (1955).

Martin, L. W., and S. C. Wiggans: Okla. State Univ. Exp. Sta., Proc. Ser. P-334 (1959).

Martin, J. P., R. B. Harding, G. H. Cannell, and L. D. Anderson: Soil Sci. 87, 334 (1959).

Matsumura, F., and G. M. Boush: Science 153, 1278 (1966).

Matsumura, F., and G. M. Boush: Science 156, 959 (1967).

Matsumura, F., and G. M. Boush: J. Econ. Entomol. 61, 610 (1968).

Matsumura, F., and G. M. Boush: Soil Biochemistry, McLaren and J. Skuins, Eds. Marcel Dekker, Inc., New York, Vol. II, p. 320 (1971).

Matsumura, F., G. M. Boush, and A. Tai: Nature 219, 965 (1968).

Matsumura, F., Y. Gotoh, and G. M. Boush: Science 173, 49 (1971c).

Matsumura, F., K. C. Patil, and G. M. Boush: Science 170, 1206 (1970).

Matsumura, F., K. C. Patil, and G. M. Boush: Nature 230, 325 (1971b).

Matsumura, F., V. G. Khanvilkar, K. C. Patil, and G. M. Boush: J. Agr. Food Chem. 19, 27 (1971a).
Menzel, D. W., J. Anderson, and A. Randtke: Science 167, 1724 (1970).
Miyazaki, S., G. M. Boush, and F. Matsumura: Appl. Microbiol. 18, 972 (1969).
Miyazaki, S., G. M. Boush, and F. Matsumura: J. Agr. Food Chem. 18, 87 (1970).
Moore, R. B.: Bull. Environ. Contam. Toxicol. 5, 226 (1970).
Otten, R. J., J. E. Dawson, and M. M. Schreiber: N.E. Weed Control Conf. Proc. 11, 120 (1957).
Petruk, G. F.: Mikrobiologiya 33, 1018 (1964).
Pfister, R. M., P. R. Dugan, and J. I. Frea: Science 166, 878 (1969).
Pimentel, D.: Ecological effects of pesticides on non-target species. Executive Office of the President, Office of Science and Technology, U.S. Government Printing Office, Washington, D.C. (1971).
Quastel, J. H., and P. G. Scholefield: Appl. Microbiol. 1, 282 (1953).
Richardson, L. T.: Can. J. Bot. 32, 335 (1954).
Rodriguez-Kabana, R., E. A. Curl, and H. H. Funderburk, Jr.: Can. J. Microbiol. 13, 1343 (1967).
Rogoff, M. H., and J. J. Reid: J. Bacteriol. 71, 303 (1955).
Rosen, J. D., D. J. Sutherland, and G. R. Lipton: Bull. Environ. Contam. Toxicol. 1, 133 (1966).
Rudd, R. L.: Pesticides and the living landscape. University of Wisconsin Press, Madison, Wis. (1964).
Slepecky, R. A., and J. V. Beck: Soc. Amer. Bacteriol., Proc. 50, 17 (1950).
Södergren, A.: Oikos 19, 126 (1968).
Tatum, W. M., and R. D. Blackburn: S.E. Ass. Game Fish Comm. Ann. Conf., Proc. 16, 301 (1962).
Teater, R. W., J. L. Mortenson, and P. F. Pratt: J. Agr. Food Chem. 6, 214 (1958).
Thompson, F. R., and C. T. Corke: Can. J. Microbiol. 15, 791 (1969).

Ukeles, R.: Appl. Microbiol. 10, 532 (1962).
Vance, B. D., and W. Drummond: J. Amer. Water Works Ass. 61, 360 (1969).
Vance, B. D., and D. L. Smith: Texas J. Sci. 20, 329 (1969).
Warren, J. R., F. Graham, and G. Gale: Phytopathology 41, 1037 (1951).
Weaver, L., C. G. Gunnerson, A. W. Breidenbach, and J. J. Lichtenberg: U.S. Public Health Rep. 80, 481 (1965).
Wershaw, R. L., P. J. Burcar, and M. C. Goldberg: Environ. Sci. Technol. 3, 271 (1969).
Wilkinson, V., and R. L. Lucas: New Phytol. 68, 709 (1969).
Winely, C. L., and C. L. San Clemente: Appl. Microbiol. 19, 214 (1970).
Woodwell, G. M., C. F. Wurster, Jr., and P. A. Isaacson: Science 156, 821 (1967).
Worth, W. A., and A. M. McCabe: Science 108, 16 (1948).
Wurster, C. F., Jr.: Science 159, 1474 (1968).

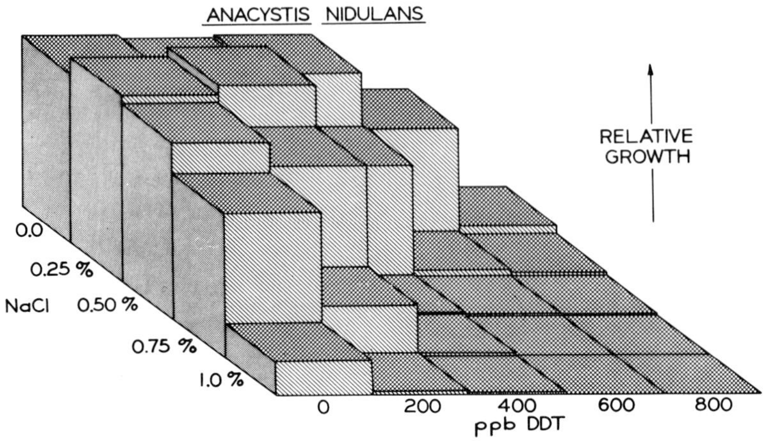

Fig. 1. Relative growth of Anacystis nidulans in response to varied DDT and NaCl concentrations. Liquid culture (15ml) in 60 mm petri dishes inoculated (120,000 cells/ml) and incubated 72 hr under 200 ft-c fluorescent lamps at 37°C. After correction for evaporation growth was measured as optical density at 660 nm. All O.D. values normalized to 1.0.

PART VI

PHOTODECOMPOSITION OF PESTICIDES

PHOTONUCLEOPHILIC REACTIONS OF PESTICIDES

Donald G. Crosby, Kenneth W. Moilanen, Masayuki Nakagawa, and Anthony S. Wong

Department of Environmental Toxicology
University of California
Davis, California

Introduction

Historically, the environmental toxicology of pesticides has concentrated upon the post-application levels of residues, especially in food, and upon the safety and toxicological significance of the applied compounds. Low residue levels were considered to represent greater safety, and "non-persistent" chemicals were strongly advocated as desirable.

However, with the recognition that as much as 90% of the applied quantity of some pesticides remains unaccounted for, scientific and practical interest in environmental transformation products has increased sharply. It now is apparent that a highly toxic material may be converted environmentally to an innocuous one, while a presumably safe pesticide can yield decomposition products which are unexpectedly lethal to some innocent form of life. The parent compound does not necessarily remain the most significant one toxicologically.

Probably all organic pesticides eventually are decomposed by the metabolic processes of living organisms. Physical forces, too, can bring about pesticide transformations, and solar radiation is among the most powerful of them. Sufficient quantum energy exists within the ultraviolet component of sunlight (290-450 nm) to permit a number of types of reactions--oxidation, reduction, isomerization, elimination, and others (Crosby, 1968; Crosby and Li, 1969). From TABLE 1 it is apparent that homolytic dissociation and subsequent free radical processes can be important in these transformations and can be effected at energies below 95 kcal/einstein (300 nm).

TABLE 1

Homolytic Bond Dissociation Energies[a]

Bond cleaved	Dissociation energy, k/cal einstein	Wavelength, nm
HO-OH	48	593
Br-C_6H_4OH	67	425
Cl-C_6H_5	80	356
H-OC_6H_5	90	317
H-CH_2OH	92	310
H-C_6H_5	104	274
H-OH	118	242

[a]From Steacie, 1954; Cottrell, 1958; Joschek and Grossweiner, 1969.

Water represents a rather general solvent, and it also is essentially transparent to sunlight. Many classes of pesticides, including the chlorobenzoic acids (Crosby and Leitis, 1969a; Plimmer, 1970), chlorophenylacetic acids (Crosby and Leitis, 1969b), halogenated phenylureas (Crosby and Tang, 1969), and chlorinated phenoxyacetic acids (Crosby and Tutass, 1966; Crosby and Wong, 1970), are photolyzed as aqueous solutions or suspensions even though they absorb very little ultraviolet light above 290 nm (TABLE 2).

TABLE 2

Light Absorption by Typical Herbicides in Water

Pesticide	Chemical name	UV absorption maxima, nm
2,4-D	2,4-Dichlorophenoxyacetic acid	283, 291
4-CPA	p-Chlorophenoxyacetic acid	289
Monuron	1,1-Dimethyl-3-(p-chlorophenyl)urea	244
Propanil	3',4'-Dichloropropionanilide	248

Photonucleophilic Reactions

Perhaps the most characteristic photodecomposition reaction of these compounds is the replacement of halogen substituents on the aromatic ring by the hydroxyl group and, to a lesser extent, by hydrogen. For example, 4-CPA (p-chlorophenoxyacetic acid, I), as its sodium salt, formed p-hydroxyphenoxyacetic acid (II) and a smaller proportion of unsubstituted phenoxyacetic acid (Fig. 1) (II). The very high dissociation energy of the H-OH bond (TABLE 1) and the continual presence of a large excess of oxygen--an efficient radical scavenger--appear to exclude abstraction of hydroxyl radicals from water as a source of hydroxyl groups, while the independence of the reaction from the presence of oxygen excludes this element as the eventual source of -OH.

The chlorine substituent appeared as chloride ion, and the rate of the reaction increased with increasing pH (Crosby and Leitis, 1969a), suggesting that a nucleophilic displacement by hydroxide ion was taking place. Such photonucleophilic reactions are known (Letzinger et al., 1965; Barltrop et al., 1967; Havinga and Kronenberg, 1968; Nijhoff and Havinga, 1968), and the relative ease with which nucleophiles attack the aromatic nucleus in its excited state could explain why the displacements proceed under environmental conditions despite the modicum of light actually absorbed. The formation of p-cyanophenoxyacetic acid (III) by irradiation of aqueous 4-CPA in the presence of potassium cyanide produced further evidence of an ionic mechanism (Crosby and Wong, 1970) (Fig. 1), and p-chlorophenol has been shown to form p-cyanophenol under similar conditions (Omura and Matsuura, 1969).

As expected, irradiation of dichloro compounds such as 2,4-D (2,4-dichlorophenoxyacetic acid, V) (Crosby and Tutass, 1966) and propanil (3',4'-dichloropropionanilide, VI) (Crosby and Moilanen, 1971) in water caused stepwise replacement of Cl by OH (and H) to provide the two monochloro- and unchlorinated products in each case. The

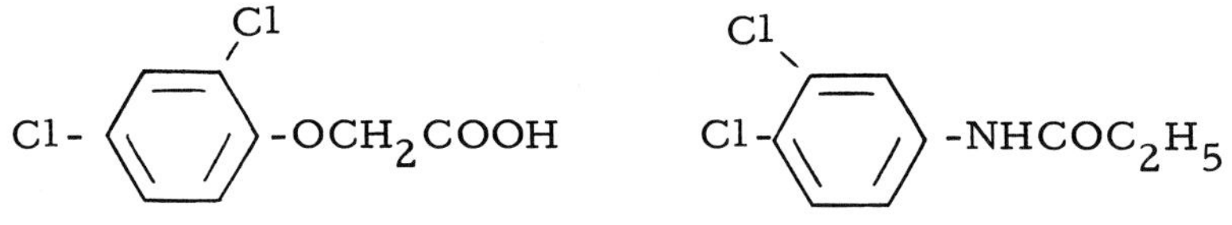

V VI

photonucleophilic displacement of chloride by chlorophenoxide ion in such <u>ortho</u>-chlorophenols introduced the intriguing possibility of the photochemical generation of chlorinated dibenzo-p-dioxins (Fig. 2). Indeed, exposure of aqueous sodium pentachlorophenate (PCP-Na, VII) to sunlight wavelengths provided identifiable amounts of 1,2,3,4,6,7,8,9-octachlorodibenzo-p-dioxin (VIII) (Crosby and Wong, 1970), but none of the corresponding 2,3,7,8-tetrachloro compound could be isolated from irradiated solutions of 2,4,5-T (2,4,5-trichlorophenoxyacetic acid) or the corresponding trichlorophenol because of the rapid photoreduction of the generated dioxins (Crosby et al., 1971).

p-Chlorophenol also appeared as a major product of 4-CPA photolysis, and the detection of small amounts of p-chlorobenzonitrile (IV) during irradiation in the presence of potassium cyanide suggested that the phenol arises in part from nucleophilic displacement of the normally stable oxyacetate sidechain by hydroxide ion (Crosby and Wong, 1971) (Fig. 1). As a further extension, it was demonstrated that the herbicide nitrofen (2, 4-dichlorophenyl 4-nitrophenyl ether, IX) was rapidly photolyzed in water to 2, 4-dichlorophenol and p-nitrophenol (Fig. 3) (Crosby and Nakagawa, 1971). The nitro group also underwent both reduction (to 2, 4-dichlorophenyl 4-aminophenyl ether, X) and nucleophilic displacement by hydroxide ions to give the corresponding phenol (XI) and nitrite ion. In the presence of cyanide, the products included 2, 4-dichlorobenzonitrile (XIII), the corresponding amide and acid, and p-nitrophenol. However, the strongly nucleophilic amine, piperidine, reacted so rapidly and extensively that there resulted only a highly complex mixture representing displacement of chlorides as well as nitrophenate; p-nitrophenol alone was identified.

Surprisingly, the rapid photolysis of dexon (sodium p-dimethylaminobenzenediazosulfonate, XIII) in aerated water also appeared to represent another example of a photonucleophilic reaction (Fig. 4) (Crosby and Moilanen, 1970). p-Dimethylaminophenol (XIII) was produced in 99% yield; irradiation in the presence of potassium cyanide provided p-dimethylaminobenzonitrile (XV), and the presence of potassium bromide resulted in good yields of p-bromodimethylaniline (XVI). Gas evolution always accompanied the reaction.

The remaining 1% of the dexon was accounted for as dimethylaniline (XIV), corresponding to the unsubstituted benzoic and phenoxyacetic acids formed upon irradiation of their respective chlorinated derivatives. The abstraction of hydrogen atoms from organic solvents by aromatic free radicals is well known (Plimmer, 1970), and this mechanism may account for many of the previously reported photoreductions. However, photoreductions produced in pure water by sunlight could not be expected to involve homolytic abstraction of hydrogen atoms from solvent, even under sensitization (Hoffman and Sher, 1966), and abstraction from dexon or its photolysis products would not permit the rather accurate accounting for reacted starting material, especially in the presence of excess oxygen.

Again, however, it might be possible to invoke a photonucleophilic mechanism in which the displacing nucleophile was hydride ion. Indeed, irradiation of dexon in the presence of a water-stable hydride (sodium borohydride) resulted in a high yield of dimethylaniline (along with a small proportion of the expected dimethylaminophenol) (Fig. 4). Analysis of the gas released during the normal photolysis in pure water revealed it to contain 5% hydrogen in nitrogen, perhaps representing the reaction of hydride ions with the hydronium ions arising from the familiar dissociation of water. The presence of hydride ions--formed, perhaps, through a reaction of hydronium ions and hydrated electrons (Hart and Ambar, 1970) (Equation 1)--also might account for the observed reduction of nitrofen's nitro group, a type of reaction which is considered to involve hydride-ion transfer (March, 1968).

$$e^-_{aq} + H_3O^+ \longrightarrow H + H_2O$$

$$e^-_{aq} + H \longrightarrow H^-$$

Equation 1

Discussion

Considering even just the examples presented here, the photonucleophilic reactions of pesticides apparently are commonplace and extensive. In the environment, they could occur wherever water and light are found together, and sunlight undoubtedly supplies sufficient illumination. In addition to the hydroxide and hydronium ions which always are present, natural waters may contain or adjoin many nucleophiles including chloride and bromide ions, sulfides, ammonia, and the amine or thiol endgroups with which biological matter is studded (Crosby, 1971, 1972). The contact and reaction of such reagents with light-activated pesticides--analogous to the action of piperidine on nitrophen--may help to account for the notable failure of pesticide scientists to provide an environmental "balance sheet" for even a single one of their compounds. The toxicological significance of the reaction products remains essentially unknown.

Conversely, simple synthetic reagents such as ammonia or sulfides might, in some cases, provide an effective means for the intentional removal or destruction of excess or unwanted pesticides (Kearney et al., 1969; Kohn, 1969) provided, of course, that the reaction products were harmless or further degraded. For that matter, the ease with which these reactions take place at ambient temperature in highly dilute aqueous media suggests that photonucleophilic displacements may be pertinent to the mechanism of herbicidal action of the diphenyl ethers, triazines, and other substances which affect light-energized processes in green plants. And it is not inconceivable that photonucleophilic transformations even had a part in the organic synthesis of a primordial planet Earth.

These few examples of a single type of environmental reaction can serve a larger cause--to bring into focus some of the fundamental but yet unanswered questions of present-day environmental toxicology. Considering the interaction of light-activated pesticides with natural substrates, where should we now look for environmental contaminants and how can we detect and identify traces of such complex unknown substances? With the variety of products available from photonucleophilic reactions alone, how are we to assess the hazards--if any--which they may present to life forms including ourselves? If we are intent upon establishing a toxicological "early warning system," just what should we monitor; is analytical evaluation of the parent pesticides really so informative of hazard? And, with both the volume of environmental toxicants and the genuine needs of mankind expanding every day, how can we train and equip people to accurately answer such questions--soon?

References

Barltrop, J. A., N. J. Bunce, and A. Thomson: J. Chem. Soc. (London) 1142 (1967).

Cottrell, T. L.: The Strengths of Chemical Bonds (2d ed.) Butterworths, London (1958).

Crosby, D. G.: Residue Rev. 25, 1 (1968).

Crosby, D. G.: "Pesticides in the soil: a symposium." Michigan State University, East Lansing (1971).

Crosby, D. G.: Advances in Chemistry Series, in press (1972).

Crosby, D. G., and E. Leitis: J. Agr. Food Chem. 17, 1033 (1969a).

Crosby, D. G., and E. Leitis: J. Agr. Food Chem. 17, 1036 (1969b).

Crosby, D. G., and M.-Y. Li: Degradation of Herbicides. P. C. Kearney and D. D. Kaufman, Eds. Marcel Dekker, New York, p. 321 (1969).

Crosby, D. G., and K. W. Moilanen: unpublished data (1970).

Crosby, D. G., and K. W. Moilanen: Abstr. 161st Meet. Amer. Chem. Soc. Los Angeles, April (1971).

Crosby, D. G., and M. Nakagawa: Abstr. 162nd Meet. Amer. Chem. Soc. Washington, D.C., September (1971).

Crosby, D. G., and C.-S. Tang: J. Agr. Food Chem. 17, 1041 (1969).

Crosby, D. G., and H. O. Tutass: J. Agr. Food Chem. 14, 596 (1966).

Crosby, D. G., and A. S. Wong: Abstr. 160th Meet. Amer. Chem. Soc. Chicago, September (1970).

Crosby, D. G., and A. S. Wong: unpublished data (1971).

Crosby, D. G., A. S. Wong, J. R. Plimmer, and E. A. Woolson: Science 173, 748 (1971).

Hart, E. J., and M. Anbar: The Hydrated Electron. Wiley-Interscience, New York (1970).

Havinga, E., and M. E. Kronenberg: Pure Appl. Chem. 16, 137 (1968).

Hoffman, M. Z., and A. Sher: Nature 210, 1039 (1966).

Joschek, H.-I., and L. I. Grossweiner: J. Amer. Chem. Soc. 88, 3261 (1969).

Kearney, P. C., E. A. Woolson, J. R. Plimmer, and A. R. Isensee: Residue Rev. 29, 137 (1969).

Kohn, G. K.: Residue Rev. 29, 1 (1969).

Letzinger, R. L., O. B. Ramsay, and J. H. McCain: J. Amer. Chem. Soc. 87, 2945 (1965).

March, J.: Advanced Organic Chemistry: Reactions, Mechanisms, Structure. McGraw-Hill, New York (1968).

Nijhoff, D. F., and E. Havinga: Tetrahedron Letters 4199 (1968).

Omura, K., and T. Matsuura: Chem. Commun. 1394 (1969).

Plimmer, J. R.: Residue Rev. 33, 47 (1970).

Steacie, E. W. R.: Atomic and Free Radical Reactions (2nd ed.) Reinhold, New York (1954).

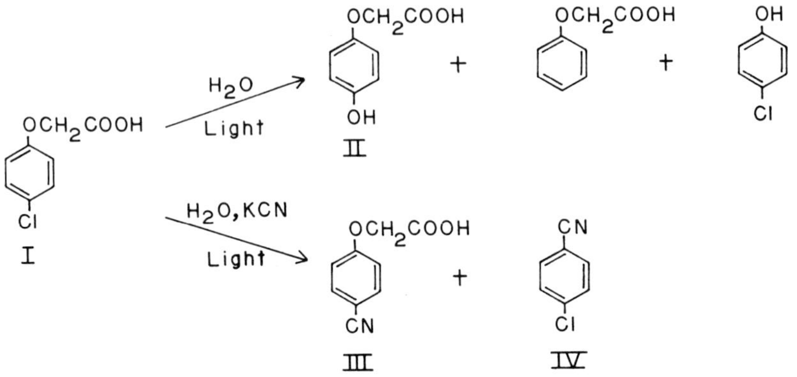

Fig. 1. Photolysis of 4-CPA.

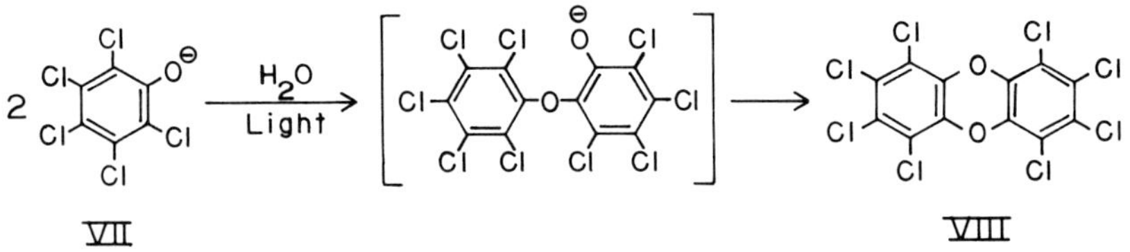

Fig. 2. Generation of dioxins from o-chlorophenates such as the sodium salt of PCP.

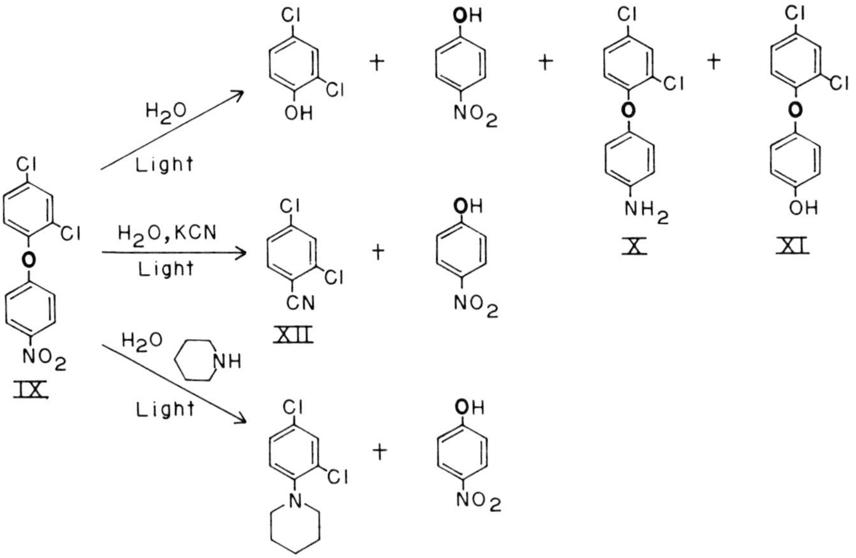

Fig. 3. Photolysis of nitrofen.

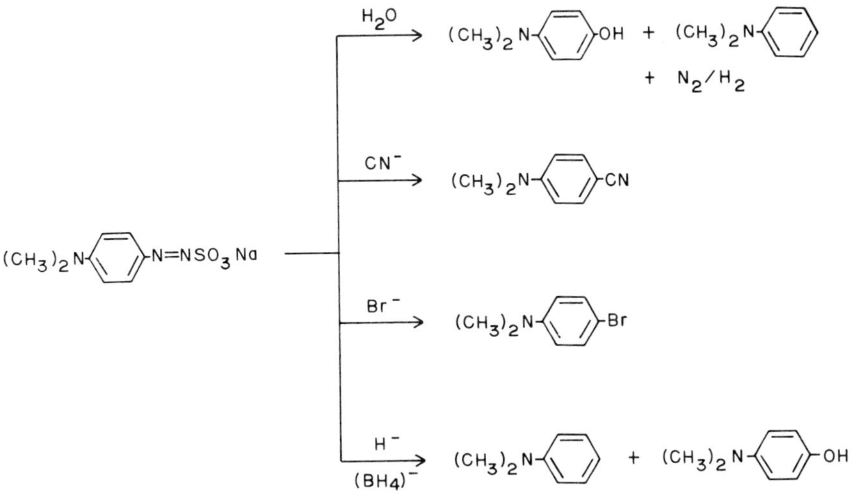

Fig. 4. Photolysis of dexon.

THE PHOTOCHEMISTRY OF SEVERAL PESTICIDES

Joseph D. Rosen

Department of Food Science
Rutgers, The State University
New Brunswick, New Jersey 08903

Introduction

Changes in the chemical structure of a pesticide released into the environment are brought about by microbiological, plant, insect, and animal metabolism as well as by hydrolysis, chemical reactions on soil surfaces, and solar energy. Because it cannot be assumed that the environmental products of a pesticide present a lesser ecological and public health hazard than the original pesticide itself, it is important to identify the products that are formed in model environmental systems, develop a method of analysis for these products to see if they are actually found in the environment, and finally, to assess their biological significance.

A common criticism of research carried out in model systems is that these systems do not reflect actual environmental conditions and the products found in the model systems may not form in the environment. The fact is, however, that unless model systems are used initially, many of the products formed could not be isolated for identification because of severe analytical problems. On the other hand, results obtained in model systems have been unfairly and unjustifiably extrapolated, resulting in misconceptions as to what will actually be the fate of a pesticide in the environment.

In this paper we will discuss the photochemical behavior of five pesticides we have been studying in simple systems during the past year. The problems of extrapolation of these results to the environment will be pointed out in some cases.

Theoretical Aspects

A molecule, upon absorbing light, acquires excess energy and is said to be in an excited electronic state. Once a molecule becomes excited, it may lose this excess energy in several ways. The simplest way is to convert the excitation energy into translational and vibrational energy and lose it to the surrounding matrix. Secondly, the excited molecule may emit a photon, a process called fluorescence or phosphorescence (depending on the nature of the excited state). Thirdly, and most important from our viewpoint, the excited molecule may undergo chemical reaction. It should be pointed out that a material need not absorb light energy directly in order to undergo photochemical processes. Thus, if an acceptor molecule is in contact with another light-excited species (donor), the donor may transfer its excitation energy to the acceptor. The acceptor then undergoes photochemical reaction just as if it had acquired the light energy directly. This process is known as photosensitization. A good example of photosensitization is the rapid conversion (77% yield) of aldrin (formula 1) to photoaldrin (formula 2) by low-energy ultraviolet light in benzene solution in the presence of the photosensitizer, benzophenone. In the absence of benzophenone, less than 1% conversion is observed (Rosen and Carey, 1968).

It is also important for purposes of our discussion to make mention of the Grotthuss-Draper law; i.e., the first law of photochemistry, which states that only radiation absorbed in a system (directly or indirectly) is effective in producing a chemical change. This law is of particular importance to those of us interested in the effects of solar radiation because the lowest recorded wavelength of sunlight to reach the earth's surface is 2863 A° (Gotz, 1931). A number of researchers have ignored the first law and have reported results based upon irradiations with lamps that emit light principally at 2537 A°. As light of this wavelength represents an energy increase of 12.8 kcal/mole greater than that at 2863 A°, results obtained solely with ultraviolet lamps emitting at 2537 A° are of academic interest only. For example, Rosen (1967) identified five photoproducts resulting from irradiation of diphenamid with high-energy light. In sunlight, the herbicide was stable.

It would appear from the above discussion that all one had to do in order to determine if a pesticide was to experience photochemical reaction in sunlight would be to simply measure the visible and ultraviolet absorption spectra of the pesticide. A pesticide which does not have significant absorbance in the wavelength region above 2863 A° would not be expected to react. Unfortunately, the situation is more complicated. While diphenamid obeys the first law of photochemistry, there are a multitude of examples in the field of pesticide photochemistry in which the law is apparently violated. A partial list of those pesticides which absorb no or very little solar energy but are significantly photolyzed by sunlight includes aldrin, dieldrin, heptachlor, carbaryl, 2,4-D, metobromuron, linuron, monuron, propanil, the pyrethroids, and paraquat.

Many ideas have been proposed for these apparent violations of the first law, among them: (a) the calculation that a small, but perhaps significant amount of solar light below 2863 A° penetrates the earth's atmosphere (Barker, 1969), (b) changes in the absorption spectrum of a material absorbed on surfaces as compared to that of the material dissolved in solvent (Slade, 1966), (c) photosensitization by air pollutants, (d) preliminary air oxidation of the substrate to small amounts of aldehydes or ketones which can then act as photosensitizers (Foote, 1971) as in the photodecomposition of polyethylene.

The moral of this discussion, therefore, is that the

best way to determine if a material will undergo photoconversion in sunlight is to expose it to sunlight.

Photolysis of SBP-1382

The synthetic pyrethroid, SBP-1382 (formula 3), attracted our attention because of its potential heavy use in New Jersey against the gypsy moth, Porthetria dispar. Forgash and Tomlin (1971) found that the LD_{50} of SP-1382 for the gypsy moth was 25 ng/larva as compared to approximately 500 ng/larva for carbaryl and over 5,000 ng/larva for DDT. An enormous amount of research has been done on the photodecomposition of pyrethroids and this has been reviewed by Chen and Casida (1969). These authors also identified the structural changes in the acid moiety of several pyrethroids by saponifying the irradiated mixture and comparing the thin-layer-separated acids to synthesized standards. Our initial approach was a slight modification of the method of Chen and Casida. We irradiated SBP-1382 absorbed on silica gel with sunlight and then we intended to separate the photoproducts by elution on the silica gel, saponify each irradiation product, and then determine the alcohol as well as the acid moieties. We found that under these conditions both the pyrethroid and its alcohol moiety (formula 4) photodecomposed very rapidly, with no detectable starting material after 5 hours. However, a small amount of decomposition was observed in the control. This control was "irradiated" in the same manner as the exposed materials (i.e., spotted on thin-layer plate, covered with a quartz plate) except that a piece of tape was placed on the quartz just over the control spot. Because of the apparent dark decomposition of SBP-1382 on silica gel, we repeated the experiment on glass plates. Again there was a very rapid decomposition of the solar-irradiated material as compared to the control, but the thin-layer chromatogram of the irradiated mixture on the glass plates was completely different than that of the silica gel irradiation. Chen and Casida did not observe such a phenomenon, at least for allethrin, where they observed qualitatively similar results for irradiations on filter paper, glass plates, and on silica gel. We now have our first example of the trouble one can get into by working with simple systems, i.e., the effect of the absorbing surface on the photoreaction. If we cannot, in some cases, extend our results from silica gel to glass, extension to soil and leaf surfaces may be even more risky.

Photolysis of Nitralin

Solar irradiation of the herbicide, nitralin (formula 5), proceeded quite rapidly on glass, silica gel, and in acetone or methanol solution. In each medium a significant quantity of material with an approximate molecular weight of 754 (by osmometry) was formed. On silica gel, significant quantities of this material were formed after only 5 hours of exposure to sunlight, detectable amounts being formed within 1 hour. This photoproduct also experienced rapid photolysis in sunlight to a material which remained at the origin of the thin-layer plate after elution with a solution of hexane and ethyl acetate (7:3).

Elucidation of the photochemical pathway, therefore, centered on identification of the material with approximate molecular weight of 754. The mass spectrum of the material exhibited no fragment higher than m/e 540, with major peaks at 498, 456, 440, 426, 393, 377, 331, 285, 269 (base peak), 255, 241, 231, 227, 226, 206, 190, 176, 162, 143, 116, and 105 m/e. The peaks at m/e 540, 498, and 456 indicated that at least two propyl groups were present. Elemental analysis indicated 26 carbons, 30 hydrogens, 8 nitrogens, 2 sulfurs, and 15 oxygens. This represents a loss of 8 hydrogens and a gain of 2 nitrogens and 3 oxygens after two nitralin molecules have combined. The infrared spectrum indicated that at least one sulfonyl group was present but the nuclear magnetic resonance spectrum indicated alteration of the hydrogen atoms of the methylsulfonyl group, suggesting "dimerization" occurred at the methylsulfonyl group.

It has previously been reported that ultraviolet irradiation of trifluralin (formula 6) resulted in successive losses of the propyl group to form α,α,α-trifluoro-2,6-dinitro-N-propyl-p-toluidine (formula 7) and α,α,α-trifluoro-2,6-dinitro-p-toluidine (formula 8). Compound 7 is also photochemically converted to α,α,α-trifluoro-2-nitro-6-nitroso-p-toluidine (formula 9) (Probst and Tepe, 1969). Our work has indicated that this is not the major pathway for nitralin photolysis, probably because of the greater susceptibility to photochemical processes of the methylsulfonyl group as compared to the trifluoromethyl group. We did, however, find a small amount of material which we have tentatively identified (by mass spectroscopy, infrared spectroscopy, and thin-layer chromatography) as 4-(methylsulfonyl)-2,6-dinitro-N-ethyl-N-methylaniline (formula 10).

Photolysis of Pyrazon

Solar irradiation of the herbicide, pyrazon (formula 11), in aqueous solution for 1 month in pyrex flasks resulted in approximately a 50% loss of starting material. Since pyrex glass effectively filters out 90% of the sunlight principally responsible for the photodecomposition of pyrazon, it may be assumed that under environmental conditions (where no pyrex glass covers the pyrazon) the rate of photodecomposition would be much faster.

The major product (22%) had an average molecular weight of approximately 9,000 and an infrared spectrum quite similar to that of another of the photoproducts obtained in 13% yield, 5-amino-4'-chloro-2,2'-diphenyl-4,5-iminodi-3(2H)-pyridazinone (formula 12). The similarities between the two infrared spectra suggest that the major product was formed by further polymerization of the dimer, compound 12. Another compound found in the reaction mixture (8%) was identified as 2,5,7,10-tetrahydro-2,7-diphenylpyrazino [2,3-d:5,6-d'] - dipyridiazine-1,6-dione (formula 13). When the solar irradiation was carried out on silica gel, only two products were formed: compound 12 and the polymer. Irradiation of compound 12 resulted in formation of polymer, not compound 13. These results can readily be explained in terms of a concentration effect. It is well known that systems of high concentration favor polymerization reactions over intramolecular cyclizations. When absorbed on silica gel, pyrazon is in a much more concentrated state than when dissolved in water.

Although we have no experimental evidence, the photolysis of pyrazon to compounds 12, 13, and polymer may proceed via photonucleophilic substitution. Another analogy for this reaction comes from previous work, where 3,4-dichloroaniline (formula 14) and 3,3',4,4'-tetrachloroazobenzene (formula 15) reacted to form 4-(3,4-dichloro anilino)-3,3',4'-trichloroazobenzene (formula 16) (Rosen et al., 1970) under the influence of light and riboflavin.

Photolysis of Abate

The larvicide, Abate (formula 17), when exposed to sunlight at a concentration of 600 ppm in methanol solution, experienced rapid decomposition as shown in Fig. 1. The major initial product was the abate-sulfoxide (formula 18) which apparently further photolyzed rapidly to at least six other products. In addition, abate-sulfone (formula 19) was formed either by further oxidation of compound 18 or directly from compound 17. The six unidentified products may have been formed either by methanolysis, oxidation, or photoisomerization of the sulfoxide directly; or by initial photoxidation of the sulfoxide to the sulfone with subsequent photodecomposition of the latter. Exposure of the sulfone to sunlight for 9 days in methanol in a quartz flask resulted in no reaction, although exposure in the photosensiting solvent, acetone, for the same period led to extensive photodecomposition with the formation of a new, as yet unidentified compound. Recalling the work of Ivie and Casida (1971), who showed that Abate was an effective sensitizer for the photodecomposition of dieldrin, we exposed the sulfone to sunlight in methanol in the presence of Abate. Within 3 days, a significant quantity of the same unidentified compound from the photolysis of the sulfone in acetone was obtained. In addition, this compound was formed by irradiating the sulfone with an ultraviolet lamp whose principal emission was at 2537 A^{o}.

However, the unidentified compound did not appear among the photoproducts of Abate or its sulfoxide, nor were any of the Abate or abate-sulfoxide photoproducts present in the reaction mixture resulting from the exposure of abate-sulfone in acetone. We concluded, therefore, that the terminal photoresidues are formed from the photolysis of the sulfoxide, without involvement of the sulfone.

Although the terminal photoresidues have not yet been fully isolated and identified, we suggest the structures 20 through 28 based upon the results of Koivistionen and Merilänen (1962) for the photolysis of parathion as well as the photoisomerizations of Kitazin P and Inezin.

Caution must be exercised in extending these results to what would occur under actual environmental conditions. Because of the insolubility of Abate in water and because it was desired to obtain enough material to facilitate product identification, the photolysis experiments were conducted in methanol at relatively high concentrations, 600 ppm. However, the solubility of Abate in water is approximately 0.5 parts per billion. The effect of this low concentration may be of particular importance, when the proposed mechanism for the photooxidation of sulfides by singlet oxygen (Foote and Peters, 1972) is considered. An examination of this proposed mechanism (Fig. 2) reveals that relative formations of sulfoxide and sulfone are concentration dependant. Thus, at very low concentrations of sulfide, the intermediate R_2S^+-O-O^- may have difficulty in finding another molecule of sulfide to react with in order to form the sulfoxide. The slower reaction, i.e., unimolecular rearrangement to sulfone, may then predominate. Therefore, under actual environmental conditions, the final photolysis product may be the light-insensitive abate-sulfone and identification of the abate-sulfoxide photoproducts may be of academic interest only.

Photolysis of Aglypt

The herbicide, aglypt (formula 29), is very rapidly converted to 3-methyl-thio-6-phenyl-1, 2, 4-triazin-5-one (formula 30) in 40 ppm aqueous solution by sunlight as recently described by Rosen and Siewierski (1971). The half-life of aglypt under these conditions is 64 minutes as compared to 220 minutes for the t-butyl analogue, BAY 94337 (formula 31). Although compound 30 was only ten times less effective than aglypt as a Hill Reaction inhibitor, field trials against a wide variety of weeds indicated it was of no value as a herbicide (Eue, 1971). During the course of the Hill Reaction studies, Dr. C. C. Still of Rutgers University observed two breaks in the slope of the oxygen evolution curve. This led to speculation that the first break was due to photosynthesis inhibition by aglypt while the second was due to rapid interaction of the photochemical by-product, imidogen ($\ddot{\underset{..}{N}}$-H), with one of the chloroplast components involved in photosynthesis. Experimental proof for such an occurrence is difficult to obtain because there are no radioactive isotopes of nitrogen available. A tritium label on the anilino group of aglypt was also considered undesirable because part of the tritium label would wind up in the phenolic position of the photoproduct, compound 30, and our results would be complicated by tritium-hydrogen exchange. In an attempt to circumvent this problem, we obtained 4-methyl-3-methylthio-6-phenyl-1, 2, 4-triazin-5-one (formula 32) from Farbenfabriken Bayer with the hope of generating a C^{14}-labeled carbene under mild photochemical conditions. Unfortunately, although compound 32 underwent extensive photodegradation, it did not form product 30, indicating that 29 and 32 photolyze by different pathways.

As organic chemists, we were also intrigued by the possibility of photolyzing compound 29 in benzene through a pyrex filter in order to trap imidogen with benzene to form the never-before-synthesized azepine (formula 33). Although compound 30 was formed quite readily under these conditions, azepine was not found and the fate of imidogen in this system remains unelucidated.

References

Barker, R. E.: Photochem. Photobiol. 7, 275 (1969).

Chen, Y.-L., and J. E. Casida: J. Agr. Food Chem. 17, 208 (1969).

Eue, L.: Farbenfabriken Bayer, Leverkusen, private communication (1971).

Foote, C. S.: University of California at Los Angeles, private communication (1971).

Foote, C. S., and J. L. Peters: J. Amer. Chem. Soc., in press (1972).

Forgash, A. J., and A. Tomlin: Rutgers University, private communication (1971).

Gotz, F. W. P.: Strahlentler 40, 690 (1931).

Ivie, G. W., and J. E. Casida: J. Agr. Food Chem. 19, 410 (1971).

Koivistionen, P., and M. Merilänen: Acta Agr. Scand. 12, 269 (1962).

Probst, G. W., and J. D. Tepe: Degradation of herbicides. P. C. Kearney and D. D. Kaufman, Eds. Marcel Decker, N.Y., pp. 259-260 (1969).

Rosen, J. D.: Bull. Environ. Contam. Toxicol. 2, 349 (1967).

Rosen, J. D., and W. F. Carey: J. Agr. Food Chem. 16, 536 (1968).

Rosen, J. D., and M. Siewierski: Bull. Environ. Contam. Toxicol. 6, 406 (1971).

Rosen, J. D., M. Siewierski, and G. Winnet: J. Agr. Food Chem. 18, 943 (1970).

Slade, P.: Weed Res. 6, 158 (1966).

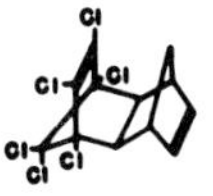

1

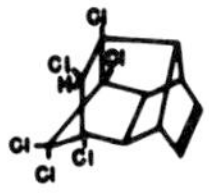

2

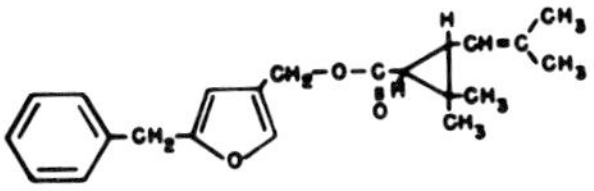

3

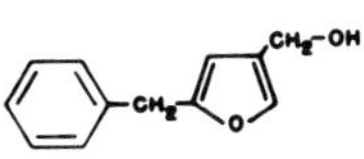

4

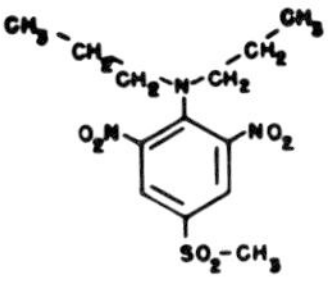

5

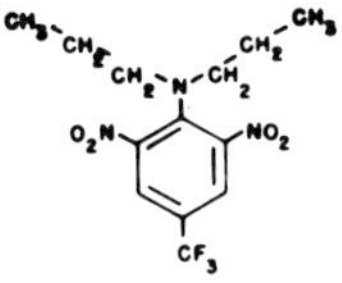

6

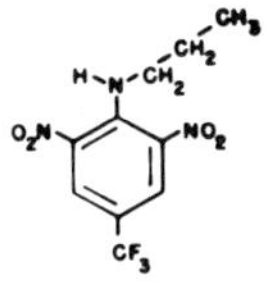

7

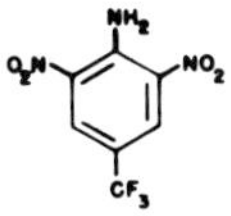

8

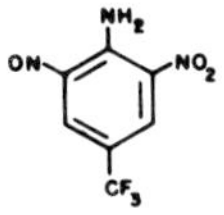

9

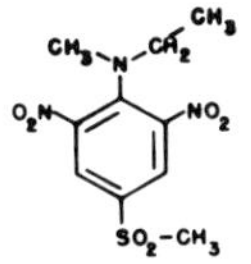

10

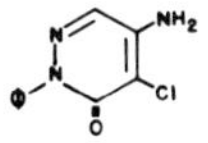

11

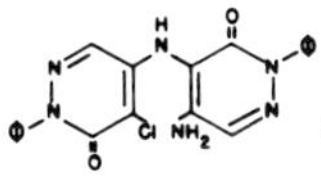

12

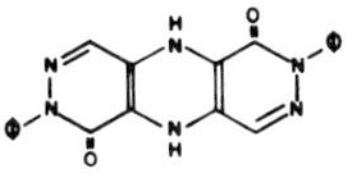

13

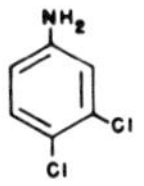

14

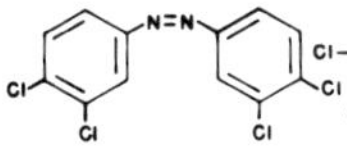

15

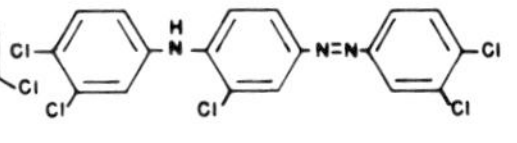

16

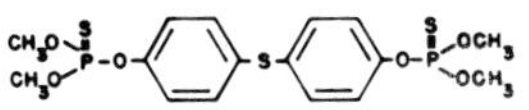

17

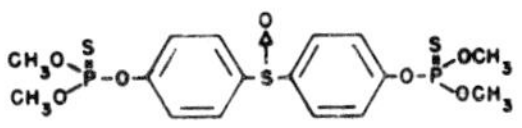

18

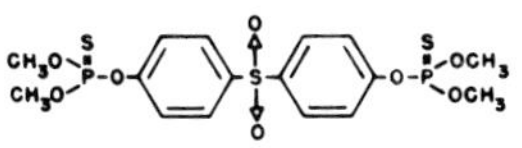

19

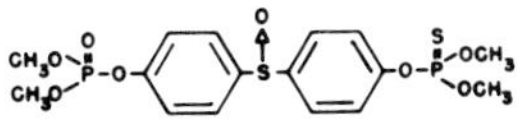

20

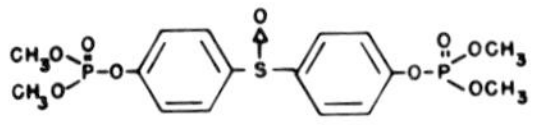

21

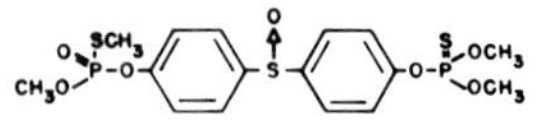

22

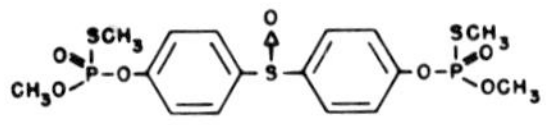

23

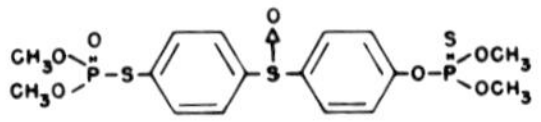

24

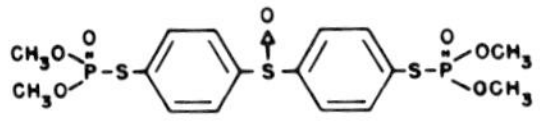

25

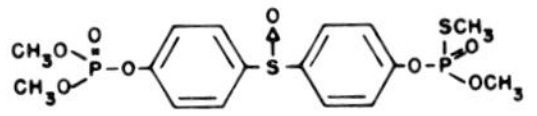

26

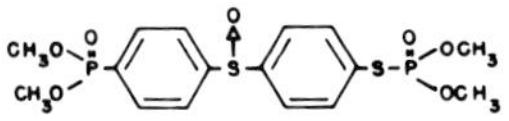

27

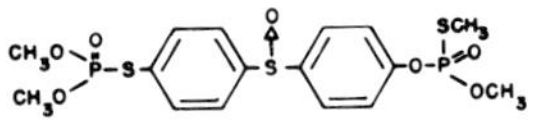

28

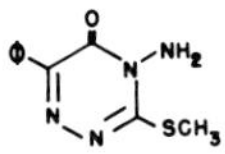

29

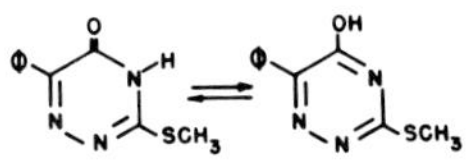

30

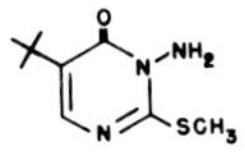

31

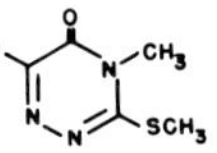

32

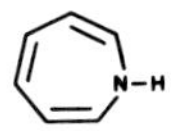

33

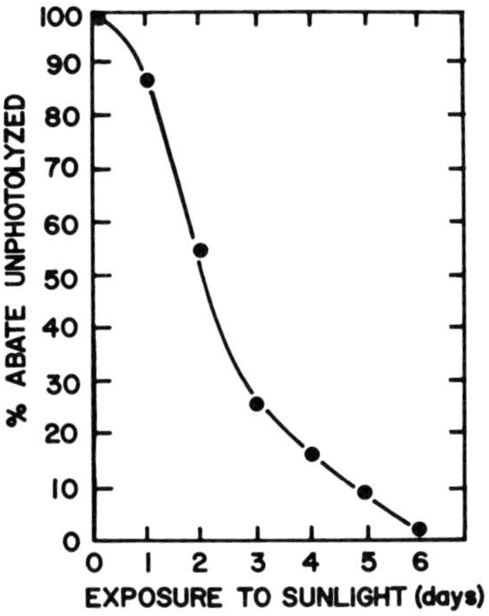

Fig. 1. Photodecomposition of Abate upon exposure to sunlight (600 ppm in methanol, quartz flasks).

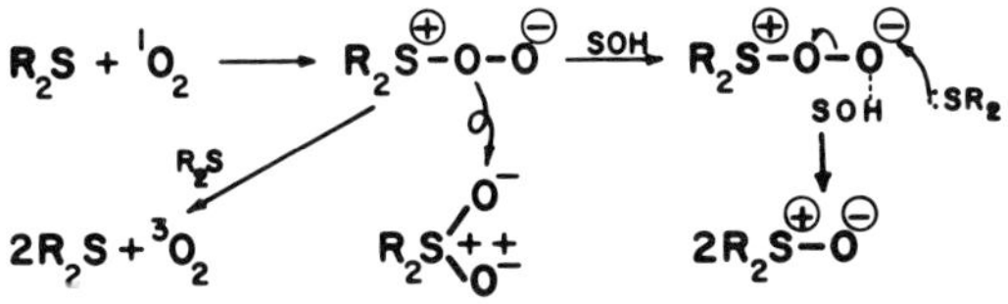

Fig. 2. Mechanism of photooxidation of alkyl sulfides (Foote and Peters).

ROLE OF PHOTOSENSITIZERS IN ALTERATION OF PESTICIDE RESIDUES IN SUNLIGHT

Louis Lykken

Division of Entomology, University of California
Berkeley, California 94720

Introduction

The use of chemicals for the control of pests on plants inevitably leaves residues in or on their foliage and the soil they grow in. In the case of systemic chemicals, much of the residue translocates from one part of the plant to another, a major portion of it being found inside the stem, leaves, etc. However, in the case of nonsystemic ones, the residues generally remain on the surface to which they are applied, at least for a short time. The surface residues generally are subject to alteration in chemical structure by biological, photochemical, and/or chemical reactions; however, some of them remain chemically intact but gradually dissipate as a result of weathering. This creates a problem because, practically without exception, the residues are toxic and, therefore, are potentially harmful to humans, useful animals, and denizens of the environment (Lykken, 1968).

Persistence is a desirable characteristic of a pesticide chemical residue from the viewpoint of an agriculturist or other persons that directly benefit from the pest control it achieves. However, from the point of view of an environmentalist and other ecologically concerned persons, this attribute is an undesirable one because of the tendency of some residues to permeate the ecosystem and to accumulate there, and because some of the residues tend to magnify to harmful levels in the food chains of certain species of predacious wildlife. As a result, controversy rages in regard to the desirability of using persistent pesticide chemicals and a movement is underway to replace such chemicals with more fugitive ones or with ones that degrade readily in the ecosystem (Lykken, 1968).

In order to achieve the greatest effectiveness and the least environmental contamination, nonsystemic pesticide chemicals should persist, on the surface to which they are applied, only for the period required for the control of the pest of interest. Ideally, the chemical should leave a toxic residue that persists only for the required period and that rapidly drops to inconsequential levels soon afterwards. It should be possible to accomplish this objective by simultaneous or subsequent application of a substance which acts to regulate, as desired, the persistence period of the residue. Unfortunately, attempts to accomplish this regulation by formulation methods and recipes, by special application methods, by use of chemical residue-breakdown agents, etc. have been universally unsuccessful (Lykken, 1968).

Widely applicable methods for accelerating the dissipation of pesticide chemical residues on plants and other surfaces, by nonphotochemical means, are still wanting. However, there is some evidence that it may be possible to control the metabolic degradation of herbicide residues within plant tissue by use of chemical additives that enhance or reduce metabolic degradation of the residue (Fox and Bingham, 1969).

It is common knowledge that the greatest loss of nonsystemic, surface residues of pesticide chemicals on plants usually is the result of the combined effect of sun, wind, and rain, a phenomenon known as weathering (Van Middelem, 1966; Crosby, 1969a). Of these weathering factors, the ultraviolet component of sunlight undoubtedly brings about the greatest amount of chemical transformation in the surface residues and, therefore, is an important factor in their degradation (Crosby, 1969b). This suggests the possibility of controlling the degree of persistence of a pesticide chemical residue by use of a chemical additive that enhances, induces, or inhibits the photoalteration of the chemical (Crosby, 1965).

Many organic chemicals sensitize, induce, or enhance photochemical reactions and others slow down or quench such reactions, especially oxidations (Foote, 1968). For example, it is known that chlorophyll is an excellent photosensitizer for all types of photooxidation and that carotenoids quench such reactions (Foote, 1968). Thus, since sunlight is an important factor in the degradation of pesticide residues, it should be possible to apply a photosensitizer or a photoinducer to the pesticide-treated surface to achieve rapid breakdown of persistent residues.

Up to the time of the first publication of Ivie and Casida in 1970 (Ivie and Casida, 1970), the literature contained little information on the use of photosensitizers to alter pesticide chemical residues on plants or other surfaces exposed to sunlight. However, some information did exist about the use of various chemical additives to sensitize photoalteration of certain pesticide chemicals in the dissolved state.

Sensitized Photoalteration of Various Pesticide Chemicals

Benzophenone, a well-known photosensitizer, accelerates the conversion of certain chlorinated cyclodiene insecticide chemicals to photoisomers when solutions of the compounds are exposed to ultraviolet light. When exposed to strong ultraviolet light in a benzone solution containing equal molar concentrations of benzophenone, aldrin and dieldrin convert in good yield to their "half-cage" isomers shown in Fig. 1 (Rosen and Cary, 1968). In the presence of the same photosensitizer, heptachlor converts under similar conditions to a "cage" isomer, as shown in Fig. 1 (Rosen et al., 1969; Rosen and Siewierski, 1970). There is evidence that acetone also photosensitizes the latter reaction (Fischler and Korte, 1969; McGuire et al., 1970). In a similar photosensitized reaction, isodrin converts to its "birdcage" isomer, as depicted in Fig. 1 (Rosen et al., 1969). Benzophenone also is an effective sensitizer in the photolysis of benzene solutions of chloroanilines in sunlight or ultraviolet light, whether or not oxygen is present (Plimmer and Kearney, 1969). Under these conditions, 3,4-dichloroaniline, a biological metabolite of certain chloroaniline-type pesticides, produces a complex mixture of photoproducts, including 3,3',4,4'-tetrachloroazobenzene when oxygen is absent, and 3,3',4,4'-tetrachloroazoxybenzene in the presence of oxygen.

In aqueous solutions, riboflavin-5'-phosphate sodium (FMN) sensitizes the photolysis, in sunlight, of 3,4-dichloroanilines, forming a number of products (Rosen et al., 1970). For example, 3,4-dichloroaniline gives 3,3',4,4'-tetrachloroazobenzene and 4-(3,4-dichloroanilino)-3,3',4'-trichloroazobenzene, among other products. Under similar conditions, 4-chloroaniline (a soil metabolite) gives 4,4-dichloroazobenzene and 4-chloro-4'(4-chloroanilino)azobenzene; 4-chloro-2-methylaniline (a plant metabolite) produces 4,4'-dichloro-2,2'-dimethylazobenzene and 4-chloro-4'-(4-chloro-2-methylanilino)-2,2'-dimethylazobenzene.

The presence of certain aromatic amines induces the photolysis, under ultraviolet light, of DDT dissolved in cyclohexane (Miller and Narang, 1970). The products formed, in the presence of diethylaniline, include DDE, DDD, dichlorobenzophenone (DDCO), and two other unidentified compounds. These reactions are thought to involve a charge-transfer from the amine to DDT.

There is evidence that chlorophyll is the active factor in catalyzing the photodegradation of pyrethroids (Glenn Jones, 1960).

From the aforementioned data, it is apparent that not much was known about the effect of sensitizing chemicals in the photoalteration of pesticide chemicals at the time when the studies by Ivie and Casida were initiated in 1968 (Ivie and Casida, 1970). Moreover, even though the phenomenon was known, no information existed in regard to the degree to which photosensitizers might actually affect the environmental residues of pesticide chemicals, in general, and persistent ones, in particular. Also, there was a paucity of information about whether or not photosensitizer reactions occur among compounds applied to plants and surfaces for pest control. Ivie and Casida hoped that their studies on the possible interactions of photosensitizers with pesticide chemical surface residue exposed to sunlight might lead to practicable methods for controlling the persistence of such residues. For example, it seemed reasonable to expect that application of a photosensitizer might rapidly reduce a persistent pesticide residue to a tolerable level (Ivie and Casida, 1970).

The remainder of this paper summarizes the results obtained in extensive studies made at Berkeley to reveal (a) the degree to which residues of important pesticide chemicals are altered by exposure to sunlight or ultraviolet light in the presence of various photosensitizers, and (b) which of the common pesticide chemicals effectively enhance the photoalteration of several important insecticide chemicals (Ivie and Casida, 1970, 1971a, 1971b). The data reported here are useful in determining whether or not the use of photosensitizers is of value in enhancing the rate of photodegradation of pesticide chemical residues under environmental conditions.

Sensitized Photoalteration and Photosensitizer Activity of Pesticide Chemicals in Silica Gel Chromatoplates

On the basis of a preliminary survey of 16 known chemical photosensitizers and related compounds, and 29 pesticide chemicals and analogs (having structural similarities to known photosensitizers), rotenone is the most effective compound for enhancing the photochemical alteration of dieldrin to photodieldrin. Measurable sensitization of the conversion of dieldrin-^{14}C residues to photodieldrin occurs when rotenone is applied to bean plants at levels as low as 0.3 ppm (Fig. 2). The potency of rotenone as a sensitizer for the photoisomerization of dieldrin is approximately 100-fold greater than that of benzophenone.

Although a few pesticide chemicals other than rotenone show some photosensitizer activity against dieldrin, they are effective only at very high concentrations (Ivie and Casida, 1970).

Results from the testing of 28 different rotenoids and related compounds indicate that the C and D rings are most critical and that the structural requirements for photosensitizer activity are met by compounds containing the chromanone moiety (Fig. 3), with aliphatic or aromatic substituents at positions 6a and/or 12a being necessary for activity. Hydroxylation at the 11 position results in potency loss, possibly due to chelation of the hydroxyl group with the ketone, and compounds with a double bond between positions 6a and 12a have greatly diminished activity. The enhanced conversion to photodieldrin also occurs with certain commercial dieldrin and rotenone formulations (Ivie and Casida, 1970).

Rotenone also catalyzes the photochemical alteration of aldrin, isodrin, endrin, heptachlor, and heptachlor epoxide but it is less effective or ineffective with DDT, DDD, lindane, and endosulfan. Many of the products appear to be photoisomers and they probably are as toxic or are less toxic than their precursors. Epoxidation does not appear to be a major photochemical reaction (Ivie and Casida, 1970).

The photosensitizing activity of rotenone is not limited to catalyzing photochemical isomerization but its major effect is its action as a photosensitizer. Its presence increases the photodecomposition of certain methylcarbamate and phosphorothionate insecticide chemicals, and of piperonyl butoxide. Photodegradation of radiolabeled samples of malathion, DDT, diazinon, Baygon, Isolan, Matacil, and carbaryl are not sensitized under similar conditions (Ivie and Casida, 1970).

Twenty-eight potential photosensitizers were screened for their effectiveness in accelerating the photoalteration of 23 pesticide chemicals radiolabeled with carbon-14 (TABLE 1). Use was made of a simple analysis technique which involved application of mixtures of 10 μg of the

TABLE 1

The Sensitizing Activity of Photosensitizers, with Varying Triplet-Energy Values, in the Alteration of Various Pesticide Chemicals Exposed to Sunlight as Deposits on Silica Gel Chromatoplates (From Ivie and Casida, 1971a)

^{14}C-pesticide chemicals studied		Name and triplet-energy value of compounds tested as photosensitizers	
		Active to some degree	
Aldrin	Diazinon	Acetophenone (74)	Trans-stilbene (50)
Dieldrin	Dyfonate	Xanthone (74)	Pyrene (49)
Endrin	Imidan	Carbazole (70)	7H-Benz(de)anthracen-7-one (47)
DDT	Malathion	Triphenylamine (70)	
	Sumithion	Dibenzothiophene (70)	Anthracene (42)
Dessin		Benzophenone (69)	
Dinoseb	Baygon	Fluorene (68)	Inactive
	Carbaryl	4, 4'-Dichlorobenzophenone (68)	Biphenyl (65)
Allethrin	Isolan	Thioxanthone (65)	Naphthalene (61)
Pyrethrin I	Landrin	Anthraquinone (62)	Chrysene (57)
	Mesurol	Phenanthrene (62)	Fluorenone (53)
Rotenone	Pirimicarb	Flavone (62)	Dicyclopentadienyliron (<43)
	Zectran	4, 4'-Bis(dimethylamino)-benzophenone (61)	
Piperonyl butoxide	2, 4-D	2'-Acetonaphthone (56)	2, 4-Pentanedione, iron III derivative (<43)
		1'-Acetonaphthone (56)	
		1'-Naphthaldehyde (56)	
		Benzil (54)	Crystal violet (39)

candidate sensitizer and 2 μg of the C^{14}-pesticide chemical as small spots on thin silica gel chromatoplates. The treated plates were exposed to sunlight, usually for 1 hour, and the photodecomposition products of the labeled pesticides were resolved by developing the plates in an appropriate solvent system, followed by radioautography. Evaluation of the sensitizing activity of test photosensitizers, whose triplet-energy values range from 74 to 39 kcal/mole, was made by visual examination of the darkened areas on the radioautograms and each compound was assigned a photosensitizer rating, based on the extent of movement of radioactivity from the original chemical to its alteration products in the presence of added sensitizer. Control experiments were made in an identical manner but the plates were held in darkness prior to evaluation (Ivie and Casida, 1971a).

In all cases where it occurs, sensitization is dependent on exposure to sunlight. Twenty-one of the 28 compounds tested as photosensitizers are active, to some degree, in sensitizing the photoalteration of one or more of the 23 radiolabeled pesticide chemicals. No direct correlation occurs between triplet-energy values (enclosed in parenthesis in TABLE 1) and photosensitizing activity, but the chlorinated hydrocarbon insecticide chemicals (particularly aldrin and dieldrin) are generally more susceptible to sensitization by the compounds having high triplet-energy values (Ivie and Casida, 1971a).

Figure 4 shows the effect of the 10 sensitizers with highest triplet-energy values on the photodecomposition of radiolabeled dieldrin on silica-gel plates. It is evident that most of these compounds are active in enhancing the conversion of dieldrin to the methylene-bridged isomer, photo-dieldrin, and that in the absence of an effective sensitizer, dieldrin is very stable in the presence of light. One sensitizer, triphenylamine, apparently acts by a different mechanism from the others because it converts dieldrin to an additional photoproduct as well as to products which remain at the origin. It is of interest that, based on a recent report, aromatic amines catalyze the photodechlorination of aliphatic halides by a photo-induced charge-transfer from the amine to the halide (Miller and Narang, 1970). Examination of the detailed data shows that only a few sensitizers have broad-spectrum activity (Ivie and Casida, 1971a).

Figure 5 is a photograph of a composite TLC radioautogram showing the effects of anthraquinone on the photodegradation of each of the 23 radiolabeled pesticide chemicals studied. For each pesticide chemical, the column marked with a minus sign shows the degree of photodecomposition that takes place in the absence of anthraquinone and the column marked with a plus sign shows the photodecomposition that occurs in the presence of anthraquinone. Examination of Fig. 5 reveals that only one compound, Baygon (a methylcarbamate), is not affected by this sensitizer. It is of interest to note that aldrin converts to three products: photoaldrin, dieldrin, and photodieldrin; that the organophosphorus compound, Dyfonate, is completely destroyed while malathion converts to a number of derivatives; and that SD-8530 (Landrin), a methylcarbamate, also converts to a number of photoproducts. Figure 5 clearly demonstrates that photosensitizers do enhance the photoalteration of certain pesticide chemicals, at least under the condition used (Ivie and Casida, 1971a).

In view of the fact that rotenone has remarkable photosensitizing activity, that many pesticide chemicals are structurally related to known photosensitizers, and that contact occurs between pesticide chemicals in the environment, there was reason to believe that one pesticide chemical might markedly alter the persistence of another by sensitizing its photoalteration. Accordingly, 175 unlabeled pesticide chemicals were tested for their photosensitizing action on each of six radiolabeled pesticide chemicals, in the manner described above. As indicated by TABLE 2, it was found that contact between most of the pesticide chemicals did not, in general, result in interactions of a photosensitizing nature. However, there were a few notable exceptions. Three compounds--the organophosphorus insecticide, abate; the diphenylsulfone acari-

TABLE 2

Photosensitizing Activity of Various Unlabeled Pesticide Chemicals in Altering the Composition of Six ^{14}C-labeled Insecticide Chemicals Exposed to Sunlight as Deposits on Silica Gel Chromatoplates[a]

^{14}C-Insecticide chemicals studied	Insecticides			Herbicides and fungicides	Miscellaneous
	Organo phosphorus products	Carbamates	Other		
Dieldrin	Abate	Carbaryl	Allethrin	Amiben	Anthraqui-
DDT	Baytex	Landrin	Genite	Carbyne (12E)	none
	Co-Ral	Mobam	Neopyna-	Dowcide A	Diphenyla-
Diazinon	Dimethoate	Temik	min	Dyrene (50W)	mine
Malathion	(44E)	Zectran	NRDC 104	MCPA (22E)	Tropital
Sumithion	Dyfonate		Pentac	Morestan (89W)	
	Imidan		Phenothi-	Pentachloro-	
Carbaryl	Methyl tri-		azine	phenol	
	thion		Rotenone	Phthalimide	
	Trithion		Sulfoxide	Phygon	
			Sulphenone	Pyramin (80W)	
			Tetradifon	Thiabendazole	
			Xanthone	Tordon (22E)	
				Vegedex	

[a]Ivie and Casida, 1971a.

cide, tetradifon; and the polycyclic compound, rotenone--were moderately active in converting Dieldrin to photodieldrin. Also, the aromatic amines, diphenylamine and phenothiazine, were active in decomposing dieldrin and, particularly, DDT (Ivie and Casida, 1971a). (As mentioned above, these reactions probably involve photodechlorination.)

Sensitized Photoalteration of Insecticide Chemicals on Bean Plant

These findings indicated a need to determine whether or not the pesticide chemical-photosensitizer relationships on silica gel hold when the chemicals are tested under more natural conditions, namely as deposits on plant foliage exposed to sunlight. This was done by applying methanol solutions of the compounds to the dorsal surface of young bean plants, exposing the plant to sunlight for 1 hour, and analyzing ether rinses of the leaves by thin-layer chromatography and radioautography. Several sensitizers were screened for their activity against dieldrin, using 10-ppm deposits of dieldrin-^{14}C and 0-, 10-, or 100-ppm levels of sensitizer (Ivie and Casida, 1971b).

It was found that, when applied to leaf surfaces exposed to sunlight, rotenone was the most active sensitizer in accelerating the photoisomerization of the radiolabeled dieldrin. While less active than rotenone, xanthone, tetradifon, triphenylamine, and flavone also were potent sensitizers of dieldrin photoalteration of bean leaves. (The uniquely high photosensitizing activity of rotenone on plant foliage is somewhat surprising because several other sensitizers showed equal or greater effectiveness compared to rotenone when tested on silica gel as previously described.) Also, it was found that, as shown in Fig. 6, the efficiency of rotenone in catalyzing dieldrin photoalteration is greater on glass than on bean leaves and that in several cases photosensitization occurs much more readily on silica gel than on leaves. (These findings possibly reflect differences in the surface characteristics of the exposure media.) (Ivie and Casida, 1971b)

On bean leaves, rotenone effectively enhances the rate of photoalteration of aldrin to photoaldrin and some photodieldrin; under the same conditions, rotenone sensitizes the photoalteration of endrin to the ketone and aldehyde "semicaged" isomers (that do not contain the epoxide ring), along with several other products.

Pesticide chemicals, other than cyclodienes, are also sensitized by rotenone on bean leaves but, in most cases, sensitization requires a high rotenone level. Tests made with a number of rotenoids and related compounds for photosensitizer activity indicated that the 2- or 3-substituted-4-chromanone moiety is apparently responsible for high sensitizing action. Although rotenone is quite active as a sensitizer against certain pesticide chemicals, it probably retains its effectiveness for only a short period in the environment because it quickly photodecomposes to non-sensitizing derivatives (Ivie and Casida, 1971b).

The photosensitizing action of certain aromatic amines on chlorinated hydrocarbon insecticide chemicals probably involves dechlorination by a photoinduced charge transfer from the amine to the halide. Certain of these photosensitizers are quite active in catalyzing the photodegradation of some chlorinated hydrocarbons; for example, DDT is quite stable on bean leaves but it breaks down rapidly after application of triphenylamine to the plant surface. The activity of triphenylamine as a sensitizer is short-lived, however, because, as shown in Fig. 7, its action is essentially complete within 30 minutes after application. The products of triphenylamine-sensitized DDT photodecomposition include DDE and dichlorobenzophenone as well as at least four unidentified products which do not include DDD (Ivie and Casida, 1971b).

Aqueous suspension of spinach chloroplasts are effective in accelerating the photodecomposition of 6 of the 23 radiolabeled pesticides studied, showing highest activity against the N-methylcarbamates, Mesurol and Zectran, and the dinitrophenol derivative, dinobuton. Chloroplast sensitization is light dependent; the reactions do not proceed in darkness. The sensitizing activity of chloroplasts is probably due largely or completely to their chlorophyll content, because purified chlorophylls *a* and *b* are sensitizers under these conditions; however, they are not as effective as an equivalent amount of chloroplasts (Ivie and Casida, 1971b).

It is of interest that the toxicity of photodieldrin to several insects and mammals is equal to or greater than that of dieldrin, and that photodieldrin is more persistent on plant foliage than dieldrin, apparently due to lower volatility (Ivie and Casida, 1970). The same toxicity relationship holds for heptachlor and its photoisomer, but the photoisomer of isodrin is less toxic than isodrin (Rosen et al., 1969).

The photodecomposition of dinobuton (2-*sec*-butyl-4, 6-dinitrophenyl isopropyl carbonate, an acaricide) in sunlight is sensitized by rotenone and xanthen-9-one, the former being the most potent one (Bandal, 1971). At least 10 photoproducts are formed on bean plants, including the hydrolysis product: dinobuton phenol (DNBP), a reduction product: 2-*sec*-butyl-4-nitro-6-aminophenol (6-NH_2-NBP), and a number of unidentified polar compounds; there is evidence that some oxidation of the *sec*-butyl side chain takes place. Similar behavior is found on thin-layer silica-gel chromatoplates in sunlight.

Summary

The use of chemicals for pest control inevitably leaves toxic residues on plant foliage which are subject to photoalteration and other reactions. It appears possible to regulate the degree of persistence of such residues by application of photosensitizers that enhance or stabilize their alteration in sunlight.

The chemical transformation of pesticide surface residues is promoted by ultraviolet light; this transformation is catalyzed by various photosensitizers, including chlorophyll. Benzophenone accelerates the production of photoproducts in solutions containing aldrin, dieldrin, heptachlor, and isodrin; it is also effective in the photolysis of chloroanilines, as is flavin mononucleotide. Certain aromatic amines induce the photolysis of DDT and chlorophyll catalyzes the photodegradation of pyrethroids. However, up to the time when Ivie and Casida reported their work in 1970 and 1971, little was known about whether or not photosensitized reactions occur when two or more residues intermingle on a surface.

Rotenone is the most effective of 45 photosensitizers, pesticide chemicals, and related compounds for conversion of dieldrin to photodieldrin on plant foliage exposed to sunlight and for the photoalteration of various other pesticide chemicals. The photodecomposition of certain pesticide chemicals is, when exposed to sunlight on silica gel TLC plates, accelerated by photosensitizing agents, including rotenone. Twenty-one of 28 known triplet-energy state compounds are active, to some degree, in sensitizing the photoalteration of one or more of the 23 radiolabeled pesticide chemicals studied. Of the 175 unlabeled pesticides tested for such activity against six labeled insecticide chemicals, 40 show some photosensitizing action. The conversion of aldrin, dieldrin, and endrin to photoisomeric derivatives is accelerated by a variety of high triplet-energy state compounds. Some aromatic amines sensitize the photodecomposition of DDT as well as several chlorinated cyclodienes, possibly by formation of charge-transfer complexes. Anthraquinone shows the highest activity of the photosensitizers tested in combination with organophosphorus compounds, methylcarbamates, pyrethroids, dinitrophenol derivatives, piperonyl butoxide, and 2,4-D. Certain of these relationships also hold true when tested on bean leaves exposed to sunlight, rotenone showing the greatest photosensitizing activity against chlorinated cyclodienes. An aqueous suspension of spinach chloroplasts effectively accelerates the photoalteration of 6 of 23 common pesticide chemicals, highest activity being obtained with Mesurol, Zectran, and Dessin. The toxicity of some of the photoproducts to insects and/or mammals and their persistence on foliage are equal to or greater than that of the precursor.

The available data suggest that photosensitizing chemicals might constitute an additional factor useful in manipulating the persistence of pesticide residues. Also, there is evidence that the use of photosensitizers uncovers photoproducts which might be overlooked without them.

Acknowledgment

Work reported herein was partially supported by AEC Contract AT(04-3)-34, Project Agreement No. 113, and by PHS Grant ES 00049.

References

Bandal, S. K.: Ph.D. Thesis, University of California, Berkeley (1971).

Crosby, D. G.: Research in Pesticides. C. O. Chichester, Ed. Academic Press, New York, p. 213 (1965).

Crosby, D. G.: Ann. New York Acad. Sci. 160(1), 82 (1969a).

Crosby, D. G.: Residue Rev. 25, 1 (1969b).

Fischler, H.-M., and F. Korte: Tetrahedron Letters 32, 105 (1969).

Foote, C. S.: Science 162, 963 (1968).

Fox, C. L., and S. W. Bingham: Residue Rev. 29, 105 (1969).

Glenn Jones, G. D.: Ann. Appl. Biol. 48, 352 (1960).

Ivie, G. W., and J.E. Casida: Science 167, 1620 (1970).

Ivie, G. W., and J. E. Casida: J. Agr. Food Chem. 19, 405 (1971a).

Ivie, G. W., and J. E. Casida: J. Agr. Food Chem. 19, 410 (1971b).

Lykken, L.: Agr. Chem. 23(8), 18 (1968).

Lykken, L.: Chemistry 44, 18 (1971).

McGuire, R. R., M. J. Zabik, R. D. Schuetz, and R. D. Flotard: J. Agr. Food Chem. 18, 319 (1970).

Miller, L. L., and R. S. Narang: Science 169, 368 (1970).

Plimmer, J. R., and P. C. Kearney: Paper No. 29, 158th Amer. Chem. Soc. Meet., New York, September (1969).

Rosen, J. D., and W. F. Cary: J. Agr. Food Chem. 16, 536 (1968).

Rosen, J. D., and M. Siewierski: J. Agr. Food Chem. 18, 943 (1970).

Rosen, J. D., M. Siewierski, G. Winnett: J. Agr. Food Chem. 18, 494 (1970).

Rosen, J. D., D. J. Sutherland, and M. A. Q. Khan: J. Agr. Food Chem. 17, 404 (1969).

Van Middelem, C. H.: Organic Pesticides in the Environment, p. 228. Adv. Chem. Series 60. Amer. Chem. Soc., Washington (1966).

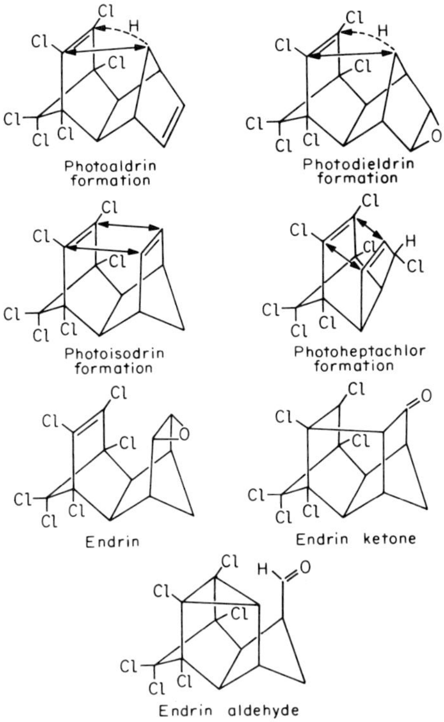

Fig. 1. Structures of aldrin, dieldrin, isodrin, and heptachlor, indicating the reactions involved in photoisomer formation, and of endrin and its two photoisomers, endrin ketone and endrin aldehyde. (From Ivie and Casida, 1971b.)

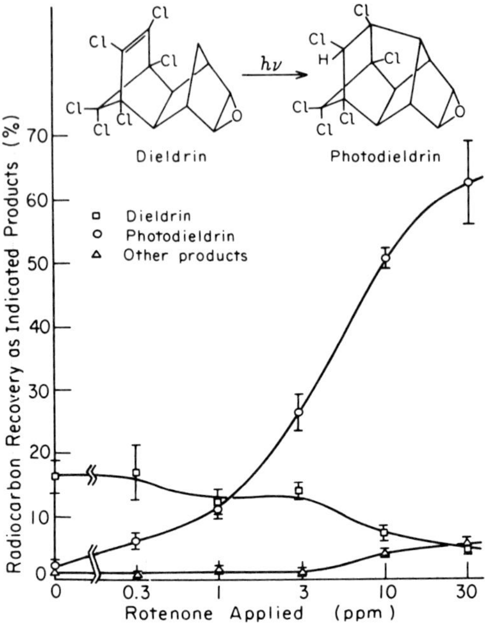

Fig. 2. The effect of concentration of rotenone on the degree of conversion of dieldrin-^{14}C (10 ppm) to photodieldrin-^{14}C under exposure to sunlight on bean leaves. (From Ivie and Casida, 1970.)

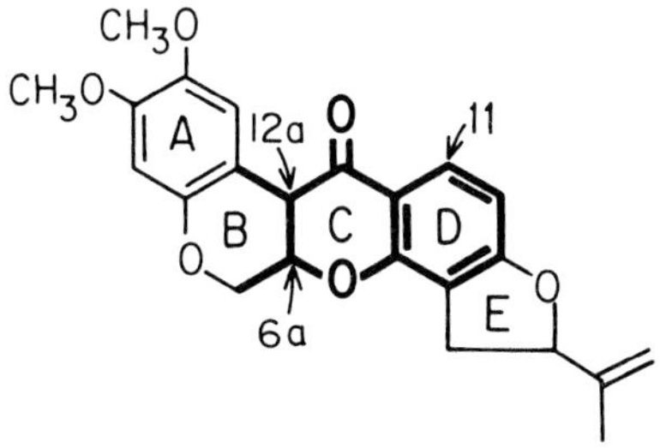

Fig. 3. The chromanone moiety (shown in heavy lines) necessary for high photosensitizing activity in rotenone. (From Ivie and Casida, 1970.)

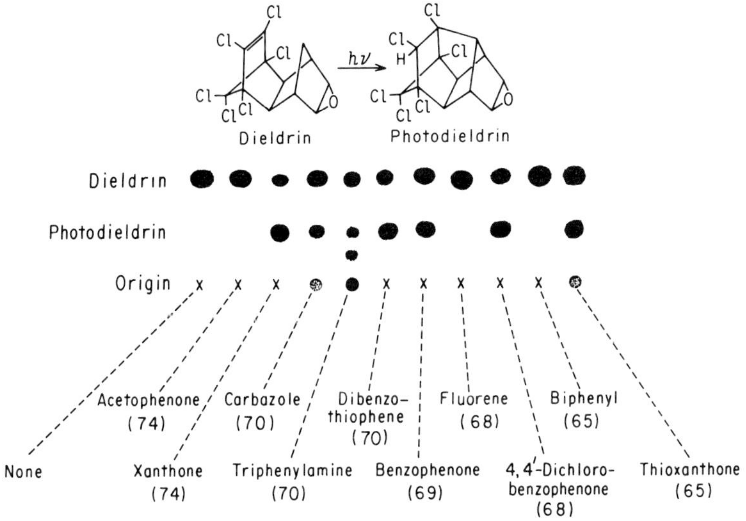

Fig. 4. The sensitizing activity of photosensitizers, with different triplet-energy values, in the alteration of dieldrin exposed to sunlight on silica-gel chromatoplates. (From Ivie and Casida, 1971a.)

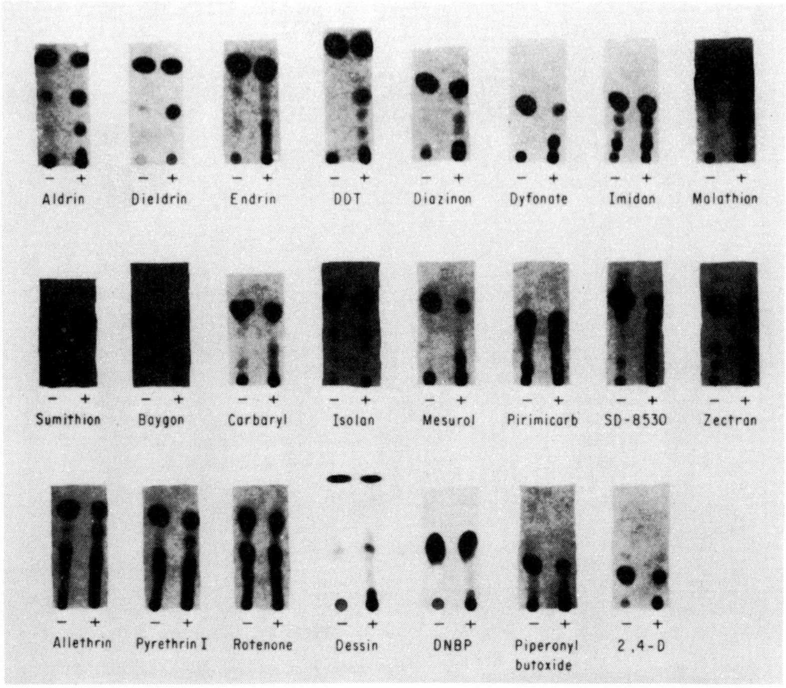

Fig. 5. Composite TLC radioautogram showing the photosensitizing effects of anthraquinone on various radiolabeled pesticide chemicals exposed to sunlight as deposits on silica-gel chromatoplates. (From Ivie and Casida, 1971a.)

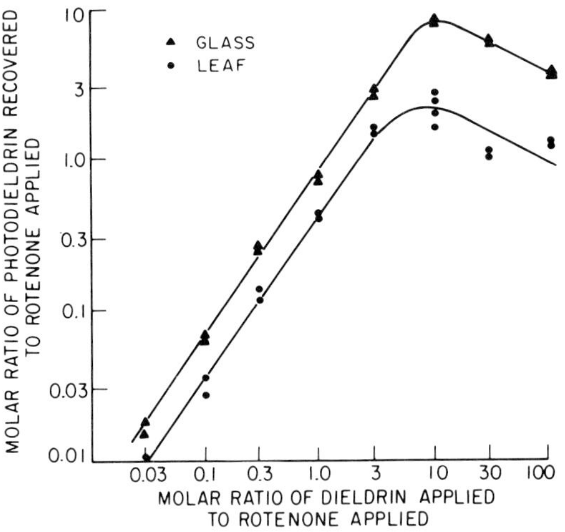

Fig. 6. The relative efficiency of rotenone in catalyzing the photoconversion of dieldrin on a glass and leaf surface exposed to sunlight. (From Ivie and Casida, 1971b.)

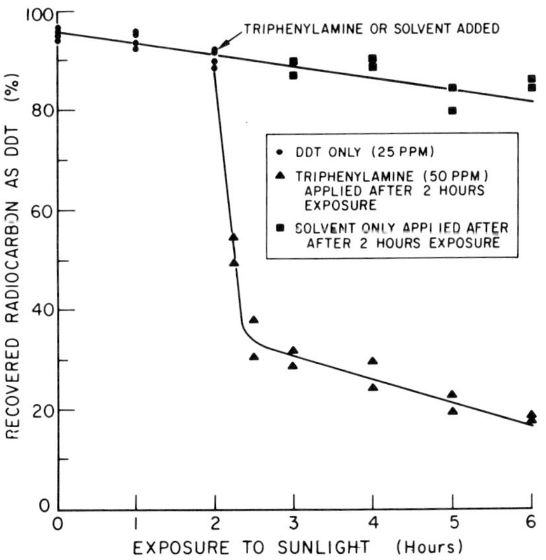

Fig. 7. Effect of Triphenylamine application on the residual persistence of DDT on bean leaves. (From Ivie and Casida, 1971b.)

PART VII

TOXIC EFFECT OF PESTICIDE RESIDUES ON WILDLIFE

ECOSYSTEMIC TRANSFERAL OF PESTICIDES RESIDUES IN AN AQUATIC ENVIRONMENT

Robert L. Rudd and Steven G. Herman

Department of Zoology, University of California
Davis, California

Introduction and History

This paper is not intended to be the usual research report. Instead, it summarizes a well-known case history, and by employing data from that case attempts a general appraisal of the current state of pesticide-ecological investigations. In the process of so doing, we will as well direct ourselves to some particular questions that indicate how the contaminant ecologist works and thinks, and the nature of the information and difficulties with which he works.

The example in question is the insecticidal treatment history of Clear Lake, California. A succinct accounting has been given by us recently (Herman et al., 1969). We quote this accounting because these background details are important in understanding our work and the attitudes of ecologists.

"Clear Lake, Lake County, California, is an eutrophic, fresh-water lake of 42,000 surface acres situated at an elevation of 1325 feet in the Inner Coast Range approximately one hundred miles north of San Francisco. It is shallow (average depth 22 feet) and does not exhibit thermal stratification. The bottom is composed of soft and highly organic ooze. Lake level is regulated by a dam. Cool, wet winters and hot, dry summers characterize the climate of the Clear Lake basin. The waters of Clear Lake are quite turbid, averaging less than three feet Secchi disk visibility throughout the year (Cook et al., 1964). The fish fauna is rich and includes eight native and twelve exotic species (Cook et al., 1966). The lake is highly productive, supporting sizeable recreational and commercial fisheries.

"Since aboriginal times, inhabitants of the Clear Lake shoreline have been aware of the presence in the summer of a small but abundant midge, *Chaoborus astictopus* (Diptera: Chaoboridae), known locally as the Clear Lake gnat. This insect does not compete with man for food or fiber, does not bite, and does not adversely affect the public health. Objection to it results from the fact that the species is highly phototropic. Its creation as a major pest might well be dated to the introduction of the electric light to Lake County.

"Clear Lake was treated with DDD (TDE) three times over a period of nine years in an attempt to control the Clear Lake gnat. From September, 1949 until September, 1957, 120,726 pounds of DDD were applied directly into the lake from moving barges. Final concentrations of the insecticide were calculated to approximate 1 part DDD to 70 million parts of water (0.0143 ppm) after the first application and 1 part DDD to 50 million parts of water (0.020 ppm) after each of the subsequent two treatments. It is now known that coverage was not always uniform, and, in fact, concentrations must have varied widely in different parts of the lake after treatments.

"The second lake treatment took place in September, 1954. One hundred Western grebes, *Aechmophorus occidentalis*, were reported dead the following December. Pathological examination revealed no infectious disease. Additional reports of grebe mortality were received in March, 1955 and December, 1957. Post-mortem examinations were made of specimens submitted to the California Department of Fish and Game; again, no infectious disease was found. Early in 1958, a composite sample of fat from two "sick" Western grebes was found to contain 1600 ppm (wet weight) of DDD. Samples of several species of fish, more grebes, and amphibians were analyzed. All contained DDD. It was concluded that DDD transferal had occurred through the food chain, poisoning the grebes, and that the nesting population of Western grebes at Clear Lake had declined as a result of the DDD treatments.

"Alden H. Miller estimated the pretreatment population of Western grebes on Clear Lake to have been in excess of one thousand pairs. Reproductive failure was evident shortly after the first treatment in 1949. No Western grebe nests were found in 1958 and 1959, in spite of intensive searches by Miller and other workers. Fewer than fifty individuals were seen on the entire lake during the breeding seasons of these years. During the 1960 season, no more than thirty grebes were seen on the lake, and neither nests nor young were seen. Department of Fish and Game personnel noted thirty-eight grebes in the vicinity of a nesting colony in 1961; sixteen nests were found on June 14 but could not be relocated. No young were observed. One juvenile was seen in 1962, and three were reported in 1963. Reproductive success remained low, and in 1966 the present study began with a reevaluation of the problem. Many nests and eggs were found, but only three young were seen, and chlorinated hydrocarbon residues in adults and eggs were high."

Clear Lake is the best studied body of water in California. From our standpoint, however, there are four ingredients of such study that, in combination, are important. These are: a long and continuing history to control the Clear Lake gnat; a more recent but equally intensive limnological and ecological set of studies; practical studies of fisheries management; and, finally, but important, the elevation of the gnat control problem and its biological consequences into the public arena as a controversial _cause celebre_ (most prominently in Rachel Carson's _Silent Spring_, but see also Rudd, 1964).

We have described in essence the "life history" of an induced residue over almost two decades. For the remainder of this paper we direct ourselves to two questions: what is the general state of pesticide ecology, using the Clear Lake example as a basal reference point? And, what are the factors explaining the "recovery" of a population of breeding birds whose reproductive performance had been inhibited for so many years?

The general case needs first to be stated.

A General Appraisal of the Current State of Pesticide-Ecological Studies

The subject of pesticidal intrusions into biological systems has been both heatedly debated and extensively reviewed in recent years. I have reviewed these matters in some detail previously (Rudd and Genelly, 1956; Rudd, 1964) and have published more general accountings recently (Rudd, 1970, 1971). Other important reviews are provided by Chichester (1965), Moore (1967), Woodwell (1967), the American Chemical Society (1969), Miller and Berg (1969), and the U.S. Department of Health, Education and Welfare--the so-called "Mrak Report" (1969). The ecologist finds most of general interest in the Mrak Report and of more theoretical interest in Moore's review. An excellent more specialized overview, based largely on aquatic systems, is by Nicholson (1967).

The number of investigations in the last decade has greatly increased. From the standpoint of the ecologist (exempting the controversial aspects) there seem to be five major categories of investigations.

Confirmation of acute effects in place: Many papers are still being published (and should be) on essentially local species-pesticide interactions.

Confirmation of the widespread occurrence of residues in both organic and inorganic systems: Extensive monitoring systems are based on the fact of occurrence and on the needs to assess changes in levels and to satisfy statutory requirements. The results of such monitoring have also extended spatially and biologically known occurrence of residues (e.g., Risebrough et al., 1968, 1969; Peterle, 1969; Cory et al., 1970; Wolman and Wilson, 1970). In addition, new kinds of contaminants have been recorded. Best known of these are lead, mercury, cadmium, and polychlorinated biphenyls (PCB). In most instances, still, although occurrences and types continue to expand, we do not yet know the biological significance of their presence.

Attempts to place both concentrations and effects in food-chain sequences: There are several earlier descriptions which essentially yield three or four linear linkages. Hickey and his associates have described these patterns in Lake Michigan (e.g., Hickey et al., 1966). We have made similar descriptions in our studies at Clear Lake, California (e.g., Herman et al., 1969). Attempts to describe residual occurrences in more complicated food-webs or in complete community studies are rarer.

More complicated food-web linkages are difficult to assess. We have attempted such descriptions of pesticide dispersal in our Clear Lake studies (Rudd et al., unpublished studies). To our knowledge the best paper of this type yet to appear describes an estuarine community on Long Island Sound in New York (Woodwell et al., 1967).

Attempts to link residue occurrence and inhibition of biological productivity: No one yet has described total community disruption due to pesticide residues. Thus far, two major areas of inhibition have been suggested: the reduction of photosynthesis in marine phytoplankton (e.g., Wurster, 1968) and interference with reproductive processes in fishes and birds. In the latter instance, studies have centered about population declines (e.g., in carnivorous birds, Cade et al., 1968; Wurster and Wingate, 1968; Anderson et al., 1969; Fyfe et al., 1969; Hickey, 1969) or the reproductive mechanisms involved in such declines (e.g., in fishes, Burdick et al., 1964; in birds--particularly egg-shell "thinning"--e.g., Hickey and Anderson,

1968; Fyfe et al., 1969; Gress, 1970; Lincer and Peakall, 1970; Peakall, 1970).

Attempts to model ecological systems: The idea of representing food-web diversity in modular form is not new but the very complexity of the system and the difficulty in acquiring proper input data have precluded proper attempts until relatively recently. We have in our Clear Lake studies constructed simple line and diversity models in graphic form. A beginning has been made on simpler computer regressions within known food-webs.

Equations of first order kinetics may be adaptable to data on pesticide ecology. Eberhardt et al. (1971) have projected such analyses based on Peterle's Ohio marsh studies of DDT uptake. These authors rightly observe that models already exist for use in tracing the fate of radionuclides in ecosystems but that little of this sort of effort exists for the fates of pesticides (see, e.g., Wiegert and Odum, 1967). Phillips (1971, in press) presents both graphic and mathematical models dealing with monitored levels, fates, and translocations of toxic chemicals in the biosphere. (See also Harrison et al., 1970.) Very recently, Metcalf et al. (1971) have developed laboratory "model ecosystems" dealing specifically with the biodegradability and ecological magnification of organochlorine insecticides.

The foregoing examples or types of study represent in our minds only the beginnings of such kinds of modular analysis.

Populations and Contamination Values of Western Grebes

In the above discussion, we delineate the earlier impact of DDD residues on the reproductive success of Western grebes. Essentially, the first phase of impact was on adult birds, followed for almost a decade by no production of young. Three young were seen in 1966. Figure 1 indicates the population of young in the 4-year period 1966-1969. The year 1969 was the first in a decade in which it was commoner to see a pair of mated birds with young than without. Still, in the years indicated, reproductive success must be related.

In September, 1971 we made aerial and surface counts to determine present population numbers. Counts along a 5-mile surface transect yielded ratios of almost precisely 1 young bird per mated pair. Total count by air revealed 615 birds, of which, by calculation, one-third or 205 should have been young. Comparison of other populations indicates that, at this time in development, we should normally expect about 1.7 young per mated pair. Some continuing effect of stored chlorinated hydrocarbon seems still to be manifesting itself.

The inverse correlation of surviving numbers of young grebes and DDD content in eggs is clearly shown in Fig. 1. We do not have residue data for 1971, but we presume that the mean value of DDD in eggs is much lower than the 117.4 ppm mean figure obtained in 1969. TABLE 1 contrasts the years 1967 and 1969, in which interval a turning point in both levels and impact of residues seems to have been reached.

TABLE 1

A Comparison of DDE-DDD Concentration Levels in Western Grebe Eggs in 1967 and 1969 (in ppm)

	DDE		DDD	
	1967	1969	1967	1969
$\overline{X}$	147.1	47.9	298.7	117.4
R	41.0-494.0	9.74-195.0	69.2-1007.0	321-257.0
N	17	28	17	28

Direct accumulation of residues in eggs followed by toxic manifestations as the yolk sac is absorbed in young birds are the presumed bases of hatchling mortality.

Other effects on eggs do occur and doubtless contribute to reduced reproductive success. Shell thickness in Clear Lake eggs (N = 24) averaged 0.33 mm with a range of 0.24 to 0.43 mm. By contrast, a sample of 24 eggs from Eagle Lake in northeastern California, not heavily influenced by pesticide applications, averaged 0.36 mm in thickness with a range of 0.29 to 0.43 mm. A comparison of means (student's t test) shows a statistical significance of difference at the 0.1% level. Curiously, a collection of 25 eggs from Tule Lake (on the California-Oregon border in an agricultural area) showed an average thickness almost precisely the same as those from Clear Lake. Yet, eggs collected at Tule Lake in the period 1893-1940 (before the chlorinated hydrocarbons were used) showed an average thickness of 0.389 mm with a range of 0.31 to 0.46 mm. A 15% reduction in thickness of grebe shells from the pre- to the post-"DDT years" seems indicated. We cannot judge population-wide results of such change in thickness.

Linkage of residues in adult birds with continuing mortality is difficult. Basically, exempting winter immigrations, the breeding population in recent years has ranged between 150-200 pairs. An essential stability is indicated but we do not have information on replacement rates. We do feel, however, that mortality of adult birds at Clear Lake due to high residue loads is likely. The best evidence is provided by two moribund adult female grebes captured at Clear Lake in 1968. One was picked up by hand; the other at the time it was shot was behaving oddly. Both were collected at the windward edge of Clear Lake; both had brood patches; and both were emaciated. Residue figures from these birds as compared with "normals" lend conviction to the view that pesticides (under the special stresses of the breeding season) were responsible for the moribund state. The following comparisons identify DDD residues (wet weight) only in two tissues.

Breast muscle	♀-1	41.1 ppm	♀-2	23.0 ppm
Brain	♀-1	47.9 ppm	♀-2	45.9 ppm
Breast muscle	"Normal" $\bar{x}$ (N=59)			13.35 ppm
Brain	"Normal" $\bar{x}$ (N=34)			3.76 ppm

The residue values in brain tissue are particularly noteworthy. They range well beyond values considered to be lethal in other species of birds.

Trophic Concentrations of DDD-DDE

Biological impoundment (particularly in lipid tissues) and subsequent magnification of pesticide residues in food-chains are now well established phenomena (see Introduction). We do not wish to belabor these points with additional confirming data as we have already described them (Herman et al., 1969; Rudd, 1964).

We do wish to add three features that seem significant to us. The first establishes that DDD and DDE are impounded in noticeable amounts in the bottom sediments of Clear Lake. Bottom samples were taken both by means of an Ekman dredge and a single deep-coring of 42 inches. Dredgings from the top 5 inches of sediment showed values ranging from 0.05 to over 1.0 ppm (wet weight). The deep-core showed, as expected, a vertical stratification of residues ranging from 31 ppb in the top 6 inches, through 19 ppb in the second 6 inches and 1.0 ppb in the third (12-18 inches deep) section. No residues were detectable in the 18-42 inch section. Average value of DDD-DDE residues in all bottom mud samples was 0.8 ppm (dry weight). Single pooled samples of oligochaete worms averaged 3.4 ppm (wet weight). Chaoborid midge larvae from these same sediments had a residue concentration (DDD-DDE) of 0.16 ppm.

Secondly, inflow waters to Clear Lake contain very little residue load, even though orchard soils in the surrounding watershed have sometimes surprisingly high levels. We conclude from points one and two that the DDD applications to the lake for gnat control (plus adventitious additions) are primarily responsible for the residues we have measured and that a continuing recycling and biological impoundment is evidenced.

Thirdly, we suggest that the "traditional" linear method of describing food-chain impoundments of pesticide residue yield to closer analysis of food-web relationships, and, further, that studies of biomass (particularly the residue-impounding lipid fractions) be intensified. Only with these means will the seemingly unusual survivability of pesticide residues in ecosystems be understood.

The Importance of the "Nutrient Shunt"; The Introduction of *Menidia*

After so many years of continuing residual effect on the reproduction of the Western grebe, the 1967-1969 lowering in residue values was startling to us (TABLE 1, Fig. 1). Among the explanations for this phenomenon, the most likely seems to be the introduction of an alien fish in the fall of 1967. This small Atherinid fish (*Menidia audens*) increased in numbers unprecedented in the history of fish introductions. Accompanying this dramatic irruption, profound changes seem to be occurring in the Clear Lake ecosystem (Cook et al., 1964, 1966; Cook, 1968; Cook and Moore, 1970).

Relative to the role that these now-abundant fishes might play in reducing chlorinated hydrocarbon residue loads in grebes, three alternatives (possibly all having influence) seem likely: (a) Grebes eating "silversides" (*Menidia*) acquire a less contaminated diet. From present residue figures, this seems unlikely in terms of fresh fish weight. Mean lipid values, however, have yet to be calculated. They may be higher, indicating fewer grams must be eaten per unit of nutrition. If so, total pesticide intake would be less than if the birds were eating less nutritious fish. (b) Possibly the tremendous biomass of fish has bound pesticides, thus making residues unavailable at initial points in the food-web. (c) The last, and very likely, possibility is that "silversides" have provided a new and important food source for introduced centrarchid fishes normally eaten by grebes, and this, in turn, has provided a more readily available food source that matures more rapidly. Smaller individuals may then be less seriously contaminated. Study continues on this aspect.

All in all, shunting of nutrients or binding of residues offer new thoughts on how to achieve dilution of residue loads and reduction of demonstrable effects in ecosystems.

Summary

One of the best known examples of trophic concentration of persistent pesticide residues occurs in Clear Lake, California. The entire lacustrine ecosystem contains chlorinated hydrocarbon residues, chiefly at the DDT series. Several sources contribute: direct application of DDD to the lake in efforts to control a gnat; inflow waters from surrounding watersheds coursing through agricultural lands; migratory birds, particularly from marine coastal environments; and (presumably) atmospheric and household origins.

The most startling effect of this contamination was total cessation of breeding in a colony of a fish-eating bird, the Western grebe (*Aechmophorus occidentalis*). This inability to reproduce continued for a period of several years following first application of DDD to the lake in 1953. By 1970 our investigations showed reproduction of grebes approaching normalcy. We have, in effect, traced over a 17-year period the life history of an induced residue and its biological effects.

This study has many aspects. We have attempted to trace survivability of residues; their distribution in the food-web, their effects on populations; the concentration ratios following biological magnification in food-chains; and, to a lesser extent, the physiological mechanisms causing the observed reproductive effects in birds. Through the course of time, other contaminants, as they became publicly known, have also been considered.

Residue levels of DDD-DDE from bottom muds average 0.8 ppm. Core samples of these sediments indicate a vertical residue gradient to a level of 18 inches below the top of substratum. Oligochaete worms from these sediments averaged 3.41 ppm DDD-DDE whole body, net weight (pooled). From these base levels follows an increasing stored residue concentration in fishes and birds, sometimes totalling up to 700 ppm in lipid tissues.

Successful introduction of a small Atherinid fish in 1967 seems to be the cause of a sudden average decline in residue levels in the tissues of fish-eating birds. This "dilution" effect may hold promise as a managerial tool in areas with known high persisting residue levels.

Although near-normal reproductive success has returned to the Western grebe colony at Clear Lake, both eggshell thickness and egg weights remain significantly reduced, compared to eggs from other Californian colonies existing in areas removed from agriculture and other contamination sources.

The concept of the "lipid pool," that is, an ecosystemic "storage" mechanism that prevents rapid degradation and residue loss in living systems, is proposed for discussion. Detailed data are presented in support of the several aspects of this paper.

Acknowledgments

Basic support for these studies stems from the Bureau of Sports Fisheries and Wildlife, U.S. Department of Interior (Contract No. 14-16-0008-881). We appreciate this support and are particularly grateful to J. O. Keith of that agency. Dr. S. F. Cook, Jr. and Ronald L. Garrett were major participants in the execution of these studies and we acknowledge them warmly. There have been, as well, contributions and effort from many other individuals and agencies, all gratefully acknowledged.

References

American Chemical Society: Cleaning our environment--the chemical basis for action. Report of the Subcommittee on Environmental Improvement; Committee on Chemistry and Public Affairs, Washington, D.C. (1969).

Anderson, D. W., J. J. Hickey, R. W. Risebrough, D. F. Hughes, and R. E. Christensen: Can. Field Nat. 83, 91 (1969).

Burdick, G. E., E. J. Harris, H. J. Dean, T. M. Walker, J. Skea, and D. Colby: Trans. Amer. Fish. Soc. 93, 127 (1964).

Cade, T. J., C. M. White, and J. R. Haugh: Condor 70, 170 (1968).

Chichester, C. O. (Ed.) Research in Pesticides. Academic Press, New York (1965).

Cook, S. F., Jr. California Vector Views 15, 63 (1968).

Cook, S. F., Jr., J. D. Conners, and R. L. Moore: Ann. Entomol. Soc. Amer. 57, 701 (1964).

Cook, S. F., Jr., and R. L. Moore: Trans. Amer. Fish. Soc. 99, 70 (1970).

Cook, S. F., Jr., R. L. Moore, and J. D. Conners: Wasmann J. Biol. 24, 141 (1966).

Cory, L., P. Fjeld, and W. Serat: Pesticide Monit. J. 3, 204 (1970).

Eberhardt, L. L., R. L. Meeks, and T. J. Peterle: Nature 230, 60 (1971).

Fyfe, R. W., J. Campbell, B. Hayson, and K. Hodson: Can. Field Nat. 83, 191 (1969).

Gress, F. Reproductive status of the California brown pelican in 1970, with notes on breeding biology and natural history. Calif. Fish and Game, Wildl. Manage. Branch Admin. Rep. No. 70-6, 21 p. (1970).

Harrison, H. L., O. L. Louchs, J. W. Mitchell, D. F. Parkhurst, C. R. Tracy, D. G. Watts, and V. J. Yannacone, Jr.: Science 170, 503 (1970).

Hazeltine, W. E.: J. Econ. Entomol. 56, 621 (1963).

Herman, S. G., R. L. Garrett, and R. L. Rudd: Chemical Fallout. Miller and Berg, Eds. C. C. Thomas, Springfield (1969).

Hickey, J. J. Ed. Peregrine falcon populations--their biology and decline. Univ. Wisconsin Press, Madison (1969).

Hickey, J. J., and D. W. Anderson: Science 162, 271 (1968).

Hickey, J. J., J. A. Keith, and F. B. Coon: J. App. Ecol. 3 (Suppl.), 141 (1966).

Lincer, J. L., and D. B. Peakall: Nature 228, 783 (1970).

Metcalf, R. L., G. K. Sangha, and I. P. Kapoor: Environ. Sci. Technol. 5, 709 (1971).

Miller, M. W., and G. G. Berg. Chemical Fallout. Current research on persistent pesticides. C. C. Thomas, Springfield (1969).

Moore, N. W.: Adv. Ecol. Res. 4, 75 (1967).

Nicholson, H. P.: Science 158, 871 (1967).

Peakall, D. B.: Sci. Amer. 222, 72 (1970).

Peterle, T. J.: Nature 224, 620 (1969).

Phillips, J. H.: Monitoring toxic chemicals in the biosphere. Amer. Sci. (1971), in press.

Risebrough, R. W., P. Reiche, D. Peakall, S. Herman, and M. Kirven: Nature 220, 1098 (1968).

Risebrough, R. W., P. Reiche, and H. S. Olcott: Bull. Environ. Contam. Toxicol. 4, 192 (1969).

Rudd, R. L. Pesticides and the living landscape. Univ. Wisconsin Press, Madison (1964).

Rudd, R. L.: California Med. 113, 27 (1970).

Rudd, R. L.: Environment--Resources, Pollution, and Society. W. W. Murdoch, Ed. Sinauer Assoc., Stamford, Conn., p. 279 (1971).

Rudd, R. L., and R. E. Genelly: Calif. Dep. Fish Game, Game Bull. No. 7, 209 p. (1956).

U.S. Department of Health, Education and Welfare. Report of the Secretary's Commission on Pesticides and their relationship to environmental health. (The "Mrak Report.") 677 p. (1969).

Wiegert, R. G., and E. P. Odum: Radionuclide tracer measurement of food-web diversity in nature. Symposium on Radioecology. Proc. 2nd Nat. Symp. Ann Arbor, Mich. (1967).

Wolman, A. A., and A. J. Wilson: Pesticide Monit. J. 4, 8 (1970).

Woodwell, G. M.: Sci. Amer. 216, 24 (1967).

Woodwell, G. M., C. F. Wurster, and P. A. Isaacson: Science 156, 821 (1967).
Wurster, C. F.: Science 159, 1474 (1968).
Wurster, C. F., and D. Wingate: Science 159, 979 (1968).

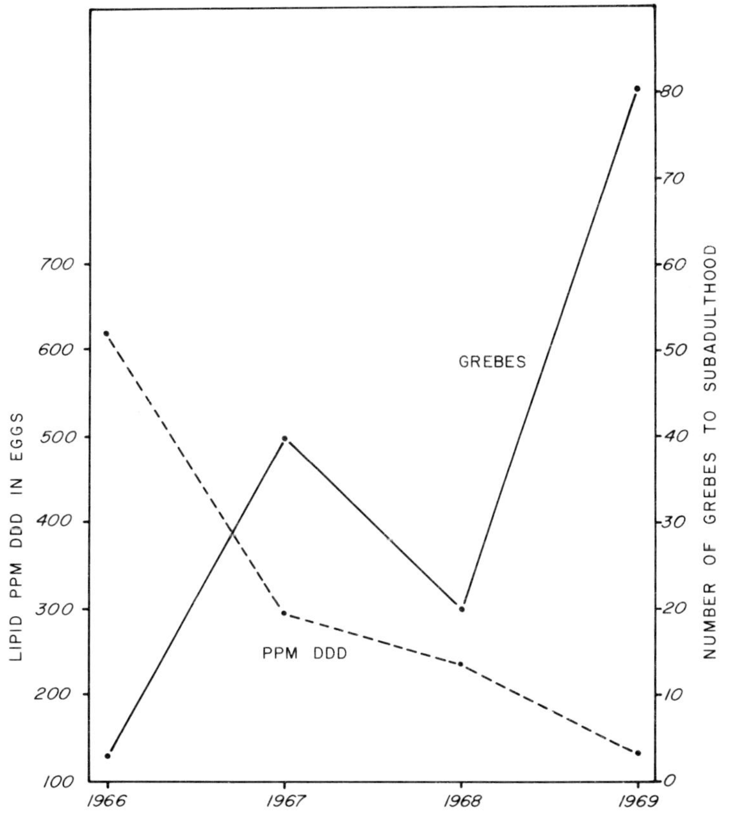

Fig. 1. Clear Lake Western Grebe reproduction and mean lipid DDD concentration in eggs compared temporally. Grebe figures for 1966 and 1967 are gross estimates, figures for 1968 and 1969 result from standard censuses.

EFFECTS OF PESTICIDES ON FISH AND WILDLIFE

Calvin M. Menzie

Office of Environmental Quality
Bureau of Sport Fisheries and Wildlife
U.S. Department of the Interior
Washington, D.C.

In the United States, following large-scale pesticide applications, birds, fish, and mammals in large numbers have been found dead and dying. At the Patuxent Wildlife Research Center, near Washington, D.C., hundreds of wildlife samples have been analyzed for residue content. The observed mortality, while of concern, could be curtailed, but the inconspicuous sub-lethal and long-lasting effects are difficult to determine and assess. We already had indications from our laboratory studies that pesticides could adversely affect reproduction, and we were concerned that the same was occurring in the wild populations. Organochlorine residues were higher in populations of many bird species, and declining populations were coincidental to the large-scale treatments.

Effects on Mammalian Species

With mammals, it has been difficult to assess long-term effects of pesticides. Residue analysis gives positive results, but lack of definitive data makes it difficult to prove more than exposure. In laboratory studies, however, adverse effects have been observed in the reproduction of raccoons fed a diet containing 2 ppm dry weight or 0.67 ppm wet weight of dieldrin. This level is frequently exceeded as residues in food organisms in areas which have been treated with aldrin or dieldrin for control of white-fringed beetle, European chafer, or corn rootworm. In these studies, adult females fed dieldrin had their estrus later than did the controls. Treated raccoons produced fewer litters than did untreated controls. Litter sizes of raccoons fed dieldrin were smaller, and embryonic survival was poorer than untreated raccoons. With males fed dieldrin, sexual development was delayed; the animals were less aggressive, and they grew more slowly and gained less weight between fall and spring than did controls (Morris, 1968, 1969).

Effects on Bird Species

Perhaps because of the large numbers, the wide distribution, and the ease of handling, we know more about birds. Brown pelican eggs were collected from 12 colonies, 10 in Florida and 2 in South Carolina--one egg from each nest. Significant differences in shell thickness and weight among several colonies, with a north-south gradient on the Atlantic Coast, was observed. Thickness increased southward. When compared with eggs collected prior to 1947, there was a decrease of 11.5% in Florida and 15.2% in South Carolina. Data for Florida is not complete, and the population trend could not be determined. In South Carolina, however, the pelican colonies show a downward population trend (Blus, 1970).

Along the Pacific Coast, brown pelicans are suffering severe reproductive failure. Eggs from Anacapa Island, California, exhibited a thinning of 34% over that of normal, and reproduction here has completely collapsed. In Mexican colonies in the Gulf of California, there is near-normal thickness of intact eggs. Residues of collapsed eggs from both areas had high levels of DDT and its metabolites--800 to 1800 ppm, lipid basis. Residues in intact eggs from the Mexican colonies averaged 72 ppm (Keith et al., 1970).

Herring gulls have also been adversely affected. Eggs collected in Wisconsin prior to 1947 showed no significant differences or geographic gradients with respect to shell thickness. In 1964, eggshell thickness was estimated to have decreased by about 32%. DDE residues averaged 202 ppm, and excessive reproductive failure was observed (Hickey and Anderson, 1968).

In Alaska, peregrine falcons have experienced large reductions in eggshell thickness since 1946. The tundra and taiga peregrines now lay eggs with shells as thin as those associated with the declining falcon populations of Great Britain, California, and eastern United States. The number of breeding pairs remained steady through 1969, but the trend in production of young has been down since 1966. In 1970, peregrines did not do well. About 60% of the known aeries failed. It would seem, therefore, that the threshhold has now been exceeded for arctic peregrines (Cade et al., 1971).

In laboratory studies, American kestrels were maintained on a diet containing 2.8 ppm p,p'-DDE. After two years, eggshells were 10% thinner than those of the controls (Wiemeyer and Porter, 1970). In other studies, they were maintained on diets containing 0.84 ppm dieldrin and 4.7 ppm DDT or 0.28 ppm dieldrin and 1.4 ppm DDT. Reproductive failure followed the same pattern among these dosed birds as had been observed in wild raptorial populations--increased egg disappearance and breakage, and eating of eggs by adults. The pesticide levels used in these studies are equal to residues found in raptor food items in the field. The eggshell thickness decreased by at least 10% (Porter and Wiemeyer, 1969).

Studies of raptor eggshells in museums and private collections indicated that colonies were stable when thinning changes of 2-3% were observed. Where the decreases observed were greater than 10%, the data indicated a downward population trend of the colony. The status of the colony could be related to the residues, as well as the shell thickness, of the eggs (Anderson et al., 1969).

As a result of field and laboratory studies, the list of birds that are subject to the eggshell thinning phenomenon has been enlarged to include the American bald eagle, American sparrow hawk, peregrine falcon, mallard and black ducks, brown pelican, cormorant, common murre, ashy petrel, herring gull, ibis, American egret, and coturnix quail. This is not a complete listing. Other birds are constantly being added as the studies are widened.

The mechanism for the phenomenon is not yet clearly understood, but some observations can be made. DDT fed to pigeons caused an increase in thyroid weight and a reduction in colloid content of the follicles. Concomitantly, liver weight increased (Jefferies and French, 1969). Relatively small amounts of DDT (10 ppm) and dieldrin (2 ppm) can induce increased avian metabolism through induction of hepatic enzymes, and the effects of the two compounds are additive (Peakall, 1967, 1969).

I would be remiss, if I did not mention the PCB's. While these materials are not pesticides, their presence is ubiquitous and a concern to all, because of their persistence and their similarity to the organochlorine pesticides. These materials have been in use since the 1930's and appear in residue analyses of birds from many parts of the world (Koeman et al., 1969; Risebrough et al., 1968). Actually, the PCB's are a family of materials marketed as Aroclors in this country, and care should be given in referring to them. Aroclor 1242, at 10 ppm, reduced egg production and hatchability and caused thinning of eggshells when fed to White Leghorn chickens. At this level, Aroclor 1254 did not induce measurable changes in eggshells or reproductive success. At 100 ppm, Aroclor 1254 did cause adverse effects, but Aroclor 1260 did not (Heath et al., 1971). Studies of pelican and cormorant colonies have produced some interesting, but not conclusive, data which suggest that DDE has a greater effect on shell thickness than PCB's. In cormorants, there was a close correlation between eggshell weight and thickness and residues of p, p'-DDE and PCB's. In pelicans, the correlation was only for p, p'-DDE (Anderson et al., 1969). Another interesting observation, but from laboratory studies, was the increased mortality of ducklings from duck hepatitis virus when previously exposed to Aroclor 1254 (Friend and Trainer, 1970).

Effects on Fish and other Aquatic Fauna

The effect of pesticides on aquatic fauna has been no less harsh than that on birds. The highest pesticide residue levels in mollusks are associated with river drainage basins supporting the most intensive agricultural use of pesticides. A similar relationship has been observed with fish. In the Laguna Madres area of Texas, adult spotted seatrout are abundant, but a progressive decline in the number of juveniles has been observed (TABLE 1). DDT and its metabolites in the ovaries of spotted seatrout reached a peak of 8 ppm prior to spawning in September 1968. There was no evidence of successful spawning later that year. One hundred miles away, the level of DDT and its metabolites was 0.2 ppm in seatrout ovaries, and a normal number of young of the year was observed. In the

TABLE 1

Spotted Seatrout Survey - Laguna Madres

Year	No. Juvenile
1964	74/hectare
1965	62
1966	30
1967	(no sample)
1968	6
1969	0.4

Laguna Madres area, seatrout feed primarily on menhaden which have high DDT residues. In other areas, the seatrout feed on shrimp, which generally die before building large residues (Butler, 1969; Butler et al., 1970). These observations are comparable to those previously experienced with lake trout in New York. Roe of lake trout contained 8 ppm DDT. No effects were observed until the yolk sac was absorbed by the fry, at which time all newly hatched fish died (Burdick et al., 1964). In laboratory studies, sub-lethal concentrations of DDT fed to brook trout caused a similar failure of reproduction (Macek, 1968).

At sub-lethal pesticide levels, laboratory studies have shown that there is impairment of mechanisms which are critical to survival. Rainbow trout accumulated residues of 19 ppb endrin, when they were fed on this material for 163 days at a level of 14.5 μg/kg body weight. When the fish were then subjected to forced swimming, the blood cortisol decreased rapidly, and liver glycogenolysis was inhibited. Additional studies yielded observations, the significance of which is not yet fully known (TABLE 2).

TABLE 2

Physiological Effects of Sub-lethal Levels of Endrin in Rainbow Trout

Endrin, μg/kg/day	Decreased significantly
up to 4.3	serum cortisol, Mg, P_i, total protein
14.5 - 43	liver glycogen
43 - 145	serum glucose, osmolality, cholesterol, and weight gain

Dieldrin yielded similar observations. These studies have shown that one of the two ammonia detoxifying systems in the brain of rainbow trout was almost completely inhibited, and high brain ammonium concentrations occurred with killing doses of dieldrin. Chronic dieldrin exposure also affected amino acid metabolism in rainbow trout. Thyroid activity was altered in Coho salmon at levels of 1/1000 the dose which caused slight mortality. Lake trout reacted similarly to chlordane or Aroclor 1248 (one of the

TABLE 3

General Levels of Persistence of Commonly Used Pesticides

Insecticide	Persistence in	
	Water	Fish
Malathion	Variable	$<$ 1 day
Dursban	Yes	$<$ 1 week
Guthion	No	$<$ 1 week
Parathion	No	$<$ 1 week
Diazinon	No	$<$ 1 week
2,4-D	No	$<$ 1 week
Dichlobenil	No	$<$ 2 weeks
Diquat	No	$<$ 3 weeks
Endothal salt	No	$<$ 3 weeks
Simazine	Yes	$<$ 3 weeks
Sodium arsenite	Yes	$>$ 16 weeks
Lindane (γ -BHC)	---	$<$ 2 days
Methoxychlor	Yes	ca. 1 week
Heptachlor	Yes	ca. 1 month
Dieldrin	Yes	ca. 1 month
Toxaphene	Yes	$>$ 6 months
DDD	Yes	$>$ 6 months
DDT	Yes	$>$ 5 months

TABLE 4

Uptake of DDT by the Eastern Oyster

Exposure level	Residue, ppm (whole body)	Bio-magnification factor
10.0 ppb	150.0 ppm	15000
1.0	30.0	30000
0.1	7.0	70000
0.01	0.72	70000
0.0001	0.07	700
Control	0.06	600

TABLE 5

Examples of Biological Magnifications

Organochlorine insecticide	Range of magnification factor for 5 aquatic mollusks
Lindane (γ -BHC)	10 - 250
Heptachlor	250 - 2500
Methoxychlor	300 - 1500
Aldrin	350 - 4500
Endrin	500 - 1250
Dieldrin	700 - 1800
DDT	1200 - 9000

PCB's). Thyroid activity of channel catfish was affected by DDT analogs or isomers, Aroclors 1248 and 1254; bluegill, by Aroclor 1248 and 1254; and rainbow trout, by DDT and dieldrin. The pattern of thyroid response to organochlorine pesticides seems to be as follows: chronic low doses stimulate activity; near-lethal doses suppress activity; intermediate doses first stimulate activity but may become inhibitory if administered long enough (Grant and Mehrle, 1969, 1970; Mehrle, 1971).

Underlying much of our problems has been the process of biological magnification. This process, the accumulation of a material in higher concentration at one trophic level than at the preceding level, is dependent only on three conditions: (1) the material must persist in the environment; (2) the material must be biologically available; and (3) the material must persist in the biological system which assimilates it (TABLE 3). That this process does occur has been amply demonstrated. The organisms that participate are many, and the degree of magnification is variable (TABLES 4, 5, 6) (Butler, 1967).

In northwest Florida, DDT was used to control the stable fly (Stomoxys calcitrans). If all the spray went into the estuary, it was calculated that the concentration would be 0.001 ppm. In this estuary, the flushing rate was slow, and DDT persisted for weeks on silt or plankton after the beach area was sprayed. Analyses showed a degree of magnification (TABLE 7) (Butler, 1969). Other field and laboratory observations have shown similar results (TABLE 8).

Summary

It is probably safe to say that of all pesticides, the organochlorine compounds constitute the most serious threat to our fish and wildlife. Their persistence in the environment and in biological systems permits their accumulation and magnification. Through this process of biological magnification, it has been possible for many

TABLE 6

Average Magnification and Loss for Seven Organochlorine Pesticides

Animal	Factor (after 5 days)	7 Day loss
Soft Clam	3000	90%
Marsh Clam	700	50
Hard Clam	500	75
Asiatic Clam	600	30
E. Oyster	1200	50

TABLE 7

Examples of Biological Magnification in Several Trophic Levels

Organism analyzed	Type of feeder	Residue
Pinfish	Plankton	0.1 - 0.5 ppm
Mullet	Plankton	3 - 10 ppm (Gonads)
Loon	Pinfish	180 ppm (Liver)
Bottlenose Dolphin	Mullet	800 ppm (Blubber)

TABLE 8

Biomagnification

Species	Compound	Exposure level	Time	Factor	Reference
Cutthroat Trout	DDT	0.3 & 0.1 ppm	30 minutes once monthly	6-8	Allison et al., 1964
Mosquito Fish	Endrin	1.0 ppm	25 minutes	10	Ferguson et al., 1966
Fathead Minnow	Endrin	0.00005 ppm	--	10000	Mount & Putnicki, 1966
Croakers	DDT	0.0001 ppm	5 weeks	20000	Hansen, 1966
Trout	DDT	--	--	200	Cope, 1966
	Hept.	--	--	300	Cope, 1966
	Dieldrin	0.0023 ppm	17-23 days	3300	SFWB, 1969[b]
Oysters	Lindane	0.001-0.05 ppm	10 days	60	Wilson, 1966
	Hept.	--	--	17600	Wilson, 1966
	DDT group	--	--	15000	Wilson, 1966
Earthworms	Organochlorines	--	--	9	SFWB, 1969[b]
Pinfish	DDT	0.001 ppm	2 weeks	12000	Hansen, 1966
Crayfish	Mirex	0.86 ppb	54 hours	16860	Ludke et al., 1971
		0.88 ppb (av.)	8 days	71358[a]	Ludke et al., 1971

[a]Digestive glands only.

[b]Bureau of Sport Fisheries and Wildlife (see Reference under Anonymous).

species of birds and some fish to accumulate sufficiently large levels of pesticides so as to threaten their very survival. Some colonies of birds are no longer sustaining themselves. Downward population trends have been noted. Similar observations have been made with fish. There is no longer doubt that many species of wildlife are faced with a serious and extensive threat to their existence. It is too soon to know whether our efforts to reverse this trend are adequate or in time. In the interim, we will continue to follow the trends through our monitoring programs.

References

Allison, D., B. J. Kallman, O. B. Cope, and C. Van Valin: U.S. Bureau of Sport Fisheries and Wildlife, Res. Rep. 64, 30 pp. (1964).

Anderson, D. W., J. J. Hickey, R. W. Risebrough, D. F. Hughes, and R. E. Christensen: Can. Field Naturalist 83, 91 (1969).

Anonymous: Bureau of Sport Fisheries and Wildlife, Division of Wildlife Research. Wildlife Research. Problems, Programs, Progress 1967. Resource Publication 74 (1969).

Blus, L. J.: BioScience 20, 867 (1970).

Burdick, G. E., E. J. Harris, H. J. Dean, T. M. Walker, J. Skea, and D. Colby: Trans. Amer. Fish. Soc. 93, 127 (1964).

Butler, P. A.: National Symposium on Estuarine Pollution, Stanford University, 107 (1967).

Butler, P. A.: BioScience 19, 889 (1969).

Butler, P. A., R. Childress, and A. J. Wilson, Jr.: FAO Technical Conference on Marine Pollution and Its Effects on Living Resources and Fishing, Rome, Italy. December 9-18, 1970, 13 pp. (1970).

Cade, T. J., J. L. Lincer, C. M. White: Science 172, 955 (1971).

Cope, O. B.: J. Appl. Ecol. 3 (Suppl.) 33 (1966).

Ferguson, D. E., J. L. Ludke, and G. G. Murphy: Trans. Amer. Fish. Soc. 95, 335 (1966).

Friend, M., and D. O. Trainer: Science 170, 1314 (1970).

Grant, B. F., and P. M. Mehrle: Pesticide effects on fish endocrine function. Progress in Sport Fishery Research, Bureau of Sport Fisheries and Wildlife. Resource Publication (1969).

Grant, B. F., and P. M. Mehrle: Progress in Sport Fishery Research, Bureau of Sport Fisheries and Wildlife. Resource Publication (1970).

Hansen, D. J.: U.S. Bureau of Commercial Fisheries, Circular 247, 10 (1966).

Heath, R. G., J. W. Spann, J. F. Kreitzer, and C. Vance: Proc. XV Int. Ornithol. Congr., 20 pp. (1971).

Hickey, J. J., and D. W. Anderson: Science 162, 271 (1968).

Jefferies, D. J., and M. C. French: Science 166, 1278 (1969).

Keith, J. O., L. A. Wood, Jr., and E. G. Hunt: Trans. North Amer. Wildl. Natural Resources Conf. 35, 56 (1970).

Koeman, J. H., M. C. T. N. DeBrauw, and R. H. DeVos: Nature 221, 1126 (1969).

Ludke, J. L., M. T. Finley, and C. Lusk: Bull. Environ. Contam. Toxicol. 6, 89 (1971).

Macek, K.: J. Fish. Res. Board Can. 25, 1787 (1968).

Mehrle, P. M.: Amino acid metabolism in rainbow trout, Salmo gairdneri, is affected by chronic dieldrin exposure. Ph.D. dissertation, University of Missouri, Columbia, Missouri (1971).

Morris, J. E.: Insecticide residues in raccoons and some effects of dieldrin on raccoon reproduction. Master's thesis, University of Missouri, Columbia, Missouri (1968).

Morris, J. E.: Division of Wildlife Research of the Bureau of Sport Fisheries and Wildlife 1969. Resource Publication 94, p. 48 (1971).

Mount, D. I., and G. J. Putnicki: Trans. North Amer. Wildl. Natural Resources Conf. 31, 177 (1966).
Peakall, D. B.: Nature 216, 505 (1967).
Peakall, D. B.: Nature 224, 1219 (1969).
Porter, R. D., and S. N. Wiemeyer: Science 165, 199 (1969).
Risebrough, R. W., P. Rieche, D. B. Peakall, S. G. Herman, and M. N. Kirven: Nature 220, 1098 (1968).
Wiemeyer, S. N., and R. D. Porter: Nature 227, 737 (1970).
Wilson, A. J.: U.S. Bureau of Commercial Fisheries, Circular 247, pp. 6-7 (1966).

STUDIES OF INSECTICIDE DETOXICATION IN INVERTEBRATES; AN ENZYMOLOGICAL APPROACH TO THE PROBLEM OF BIOLOGICAL MAGNIFICATION

Tsutomu Nakatsugawa and Peter A. Nelson

Department of Forest Entomology
State University College of Forestry
at Syracuse University
Syracuse, New York

Introduction

Some insecticides as well as other synthetic organic chemicals introduced into the environment are accumulated and concentrated in various species of organisms, causing delayed disturbances of the ecosystem. Cases of biological magnification have been documented and reviewed (Dustman and Stickel, 1969; Upholt and Kraybill, 1969; Edwards, 1970; Pimentel, 1971). While the phenomenon has drawn attention because of its threat to the survival of conspicuous vertebrates including man, the ecosystem involves a large number of inconspicuous invertebrates. A few experimental data indicate the importance of invertebrates in biological magnification of insecticides even though much of our current concept is based on indirect evidence. Proper understanding of the phenomenon would thus require pharmacological information on all the major animal species involved. In this paper we shall consider the phenomenon of biological magnification in relation to the phylogeny of detoxication systems and report briefly on our first attempt at studying detoxication enzymes in non-arthropod invertebrates.

The presence of a higher amount of insecticide

residues in animals of higher trophic levels, together with other circumstantial evidences do indicate occurrence of biological magnification (Hunt and Bischoff, 1960; Meeks, 1968). But, magnification does not necessarily occur at each step of the trophic chain, and insecticide residues may indeed decrease at a higher trophic level. For example, concentration of dieldrin in cod, Gadus morrhua (0.006 ppm), was less than that of its main food, the sand eel (0.016 ppm); dieldrin concentration in planktonic crustacea (0.16 ppm) was higher than in any of the fish examined (Robinson et al., 1967). Dieldrin was not concentrated in the food chain of soil--earthworms--thrushes (Jefferies and Davis, 1966). Stronger experimental evidence for biological magnification is available for some aquatic animals (TABLE 1). Biological magnification is a balance of intake and elimination of a chemical and is a function of various factors. It is mostly chlorinated hydrocarbons among various insecticides that have been associated with this phenomenon. This is the case presumably because these chemicals usually have high affinities toward biological constituents, are not readily biodegradable or yield similarly recalcitrant metabolites and thus are not easily eliminated from animals. It is interesting, however, that parathion and diazinon, usually considered biodegradable, do show a low degree of accumulation. Our understanding of detoxication phenomena and biodegradability of various insecticides is, unfortunately, based on a very limited number of animal species. Although several reviews have appeared dealing with comparative detoxication (Brodie and Maickel, 1962; Adamson, 1967; Williams, 1967; Terriere, 1968), the literature covers only 2 out of the 30 phyla in the animal kingdom, namely, Chordata and Arthropoda. Even within these two phyla, species examined for detoxication are limited. As for Arthropoda, almost all work has been done with insects. Are the so-called biodegradable insecticides really degradable throughout the food web? Information on the distribution of detoxication enzymes in invertebrates is necessary to

TABLE 1

Biological Magnification of Insecticides by Aquatic Animals

Animal	Insecticide	Ambient concentration (ppb)	Magnification factor	Reference
Oyster	DDT	0.1	70,000	Butler, 1964
	chlordane	10	7,300	Wilson, 1965
	lindane	50	60	Wilson, 1965
	methoxychlor	50	5,780	Wilson, 1965
	toxaphene	50	2,920	Wilson, 1965
Fishes	DDT	1	12,000	Hansen, 1966
	DDT	0.1	40,000	Hansen, 1966
Fish (*Fundulus heteroclitus*)				
	diazinon	320	10	Miller et al., 1966
	parathion	120	80	Miller et al., 1966
Mussels	parathion	120	50	Miller et al., 1966

answer this question.

Biological Magnification and the Origin of Detoxication Enzymes

It would be useful here to briefly consider our current concept of detoxication enzymes and their phylogenetic origin in relation to biological magnification. The role of detoxication enzymes is to increase the polarity of xenobiotics. Brodie and Maickel (1962) presented their thought on the origin of detoxication enzymes, particularly microsomal oxidases. With experimental evidence they argued that detoxication enzymes were needed as the animals evolved from aquatic to terrestrial forms. Presumably terrestrial animals had to develop a means of making toxicants more polar to eliminate them with a limited amount of water. Aquatic forms, on the other hand, lacked detoxication enzymes because the drugs could be flushed away. They showed that lipid-soluble drugs are readily eliminated from fishes or frogs to a large volume of water by simple diffusion without detoxication (TABLE 2). After equilibrium there was little difference for each drug between the concentration in the animal and that in the ambient water. The same distribution was obtained if the animals were placed in drug solutions. The latter experiment might be considered a case of "biological non-magnification."

Cases of biological magnification in aquatic animals attest, on the other hand, to the inability of aquatic forms to rid themselves of lipid-soluble molecules. Biological magnification is, however, not an irreversible process. For example (Butler, 1966), when oysters that had accumulated DDT were placed in clean sea water, the DDT level gradually decreased over several days' period. Whether the elimination was due to detoxication was not known. This does indicate, however, that the lipid-soluble molecule could not be quickly dialyzed out of the aquatic animal. Data on plasma half-times of various drugs in a fish have been discussed in relation to gill excretion (Adamson, 1967). The reason for the apparent contradiction among these results might be found in the difference of the so-called lipid-solubility in the two cases.

TABLE 2

Distribution of Drugs between Water and Aquatic Animals (Calculated from Brodie and Maickel, 1962)

Compound	Concentration ratio Animal/Ambient water	
	Fish (*C. auratus*)	Frog (*R. pipiens*)
Aminopyrine	1.00	1.15
Hexobarbital	1.74	1.50
Acetanilide	1.23	--
Phenacetin	1.10	1.19
Thiopental	1.73	1.56
4-Aminoantipyrine	0.97	1.00
Amphetamine	--	1.27

The drugs used by Brodie and Maickel, although lipid-soluble, were much more water-soluble than chlorinated hydrocarbon insecticides (TABLE 3). Affinity of insecticides toward biological constituents as represented by the so-called lipid solubility may thus be a very important factor in biological magnification even in aquatic organisms. It might be that the drugs used by Brodie and Maickel (1962) were lipid-soluble enough to require detoxication by terrestrial animals, but had enough water-solubility to be eliminated through diffusion by aquatic animals.

The work by Brodie and Maickel showed that in those aquatic forms body surface and/or respiratory surface were permeable to lipid-soluble molecules. At least in the cases of many fishes, frogs, and aquatic invertebrates

TABLE 3

"Lipid Solubilities" of some Drugs and Chlorinated Insecticides

Chemical	Solubility[a] in water g/100 ml	Solubility[a] in chloroform g/100 ml	Partition[b] coefficient hexane/water
Aminopyrine	5.6	100	--
Hexobarbital	0.033	soluble	--
Acetanilide	0.54	27	--
Phenacetin	0.076	7.2	--
Chlorpromazine	40	soluble	--
Lindane	--	--	1,730
Aldrin	--	--	100,000
Dieldrin	--	--	36,000
Heptachlor	--	--	111,000
Heptachlor epoxide	--	--	40,000
DDT	--	--	91,000

[a]Calculated from data in Merck Index 8th Ed. Merck & Co., Inc., Rahway, N. J.

[b]Voerman, 1969.

that have a respiratory surface in contact with water, accumulation of insecticides by this mechanism of absorption might be more significant than via food intake when ambient water is contaminated. This is likely when the amount of water that has to come in contact with such surfaces is large due to a high demand for gas exchange. If an insecticide attains a high enough vapor pressure under certain conditions, terrestrial animals would also accumulate the insecticide via respiratory surface. If a chemical has a high affinity to a body constituent, absorption from respiratory surfaces can be very efficient. Such is apparently the case with mercury vapor (Kudsk, 1969). In the absence of water contamination or in the case of terrestrial animals, food is likely to be the most important source of absorption. The actual route of biological magnification at each trophic step has never been thoroughly established. Factors relating to the phenomenon of biological magnification might be summarized as shown in Fig. 1. Since a number of insecticides are highly lipid soluble, detoxication would have a decisive role in governing biological magnification.

Lipid-solubility of insecticides might also be important in the facilitation of the elimination process by detoxication. Earlier, Gaudette and Brodie (1959) showed that microsomal oxidases detoxified only lipid-soluble molecules. Although counter evidence was presented by Mazel and Henderson (1965), the recent work of Martin and Hansch (1971) showed that lipid solubility of substrates is negatively correlated with Km's of the microsomal oxidases.

Although Brodie and Maickel (1962) considered that aquatic animals did not need and, hence, lacked detoxication enzymes, Potter and O'Brien (1964) have shown that parathion can be activated to paraoxon by fish and frog livers as well, thus showing the presence of microsomal oxidases in aquatic vertebrates. Further work has amply demonstrated the presence of microsomal oxidases such as aniline hydroxylase and aminopyrine demethylase in a number of fish (Adamson, 1969). This indicates that the origin of microsomal oxidases is in a lower phylum than Chordata or Arthropoda.

Studies of Detoxication in Non-arthropod Invertebrates

Principles of approach

It is likely that the spectrum of insecticides used in our environment will keep changing in the future. A vast number of insecticides would make it difficult to select a few for studies of detoxication in non-arthropod invertebrates. Fortunately, however, a considerable amount of information is available regarding the patterns of detoxication in vertebrates and insects. Particularly useful is the knowledge on enzymological types of detoxication of various insecticides. Such information might be utilized, at least initially, for the study of invertebrate detoxication.

The detoxication of toxicants has been classified into two phases; the primary phase usually involves oxidative, reductive, and hydrolytic reactions and the secondary phase involves the synthesis of water-soluble conjugates (Williams, 1959). From the enzymological point of view, however, three types of enzymes can be recognized as important mechanisms in the primary metabolism of insecticides (Fig. 2):

Hydrolases - As a class of detoxication enzymes hydrolases have been studied for the longest time. These enzymes are soluble proteins or can be easily solubilized and do not generally require a specific cofactory for activity. They have been studied under such names as esterases, aliesterases, carboxyesterases, amidases, aromatic esterases, A-esterase, phosphatases, etc. Their implications in enzyme induction (Ball et al., 1954) and insecticide resistance (Oppenoorth and Van Asperen, 1960; Matsumura and Brown, 1961) have been shown or postulated.

Glutathione S-transferases - One of the first insecticide detoxication enzymes to be studied was DDT-dehydrochlorinase from the house fly (Sternburg et al.,

1954). Recent work seems to indicate that this enzyme is only one of a group of enzymes that require GSH for activity. Related enzymes metabolize BHC isomers (Ishida and Dahm, 1965), methyl parathion (Fukami and Shishido, 1966), and parathion (Nakatsugawa et al., 1969). These reactions yield, in some cases, stable immediate products by conjugation with GSH, and may be considered as the second phase of metabolism. The enzymes involved, however, do not seem to form activated intermediates that are necessary in other detoxication conjugations (Cohen et al., 1964) and may be treated as the enzymes of primary detoxications. Since the substrate spectrum has not been fully explored, more foreign compounds (xenobiotics) are likely to be found susceptible to this group of enzymes. A recent review covers the related enzymes for detoxication of non-insecticide xenobiotics (Boyland and Chasseaud, 1969).

Microsomal oxidases - Studies by several groups of investigators have shown that a majority of insecticides is subject to metabolism by oxidative enzymes that require $NADPH_2$ and oxygen *in vitro*. The microsomal enzymes metabolize insecticides such as DDT, cyclodienes, carbamates, organophosphates, rotenoids, and pyrethroids, and insecticide synergists or modifiers such as sesamex and SKF 525-A, as well as a variety of pharmaceutical chemicals (Wilkinson, 1968). They have been found important in such phenomena as synergism and antagonism, enzyme induction (Ernster and Orrenius, 1965), and variation of toxicity with species, sex, age, and diet (Kato and Gillette, 1965). A number of reviews are available on this group of enzymes (Hodgson, 1968; Gillette et al., 1969).

Other types of detoxication enzymes may be found in the future. In this connection, a few cases of reductive metabolism are known, but the enzymes have not been well-defined. Thus, the types of detoxication of a particular insecticide in vertebrates and insects can be predicted at least qualitatively. A survey of non-arthropod invertebrates for the types of detoxication enzymes using a few model compounds would similarly be useful in predicting the fate of various insecticides through the food web organism.

The selection of animals for the survey is another problem. Although studies of detoxication have traditionally been conducted with species that are economically or hygienically important, it is probably more meaningful, from an ecological viewpoint, to study the major phyla as defined by Russel-Hunter (1969), i.e., Molusca, Annelida, Arthropoda, and Echinodermata. The definition is based primarily on the relative biomass that each phyla occupy, reflecting the relative flow of solar-energy in the ecosystem.

Detoxication in the earthworm

For our experiments we have chosen the earthworm, Lumbricus terrestris, because of the close phyletic relation of Annelida to Arthropoda. Our primary objective was to find out if the three types of detoxication enzymes are present. Particularly significant in this respect would be the microsomal oxidases because of their seemingly specialized role as detoxication mechanism. For this purpose parathion was selected as a model substrate. Parathion metabolism in vertebrates and insects are quite similar and may be summarized as in Fig. 3. Its initial metabolism is almost exclusively carried out by microsomal oxidases, but further metabolism of one of the initial metabolites, paraoxon, involves the other two types of enzymes (Lewis, 1969; Nakatsugawa et al., 1969; Appleton and Nakatsugawa, unpublished results).

Adult earthworms, weighing 3 to 4 grams, were obtained locally. Parathion was dissolved in a worm Ringer solution and 25 μl was injected into coelomic cavity near the crop and gizzard area. Schematic anatomy of the earthworm is shown in Fig. 4. The worm could tolerate doses up to 10 $\mu g/g$ body weight without noticeable effects. Ethoxy-1-^{14}C or phenyl-1-^{14}C parathion was used to analyze the metabolism. Injected worms were placed individually in 250-ml Erlenmeyer flasks in a humid dark environment. After an appropriate time worms were processed as shown in Fig. 5. There was a great fluctuation in response among individual worms. This might be expected since the age and conditions were not closely controlled. Variable amounts of excretion seemed to occur through the nephridia. The feces also seemed to carry some water-soluble metabolites. Recovery of hexane solubles decreased over several hours and aqueous metabolites in worm extracts increased (Fig. 6). Most metabolism occurred within 6 to 8 hours. Hexane solubles contained only parathion and no paraoxon was detected by thin-layer chromatography. Anion-exchange chromatography (Plapp and Casida, 1958; Nakatsugawa et al., 1969) of aqueous metabolites was run on a reduced scale using 6 mm (I.D.) columns. Recovery of radioactivity through the columns was nearly complete. The chromatograms were different from those obtained with rat urinary metabolites (Fig. 7 and 8). Monoethyl paraoxon was absent. Chromatograms of ethoxy-labeled metabolites contained three peaks. The first (E-1) was identified as diethyl phosphoric acid by extraction, methylation with diazomethane, and gas chromatography on a Carbowax-20M column. The second peak (E-2) was perhaps diethyl phosphorothioic acid; the metabolite co-chromatographed with the authentic standard on the ion-exchanger and also on a thin-layer plate. The third metabolite (E-3) was not identified. Although these results indicated that 4-nitrophenol was produced, chromatograms obtained with phenyl-labeled metabolites were quite different from those for rat metabolites. Since the second peak (P-2) eluted at the same position as diethyl phosphorothioic acid (E-2), both P-2 and E-2 could have been the same metabolites that contained both ethoxy and phenyl groups. Reduction of parathion or paraoxon to amino derivatives by symbiotic microorganisms was considered, but the metabolites did not co-chromatograph with standard amino derivatives on chromatographic plates. Since they could be extracted into ether better from strongly acidic aqueous solutions, it might be an acidic conjugate. This leaves the identity of metabolite E-2 uncertain. Metabolite P-3 might be 4-nitrophenol judged by the position of the peak.

Production of diethyl phosphoric acid *in vivo* hinted the presence of microsomal oxidases since the metabolism must involve a desulfuration step. A clearer picture might emerge from *in vitro* experiments using a liver-like detoxication tissue of the earthworm, if any such organ is present. Various tissues were dissected out and each was incubated for 1 hour in 0.1 ml of an earthworm Ringer solution containing 10^{-5} M ethoxy-labeled parathion. Acidic metabolites unextractable with chloroform at a neutral pH were measured. Low activities were distributed among various tissues. TABLE 4 summarizes the results. Specific activities for low weight tissues are not very reliable due to errors in tissue weights. Comparison of the results with similar data for the housefly clearly shows lower and less localized distribution of detoxication systems in the earthworm.

Front parts of the digestive tract were chosen for further experiments because of ease of collecting and handling. A 20% homogenate of esophagus, crop, and gizzard was prepared in 0.25 M sucrose and 0.1 ml aliquots were incubated with 5×10^{-6} M ethoxy-labeled parathion with or without 5×10^{-4} M $NADPH_2$ in a total volume of 0.4 ml. $NADPH_2$ enhanced the production of acidic metabolites although the activity was low. To examine if endogenous inhibition was suppressing the enzyme activity, the homogenate of the earthworm tissues was added to a rat liver microsomal system. Activity of the rat enzyme, which was about 30 times as high as the worm enzyme on a weight basis, was suppressed by the worm homogenate by 57%. The observed activity of the worm homogenate might, therefore, have been lowered similarly.

Because these experiments did indicate the presence of microsomal oxidases in the earthworm, other known

TABLE 4

Tissue Distribution of Detoxication Enzymes for Ethoxy-1-^{14}C parathion Metabolism

Tissue	Tissue weight (mg) per incubation	Activity (10^{-12} mole acidic metabolites/mg tissue/hr)
Earthworm, *Lumbricus terrestris*		
Crop	11.4	0.40
Gizzard	44.7	0.15
Intestine	72.8	0.06
Brain	1.2	3.14
Nephridia	15.3	0.34
Calciferous glands	2.2	1.88
Gonads	11.5	0.42
Integument	7.2	0.84
Esophagus	28.8	0.43
Housefly, *Musca domestica* (♀)		
Whole gut	5.3	3.2
Foregut	3.0	5.8
Midgut with Malpighian tubules	3.1	22.1
Hindgut	0.45	2.1
Malpighian tubules	0.48	72.9
Ovary	11.7	0.07
Muscle	1.6	0.1
Carcass (scraped)	1.2	1.0
Carcass (unscraped)	3.8	0.4

substrates for microsomal oxidations were examined. Little 0-demethylation of 1-^{14}C-nitroanisole was observed. Oxidative "hydrolysis" of n-propyl paraoxon could not be assessed due to a relatively large amount of non-oxidative hydrolysis, which indicated the presence of phosphatases, however. In contrast, epoxidation of aldrin to dieldrin was surprisingly rapid. The metabolite was measured by using a gas chromatograph equipped with an electron-capture detector. A 120 cm x 2 mm glass column contained chromosorb G coated with 3% OV-17. The epoxidation was greatly enhanced by $NADPH_2$, giving a further indication of microsomal oxidases in the earthworm (TABLE 5).

TABLE 5

Oxidative Metabolism in Earthworm Tissue Homogenates (esophagus + crop +gizzard)

Substrate (products) / Cofactor	Metabolites produced 10^{-12} mole/hr	
	5×10^{-6} M ethoxy-1-^{14}C parathion (acidic metabolites)	2.5×10^{-5} M aldrin (dieldrin)
NADPH	47	1,920
none	10	300

Discussion

In retrospect, use of subcellular fractions of tissue homogenates might have expedited detection of microsomal enzymes in the earthworm. Such an approach might not always be justified for detoxication studies with invertebrates because of interferences by endogenous inhibitors in such homogenates. Assessment of significance *in vivo* of each enzyme would be important also because of other competing pathways.

Our results indicate the presence of microsomal oxidases in Annelida. To the best of our knowledge no enzymological work has been done on detoxication in other non-arthropod invertebrates. Besides their lack of economic and hygienic importance, many factors have probably contributed to the neglect of these phyla. The small size of many invertebrates as well as a low degree of organ differentiation for detoxication may have made detection of enzyme activities difficult. Kapoor et al. (1970) found only a very slow metabolism of methoxychlor in a snail (*Physa* sp.). Use of micro-dissection of tissues, combined with micro *in vitro* incubation systems, and sensitive analytical techniques would be a necessity in work with many invertebrates. Paucity of supporting physiological information on non-arthropod invertebrates is another factor limiting such studies. For example, our preliminary work showed a salt water clam, *Venus mercenaria*, is paralyzed and killed by parathion. Without information on neurotransmitter in the clam, the result could not be used as evidence for microsomal activation.

Only speculation can be made as to the origin of detoxication enzymes in the animal. Similarities of microsomal oxidases for detoxication with enzymes of steroid hormone metabolism (Kuntzman et al., 1966) might indicate a close relation of origin of the detoxication enzymes to the utilization by animals of steroids as pharmacologically active hormones. Although steroids are metabolized in invertebrates there is little evidence for hormonal steroids in non-arthropod invertebrates (Lehoux and Sandor, 1970). Annelids do have sex-related hormones, but their chemical nature is unknown (Clark, 1969).

It is also possible that microsomal oxidases are ubiquitous in all animals. Cytochrome P-450, the terminal component of microsomal oxidases, has been found even in microorganisms (Ambike et al., 1970; Yu and Gunsalus, 1970). Evolution of detoxication enzymes might have involved reorganization of already available components for the specific purpose of detoxication.

Acknowledgment

We gratefully acknowledge assistance of M. Morelli, H. Appleton, R. Smith, and K. Takaki. Part of the work described was supported by grants from the Research Foundation of State University of New York No. 10-7105-A and 10-7105-B.

References

Adamson, R. H.: Fed. Proc. 26, 1047 (1967).
Ambike, S. H., R. M. Baxter, and N. D. Zahid: Phytochem. 9, 1953 (1970).
Ball, W. L., J. W. Sinclair, M. Crevier, and K. Kay: Can. J. Biochem. Physiol. 32, 440 (1954).
Boyland, E., and L. F. Chasseaud: Adv. Enzymol. 32, 173 (1969).
Brodie, B. B., and R. P. Maickel: Proc. Int. Pharmacol. Meet., 1st. 6, 299 (1961, Pub. 1962).
Butler, P. A.: Commercial fisheries investigations, 65-77, U.S. Fish Wildl. Ser. Cir. 226 (1964).
Butler, P. A.: Pesticides in the marine environment, Pesticides in the Environment and their Effects on Wildlife. N. W. Moore, Ed. Suppl. Vol. 3. J. Appl. Ecol. Blackwell Sci. Pub., Oxford (1966).
Clark, R. B.: Gen. Comp. Endocrinol. Suppl. 2, 572 (1969).
Cohen, A. J., J. N. Smith, and H. Turbert: Biochem. J. 90, 457 (1964).
Dustman, E. H., and L. F. Stickel: Biological Effects of Pesticides in Mammalian Systems. H. F. Kraybill, Ed. Ann. N.Y. Acad. Sci. 160, p. 162 (1969).
Edwards, C. A.: Crit. Rev. Environ. Control 1, 7 (1970).
Ernster, L., and S. Orrenins: Fed. Proc. 24, 1190 (1965).

Fukami, J., and T. Shishido: J. Econ. Entomol. 59, 1338 (1966).

Gaudette, L. E., and B. B. Brodie: Biochem. Pharmacol. 2, 89 (1959).

Gillette, J. R., A. C. Conney, G. J. Cosmides, R. W. Estabrook, J. R. Fouts, and G. J. Mannering, Eds. Microsomes and Drug Oxidations. Academic Press, New York, N.Y. (1969).

Hansen, D. J.: Indicator organisms--fish, p. 10-11, Annual Report of the Bureau of Commercial Fisheries Biological Laboratory, Gulfbreeze, Fla. The Fiscal Year Ending June 30, 1965. U.S. Bur. Comm. Fish. Cir. 247 (1966).

Hodgson, E., Ed.: Enzymatic Oxidations of Toxicants. North Carolina State University, Raleigh (1968).

Hunt, E. G., and A. I. Bischoff: Calif. Fish Game 46, 91 (1960).

Ishida, M., and P. A. Dahm: J. Econ. Entomol. 58, 383 (1965).

Jefferies, D. J., and B. N. K. Davis: J. Wildl. Manage. 32, 441 (1966).

Kapoor, I. P., R. L. Metcalf, R. F. Nystrom, and G. K. Sangha: J. Agr. Food Chem. 18, 1145 (1970).

Kato, R., and J. R. Gillette: J. Pharmacol. Exp. Ther. 150, 279 (1965).

Kudsk, F. N.: Acta Pharmacol. Toxicol. 27, 149 (1969).

Kuntzman, R., R. Welch, and A. H. Conney: Advances in Enzyme Regulation. G. Weber, Ed. Pergamon Press, N.Y. 4, 149 (1966).

Lehoux, J-G., and T. Sandor: Steroids 16, 141 (1970).

Lewis, J. B.: Nature 224, 917 (1969).

Martin, Y. C., and C. Hansch: J. Med. Chem. 14, 777 (1971).

Matsumura, F., and A. W. A. Brown: J. Econ. Entomol. 54, 11176 (1961).

Mazel, P., and J. F. Henderson: Biochem. Pharmacol. 14, 92 (1965).

Meeks, R. L.: J. Wildl. Manage. 32, 376 (1968).

Miller, C. W., B. M. Zuckerman, and A. J. Charig: Trans. Amer. Fish. Soc. 95, 345 (1966).

Nakatsugawa, T., N. M. Tolman, and P. A. Dahm: Biochem. Pharmacol. 18, 1103 (1969).

Oppenoorth, F. J., and K. VanAsperen. Science 132, 298 (1960).

Pimentel, D.: Ecological Effects of Pesticides on Non-Target Species. Executive Off. of the President. Off. Sci. Technol. (1971).

Plapp, F. W., and J. E. Casida: Anal. Chem. 30, 1622 (1958).

Potter, J. L., and R. D. O'Brien: Science 144, 55 (1964).

Robinson, J., A. Richardson, A. N. Crabtree, J. C. Coulson, and G. R. Potts: Nature 214, 1307 (1967).

Russel-Hunter, W. D.: A Biology of Higher Invertebrates. Macmillan, New York (1969).

Sternburg, J., C. W. Kearns, and H. Moorefield: J. Agr. Food Chem. 2, 1125 (1954).

Terriere, L. C. Enzymatic Oxidations of Toxicants. E. Hodgson, Ed. North Carolina State University, Raleigh, p. 175 (1968).

Upholt, W. M., and H. F. Kraybill. Biological Effects of Pesticides in Mammalian Systems. H. F. Kraybill, Ed. Ann. N.Y. Acad. Sci. 160, p. 55 (1969).

Voerman, S. Bull. Environ. Contam. Toxicol. 4, 64 (1969).

Wilkinson, C. F. Enzymatic Oxidations of Toxicants. E. Hodgson, Ed. North Carolina State University, Raleigh, p. 113 (1968).

Williams, R. T. Detoxication Mechanisms (2nd Ed.) Chapman & Hall, London (1959).

Williams, R. T.: Fed. Proc. 26, 1029 (1967).

Wilson, A. J. Chemical assays 6-7. Annual Report of the Bureau of Commercial Fisheries Biological Laboratory, Gulfbreeze, Fla. The Fiscal Year Ending June 30, 1965. U.S. Bur. Comm. Fish. Cir. 247 (1965).

Yu, C-A., I. C. Gunsalus. Biochem. Biophys. Res. Comm. 40, 1431 (1970).

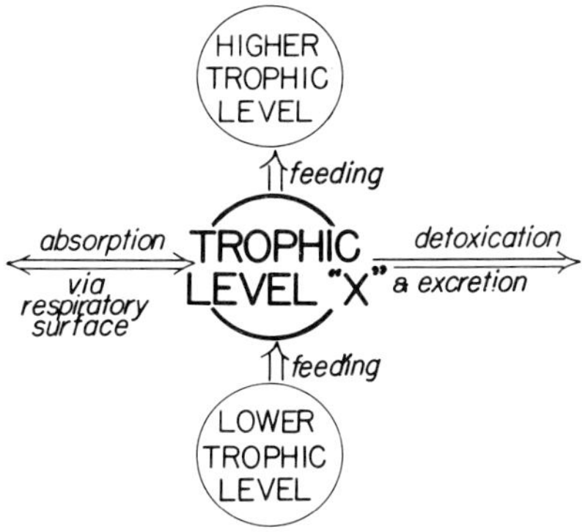

Fig. 1. Factors involved in biological magnification.

HYDROLASES	esterases, carboxyesterases, amidases, aliesterases, A-esterase, aromatic esterases, phosphatases
GLUTATHIONE S-TRANSFERASES	DDT-dehydrochlorinase, BHC(dechlorination), methyl parathion (demethylation)
MICROSOMAL OXIDASES	DDT, carbamates, rotenoids, pyrethroids,(hydroxylations); cyclodienes (epoxidations); O-P's (activation & "hydrolysis")

Fig. 2. Three types of enzymes for insecticide detoxication.

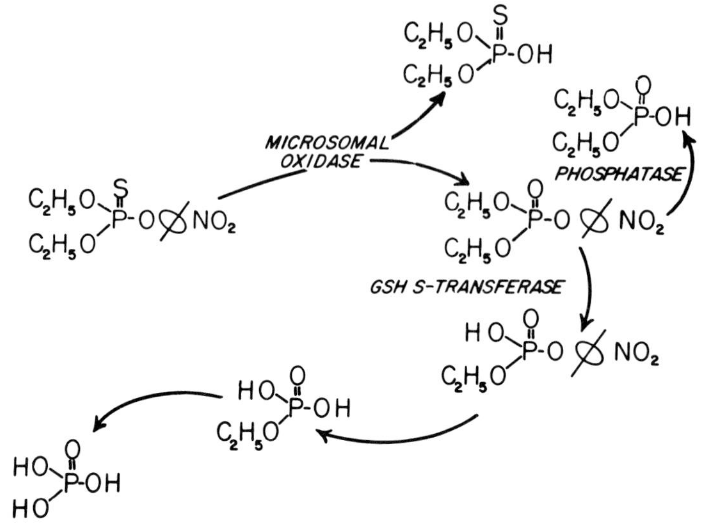

Fig. 3. Parathion metabolism in mammals and insects.

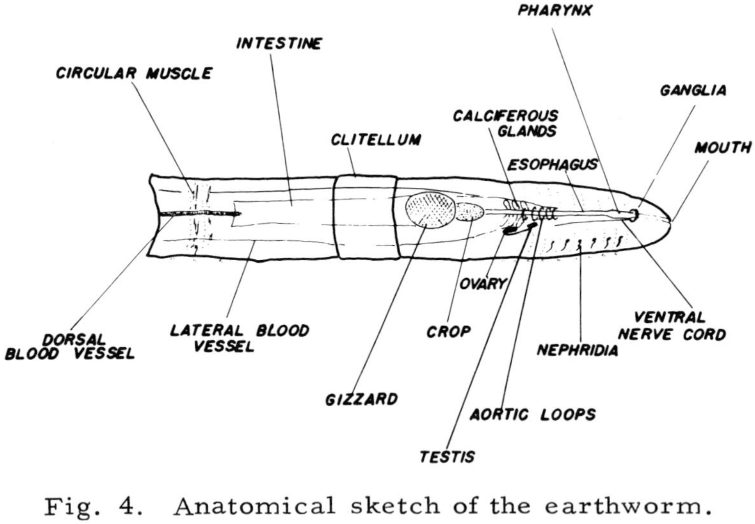

Fig. 4. Anatomical sketch of the earthworm.

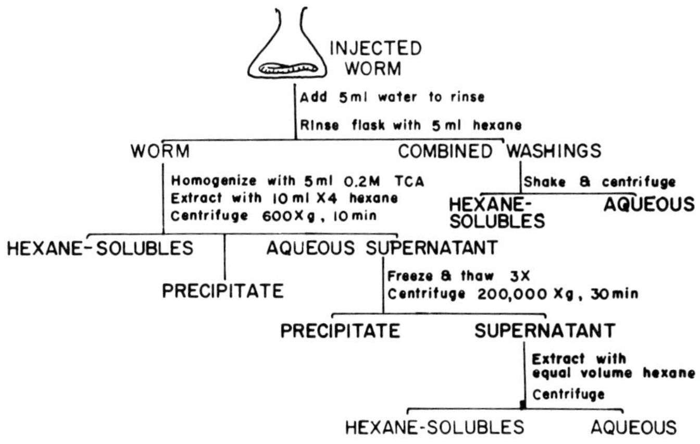

Fig. 5. Flow chart for analyses of parathion metabolism in the earthworm.

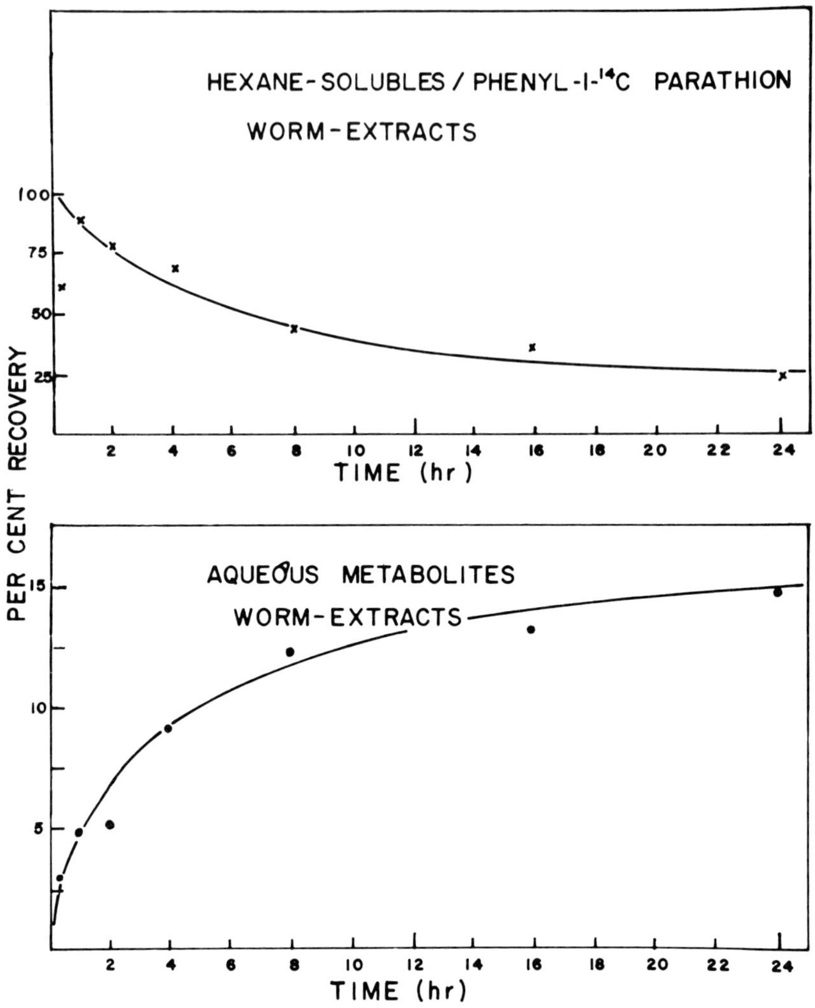

Fig. 6. Time course of phenyl-1-^{14}C parathion metabolism in the earthworm.

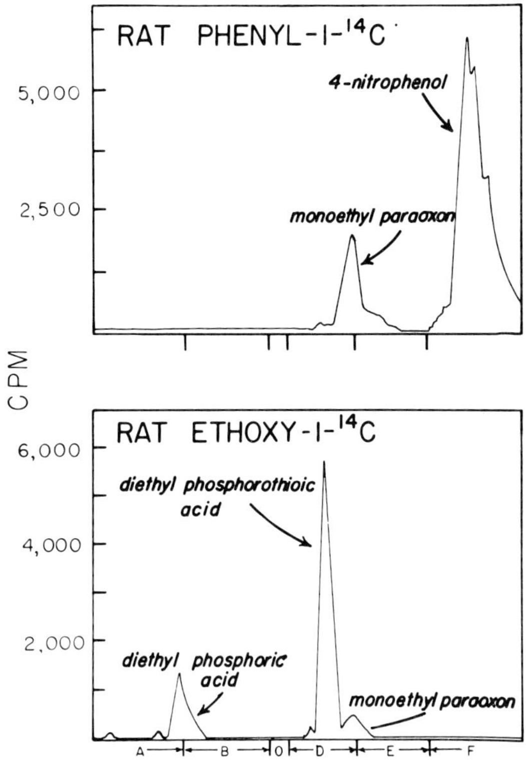

Fig. 7. Anion-exchange chromatograms of parathion metabolites from rat urine.

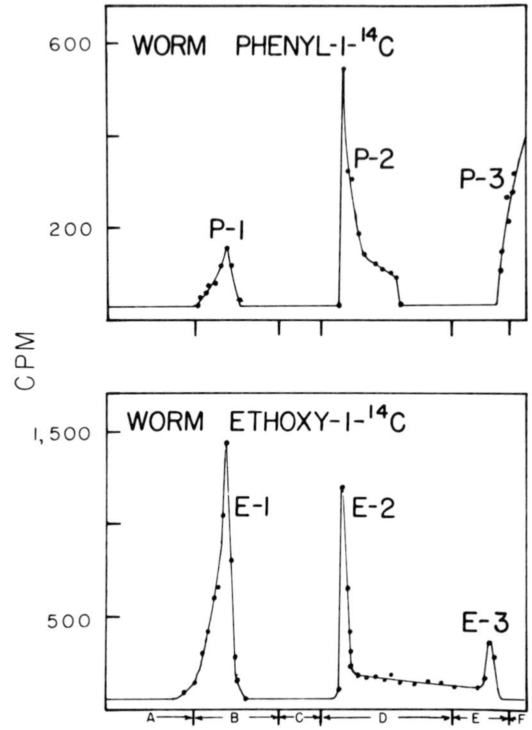

Fig. 8. Anion-exchange chromatograms of parathion metabolites from the earthworm.

BIOLOGICAL EFFECTS OF TOXIC PESTICIDAL CONTAMINANTS AND TERMINAL RESIDUES

Fumio Matsumura

Department of Entomology, University of Wisconsin
Madison, Wisconsin U.S.A.

Introduction

Pesticides introduced into the environment undergo various translocation and transformation processes. The most important factor controlling these processes is the chemical nature of the pesticide itself. Chemical stability, affinity to lipids and other biological matter, volatility, water solubility, etc. are examples of such chemical characteristics of pesticides which play important roles in determining the fate of such compounds in the environment. Since the main hazards of pesticides as environmental contaminants are through their action upon biological systems, it becomes increasingly important to study the properties of pesticides which result in biological reactivity, including the mechanisms of biological magnifications.

Among the chemical characteristics of pesticides, there seem to be three factors which are extremely pertinent to the long-term environmental hazard: the biological affinity, stability, and biological reactivity of the chemicals. It is important to mention here that the pesticides which give rise to environmental conversion products with such characteristics as mentioned above should also be included in this category.

For example, one often finds an equally or more toxic metabolic product appearing after the original compound has disappeared. Formation of dieldrin from aldrin, or photodieldrin from dieldrin by the action of both sunlight (Robinson et al., 1966; Rosen et al. 1966) and microorganisms (Matsumura et al., 1970) are good examples. Whatever the processes involved in changing the chemical structure of pesticides, as long as the final stable form happens to cause eventual biological hazards, the pesticide in question must be considered as a potential environmental contaminant.

"Terminal residues" can be defined as the chemicals which accumulate in biological material in the environment as a result of the introduction of pesticides to the environments. These include the original stable pesticide, conversion products, and the chemical impurities in the crude pesticidal preparations that remain in the environment longer than the pesticidal principles (e.g., beta-BHC, chlorinated dibenzo-p-dioxins, etc.). While all the general criteria of biological affinity, stability, and biological activities are applicable in determining any terminal residues, it is still not easy to assess the importance of such terminal residues in relation to the final magnitude of environmental hazardousness, for biological reactivity is an extremely difficult item to define.

Commonly Occurring Terminal Residues in the Environment

DDT analogs - The most important residues arising from the use of commercially available DDT formulations are p,p'-DDT and o,p-DDT, the latter being present as an impurity. Also important is DDE which is the major metabolic product in animals. There is a report that a marine diatom species (*Cylindrotheca closterium*) exclusively produced DDE from DDT (Keil and Priester, 1969). In soils and water, TDE forms from DDT (Fig. 1) as the major conversion product (Finley and Pillmore, 1963), mainly as a result of microbial action, while the presence of reduced porphyrins aqueous solutions alone are known to perform this reaction (Miskus et al., 1965). Other DDT metabolites which are produced in abundance are DDNS and DDOH (by microorganisms from soil and water--Matsumura et al., 1971a). Three marine and fresh water algae species produced DDOH as the major metabolic product (Patil et al., 1972). The compound 4,4'-dichlorobenzophenone is also a commonly occurring residue in the environment. It can form as a result of photochemical reaction in the presence of atmospheric oxygen (Fleck, 1949) from DDT, or as a result of microbial action from chlorobenzilate and chloropropylate, two widely used acaricides.

BHC - Gamma-BHC (lindane) is the actual insecticidal principle of BHC. Aside from gamma-BHC perhaps the most important terminal residue arising from the use of BHC is beta-BHC (Goto, 1971). This isomer appears to be the most stable one, among others, and is the factor causing the eventual increase of beta-BHC in the environment, in comparison to other isomers. By using a housefly enzyme, Ishida and Dahm (1965), for instance, found that the order of metabolic susceptibility among the BHC isomers and derivatives was delta-PCCH (pentachlorocyclohexene) > gamma-PCCH > alpha-BHC > gamma-BHC > delta-BHC. Beta-BHC was nondegradable by this system. PCCH isomers can also be important metabolic products (Fig. 2). Yule et al. (1967) found gamma-PCCH to be the major metabolic product of gamma-BHC by microorganisms.

There are possibilities that other conversion products are also important. Sethunathan et al. (1969) found that a *Clostridium* sp. from rice fields produced less apolar products than PCCH as the major metabolic product. In our laboratory we found that a strain of *Pseudomonas putida* quickly converted gamma-BHC to tetrachlorocyclohexene along with gamma-PCCH (Matsumura et al., unpublished data). Allan (1955) also detected the presence of 1,2,3,5-tetrachlorobenzene among the metabolic products of BHC. Tsukano and Kobayashi (1971) recently found gamma-BTC (r-3,4,5,6-tetrachlorocyclo-1-cyclohexene) among the degradation products of r-BHC in soil.

<u>Cyclodiene insecticides</u> - The insecticides belonging to this group of chemicals are generally stable. The most commonly occurring reaction in the environment to these insecticides are epoxydation reactions to produce, for instance, dieldrin from aldrin, heptachlor epoxide from heptachlor, and endrin from isodrin. Although the reaction has been known to take place for the unsaturated bonding of cyclodiene molecules, the recent findings by two independent research groups (Schwemmer et al., 1970; Lawrence et al., 1970) indicate that it can occur with saturated and chlorinated bondings such as formation of oxochlordane from chlordane. The expoxidation process of cyclodienes is an exceedingly common oxidation process in many biological systems. Some microorganisms, in addition, appear to be capable of reducing dieldrin to aldrin (Matsumura et al., 1968). These oxidation products are generally more toxic to biological systems. The basic metabolic route of dieldrin in microorganisms appears to be through oxidation, hydrolysis of the epoxy ring, and dechlorination (Fig. 3).

Another characteristic reaction associated with cyclodienes is intramolecular rearrangement reactions to form isomers. Thus dieldrin can be rearranged by the action of sunlight (Robinson et al., 1966; Rosen et al., 1966) and microorganisms (Matsumura et al., 1970) to form an intermolecular bridge to become photodieldrin. Rosen and Sutherland (1967) also showed that a similar photochemical reaction can take place with aldrin to form photoaldrin. These reaction products can be found in nature where action of sunlight is expected; i.e., such as the plant surfaces (Robinson et al., 1966; Weisgerber et al., 1970) and soils (Lichtenstein et al., 1970). Other rearrangement products formed by the action of sunlight are ketoendrin (Rosen et al., 1966), endrin aldehyde and alcohol (Zabik et al., 1971), photoheptachlor and photoisodrin (Rosen et al., 1969). Microorganisms are also capable of forming ketoendrin and aldehydes from endrin

(Matsumura et al., 1971b). A strain of soil microorganism, *Pseudomonas sp.*, from the loading dock soil of a pesticide formulator can convert dieldrin to bridged and non-bridged ketodieldrin (Fig. 4; Matsumura et al., 1968). All the above conversion products can be potential terminal residues as they are physically and chemically similar to the parent molecules. In addition, cyclodienes are known to be degraded further, sometimes through different routes. In most cases such degradation reactions yield more polar compounds than the parent molecules. The epoxy ring can be opened by the action of microorganisms; e.g., formation of aldrin diol from dieldrin (Matsumura et al., 1968; Wedemyer, 1968). It has also been found that at least one species of fungus, *Trichoderma koningi*, can degrade the chlorinated ring of dieldrin to carbon dioxide (Bixby et al., 1971), as shown in Fig. 5. 1-Hydroxy-2,3-epoxy-chlordene can be produced in the soil by the combined action of physiochemical and microbial reactions (Miles et al., 1969).

Organophosphate and carbamate insecticides - This group of pesticides are generally less stable, the most common biochemical degradation processes being through hydrolytic and oxidative systems. In the case of microorganisms, the oxidative processes which are important in higher animals in degrading these compounds are less frequently observed. This is probably because of the lack of defined mixed function oxidase systems in microorganisms. While the processes of oxidation of thiophosphates to corresponding phosphates (e.g., parathion to paraoxon, malathion to malaoxon) represent true activation of the organophosphate insecticides (the latter being more toxic than the parent compounds), the products tend to be less stable than the corresponding thiophosphates themselves. Generally speaking, most of the metabolic products of organophosphates are much more polar than the parent compounds. Some stable metabolic or conversion products are known. For instance, one of the major microbial degradation products is amino-parathion (*O*,*O*-diethyl-*O*-

p-nitrophenyl thiophosphate) (Cook, 1957; Ahmed et al., 1958). Photolytic degradation reaction of azinphosmethyl (Guthion), O,O-dimethyl S-(4-oxo-1,2,3-benzotriazin-3-(4H)-yl methyl)) phosphorodithioate gives rise to 4-oxo-3,4-dihydro-1,2,3-benzotriazin upon irradiation in hexane, indicating utilization of two hydrogens from the solvent (Kurihara et al., 1966).

Hydrolytic end products of carbamate insecticides are often more apolar than the original carbamates themselves. Thus photolytic reaction of carbaryl (Sevin) in the presence of water produces 1-naphthol which is less polar than carbaryl (Crosby, 1969). A similar reaction has been reported by Pape et al. (1970) with an N-methyl phenylcarbamate (C-8353, Ciba and Geigy Corp.) which yields the corresponding substituted phenol upon photolysis. Unlike organophosphates, however, substitutes on the phenol rings of the insecticidal carbamates are not electrophilic (e.g., unlike chlorine and nitro groups for organophosphates). These carbamate-derived phenols have not been known to cause environmental problems.

Aromatic herbicides - Phenylureas and phenylcarbamates do not persist long in the soil. The rate of degradation appears to be favored in relatively organic-poor soils which do not remove the pesticides from circulation for microbial attack (Geissbuhler et al., 1963). The general microbial degradation pathway is through hydrolytic N-demethylation reactions for phenyl ureas and hydrolysis reactions for phenylcarbamate (Kearney, 1965). Ring-substituted anilines are the resulting end products of such microbial attacks.

Plimmer and Kearney (1969) showed that 3,3',4,4'-tetrachloroazobenzene can form as a result of UV irradiation of 3,4-dichloroaniline. Rosen et al. (1970, 1971) demonstrated that in the presence of a photosensitizer (FMN or riboflavin-5-phosphate salt) and ultraviolet substituted anilinoazobenzenes (a trimer) can form. The latter authors, for instance, synthesized 4-(3,4-dichloroanilino)-3,3',4'-trichloroazobenzene which is to be formed from 3,4-dichloroaniline. The reaction for 4-chloroaniline can proceed much faster to form 4-chloro-4'-(4-chloroaniline) azobenzene.

Phenoxyalkanoate herbicides - This group of chemicals are also readily degraded in the soil. Generally persistence ranges from 1 (2,4-D) to 2 months (MCP) to 5 months for 2,4,5-T. For the herbicides themselves, the ether linkage between the ring and to the fatty acid is susceptible to microbial attack (MacRae and Alexander, 1963). Thus the end products are substituted phenols which can be further ring hydroxylated. Similar reactions have been observed when 2,4-D was irradiated by UV light (Crosby and Tutass, 1966).

Polymerization of substituted aromatics by the aid aid of ultraviolet radiation along with other initial structural changes can result in complex and often unexpected reaction products. An excellent example of the interaction of different photochemical reactions on a single pesticidal compound can be found in the work by Munakata and Kuwahara (1969), who discovered numerous reaction products including three dimers and one trimer from pentachlorophenol upon sunlight irradiation.

Fungicides - The most recent event in the search for toxic terminal residues is the finding that ethylene thiourea (ETU) can be carcinogenic to the thyroid gland or rats. ETU can be formed from ethylene bis(dithiocarbamate) fungicides such as Maneb, Nabam, and Zineb. For instance, Czegledi-Janko and Hallo (1967) found that when stored in the dark Zineb and Maneb underwent decomposition to ethylenethiourea ethylene thiuram mono- and disulfide, and other degradation products (Fig. 6). Engst and Schnaak (1967) also found the above two compounds among the decomposition products of Maneb and Zineb. The carcinogenicity of ETU has recently been found by the Food and Drug Administration (Wodicka, 1971). It is not certain, however, whether ETU poses a health threat at present.

The danger and problems associated with the use of methylmercuries as fungicides are well known. Cases of accidental poisoning by the consumption of grain which had been treated with methylmercury fungicides have been reported in the United States. Less known is the fate of other organomercury pesticides. Phenylmercuric acetate, for instance, has been used extensively in the United States as a fungicide and slimicide. It has also been used in Japan in the past as an agent to control rice blast, the major disease of rice in Japan. The most crucial question is the extent to which phenylmercuric compounds contribute to the formation of methylmercury derivatives. In the U.S. the source of contamination of many rivers has been suspected to be by phenylmercuries, because of the extensive use of phenylmercuric acetate as a slime treatment agent in paper mills. Studies in our laboratory (Matsumura et al., 1971c) have shown that soil and aquatic microorganisms can convert phenylmercuric acetate into di-phenylmercury and another apolar phenylmercury compound. In none of the cases has methylmercury been observed to be produced from phenylmercuric acetate by microbial action (Fig. 7). Diphenylmercury, which was the most apolar member of the metabolites, can accumulate in biological systems.

Biological Effects and Implications

Stable pesticidal residues and contaminants become environmental problems when they show strong biological affinities and adverse effects upon one or more biological systems. While there is no set rule to estimate the degrees of biological affinity, it can be said that those which have been frequently encountered in animals representing a high trophic level, representing enormous degrees of concentration above the environmental concentration (e.g., sea water and DDT or mercury concentrations in marine fish), should be regarded as the ones with high affinities.

Acute toxicities - Although acute toxicities may not be the best and most accurate index of toxicities and long-term potential hazards of these terminal residues, they serve as the logical assessment for the initial toxicity differences among various biological systems. For instance, the results of extensive susceptibility tests of nine insecticides on 12 fish species indicated that the salmonids are the most susceptible fish to all insecticides tested. The ictalurids and cyprinids were, on the other hand, the least susceptible fish (Macek and McAllister, 1970).

The results shown in Fig. 8 indicate the general susceptibilities of brine shrimp, *Artemia salina*, to various terminal residues and analogs of DDT (Tsai et al., 1971). It can be seen that these analogs, though many of them have been regarded as non-insecticidal, are indeed toxic to this species. Menzel et al. (1970) examined four different species of marine phytoplankton for their response to DDT, endrin, and dieldrin and found that a diatom species, *Cyclotella nana*, was very sensitive to all these chlorinated hydrocarbon insecticides. The effects of dieldrin and endrin appear to occur as low as 0.01 ppb as judged by the retardation of the growth of the diatom. On the other hand, *Dunaliella tertiolecta*, a green flagellate typical of tidal pools and estuaries, was extremely resistant to all these chemicals, showing no inhibitory effects at 1,000 ppb. Both *Skeletoneura costatum*, a costal centric diatom, and *Coccolithus huxleyi*, a cocolithaforid, showed intermediate responses to the insecticides.

An investigation in our laboratory on two blue-green algae species (Batterton et al., 1971) indicated that some of the terminal residues of cyclodiene insecticides are as toxic to these blue-green algae species as the parent chemicals, i.e., aldrin, dieldrin, and endrin (Fig. 9).

Chronic and subacute effects - To assess the chronic and subacute toxicities of stable pesticidal residues is not an easy matter. First, rules and methods have not been established to express the long-term effects of pesticides on many animals; and second, even if rules were established, the definition of toxic effects could vary among experts. Nevertheless, most researchers today would agree that chronic effects are the criteria that are most useful in assessing the environmental hazards of any pesticide. For instance, by the data shown in TABLE 1, one may conclude that DDT is the safest insecticide for birds, and that Zectran is extremely dangerous. One of the reasons for such an erroneous conclusion is that

TABLE 1

Acute Oral LD_{50} Values[a] of Insecticides for Selected Birds

Insecticides	Mallards ♀	Pheasant ♀	Coturnix ♂ (Quail)
Chlorinated hydrocarbons			
DDT	2240	1296	841
dieldrin	381	79	70
endrin	5.64	1.78	--
Organophosphates			
parathion	1.90	12.4	5.95[b]
Dursban	75.6	17.7	17
Abate	90[c]	31.5	84.1
Carbamates			
arprocarb	11.9	20[b]	28.3[b]
Zectran	3.0	4.5	3.2
carbaryl	2179	707	2290

[a]From Tucker and Crabtree, 1970.

[b]The sex different from the one indicated at the heading.

[c]Average value.

chronic toxicity data have been completely neglected from consideration. Tucker and Crabtree (1970) used a criterion that the lowest daily oral dosage was one that produced death to one or two animals (out of 6) by the end of the 30-day experimental observation period. As one can easily see, the chronic values for parathion, arprocarb, and Zectran are only slightly below those of the acuts LD_{50}'s. That means the effects of these insecticides at sublethal doses disappear after each day of administration and that the effects do not accumulate or are not carried over to the next day. The opposite is true for the insecticides which show a high ratio for acute vs. chronic values. Endrin, for instance, is toxic to mallards at a daily dose of 0.125 mg/mg (TABLE 2).

Chronic effects of stable pesticides are also difficult to obtain. Often-cited chronic effects are, for instance, hormonal changes, reproductive damage including eggshell thickness and hatchability, behavioral abnormalities, growth retardation or stimulation, etc. These are more subtle, yet very important aspects of environmental effects of pesticides. Since there are a number of excellent review papers (e.g., Robinson, 1969; Cope, 1971; etc.) on this subject, I will try to summarize in this paper one aspect of such subtle biological effects which has not been described.

The major obstacle to the understanding of the effects of stable pesticides, chlorinated hydrocarbon insecticides in particular, is that the basic mechanisms of their action is not well understood. While the site of actual attack of DDT, for instance, has long been known to be the nervous system, the biochemical cause by which the nervous systems are disrupted from normal functioning has remained unknown. Recently two groups of workers independently discovered that DDT and other chlorinated hydrocarbon insecticides can inhibit the action of ATPases (Koch, 1969; Koch et al., 1969; Matsumura and Patil, 1969; Matsumura et al., 1969). This inhibitory property of chlorinated hydrocarbons appears to extend to most of the ATPases including Mg^{++}-ATPases, Na^{+},K^{+}-ATPases, actomyosine-

TABLE 2

Ratio between Acute Oral LD_{50}'s and Chronic,[a] Minimum Lethal Dosages on Mallards for Several Insecticides

Insecticide	Chronic minimum lethal doses[a] mg/kg/day	Ratio for cumulativeness acute/chronic
DDT	50	44.8
dieldrin	1.25	76[c]
endrin	0.125	45
Abate	2.5	32 - 40
Dursban	2.5	30
parathion	3 - 6[b]	2.7 - 5.3[b]
carbaryl	125	17.4
arprocarb	2.0	2.0
Zectran	1.25	2.4

[a]From Tucker and Crabtree (1970). The lowest daily oral dosage that produced death to one or two animals (out of 6) by the end of 30 days.

[b]Data taken from those for gray partridges.

[c]Similar values for lindane: 67.

ATPases of nerves, muscles, and other parts of animal bodies. It is clear that a specific ATPase which should be inhibited particularly by DDT, or any other active insecticides (Matsumura and Patil, 1969), must be sought in the nervous system to answer the question of the actual action mechanisms of these chemicals. Nevertheless, the extent to which they exhibit broad inhibitory properties on many different kinds of ATPases raises a possibility that there may be many side effects of chlorinated hydrocarbon insecticides which have been as yet undetected.

Since Na^+,K^+-ATPases have been known to serve as the enzyme responsible for Na^+ and K^+ exchanging across many biological membranes, we have decided to study first the effect of DDT on the salinity regulatory mechanism of a blue-green algae (Batterton et al., 1972). The initial experimental results indicate that the susceptibility of a blue-green alga, Anacystis nidulans, against DDT varies greatly under different salt concentrations. At high NaCl concentrations the blue-green alga becomes extremely sensitive to DDT. It is clear from the result that this fresh water species loses its NaCl-tolerance capability in the presence of a low level of DDT: the level normally would not affect the species. A separate experiment in vitro showed also that DDT was indeed an inhibitor of Na^+, K^+-ATPases of A. nidulans, blocking all ouabain sensitive ATPase activities. The most important indication that the ATPase is related to the NaCl-tolerance mechanism comes from the in vivo finding that Ca^{++}, when added externally to the medium, can antagonize the effects of DDT.

In another set of experiments, the brine shrimp, A. salina, was subjected to various chlorinated hydrocarbon insecticides under different salt concentrations (Fig. 10). The results clearly indicate that the effects of these chlorinated hydrocarbons are strongest at either extremely low or high salt concentrations. The brine shrimp is noted for its great capabilities of tolerance on different salt concentrations. It is often found in abundance in inland salt lakes (e.g., salt ponds and lakes in Utah) where the salt concentration is so high that no other organism can survive. The loss of salt tolerance mechanisms for this species by the presence of these insecticides is, therefore, quite a surprising phenomenon.

The above examples illustrate only one aspect of pesticidal pollution. It is important to note, however, that such a finding comes from fundamental knowledge of the chemical interactions with biological materials. It is also necessary to stress that those stable terminal residues and contaminants would not have been detected from the environments if not for the specific knowledge accumulated through basic researches in laboratory as to their chemical characteristics and behavior. Factors involved in the interactions of pesticides with various ecosystems are numerous and complicated, but it certainly is hoped that there are a number of rate-limiting, key factors that can be analyzed through controlled laboratory experiments.

References

Ahmed, M. K., J. E. Casida, and R. E. Nichols: J. Agr. Food Chem. 6, 740 (1958).

Allan, J.: Nature 175, 1131 (1955).

Batterton, J. C., G. M. Boush, and F. Matsumura: Bull. Environ. Contam. Toxicol. 6, 589 (1971).

Batterton, J. C., G. M. Boush, and F. Matsumura: DDT, inhibition of NaCl tolerance by the blue-green alga, Anacystis nidulans. In preparation (1972).

Bixby, M., G. M. Boush, and F. Matsumura: Bull. Environ. Contam. Toxicol. 6, 491 (1971).

Cook, J. W. J. Agr. Food Chem. 5, 859 (1957).

Cope, O. B. Ann. Rev. Entomol. 16, 325 (1971).

Crosby, D. G.: Residue Rev. 25, 1 (1969).

Crosby, D. G., and H. O. Tutass: J. Agr. Food Chem. 14, 596 (1966).

Czegledi-Janko, G., and A. Hallo: J. Chromatog. 31, 89 (1967).

Engst, R., and W. Shnaak: Zeitschrift für Lebensmittel Untersuchungs Forschung 134, 216 (1967).

Finley, R. B., and R. E. Pillmore: Amer. Inst. Biol. Sci. Bull. 13, 41 (1963).

Fleck, E. E.: J. Amer. Chem. Soc. 71, 1034 (1949).

Geissbuhler, H., C. Haselback, H. Aebl, and L. Ebner: Weed Res. 3, 277 (1963).

Goto, M.: Organochlorine compounds in the environment in Japan. Environmental Safety and Quality, Vol. I. F. Coulson and F. Korte, Eds. Georg Thieme Verlag, Berlin, in press (1971).

Ishida, M., and P. A. Dahm: J. Econ. Entomol. 58, 383 (1965).

Kearney, P. C.: J. Agr. Food Chem. 13, 561 (1965).

Keil, J. E., and L. E. Priester: Bull. Environ. Contam. Toxicol. 4, 169 (1969).

Koch, R. B. J. Neurochem. 16, 269 (1969).

Koch, R. B., L. K. Cutkomp, and F. M. Do: Life Sci. 8, 289 (1969).

Kurihara, N. H., D. G. Crosby, and H. F. Beckman: Abstr. 152nd Nat. Meet. Amer. Chem. Soc., N. Y. Sec. A, p. 47 (1966).

Lamberton, J. G., and R. R. Claeys: J. Agr. Food Chem. 18, 92 (1970).

Lawrence, J. H., R. P. Barron, J. -Y. T. Chen, P. Lombardo, and W. R. Benson: J. Assoc. Offic. Anal. Chem. 53, 261 (1970).

Lichtenstein, E. P., K. R. Schulz, T. W. Fuhremann, and T. T. Liang: J. Agr. Food Chem. 18, 100 (1970).

Macek, K. J., and W. A. McAllister: Trans. Amer. Fish. Soc. 99, 20 (1970).

MacRae, I. C., and M. Alexander: J. Bacteriol. 86, 1231 (1963).

Matsumura, F., G. M. Boush, and A. Tai: Nature 219, 965 (1968).

Matsumura, F., and K. C. Patil: Science 166, 121 (1969).

Matsumura, F., T. A. Bratkowski, and K. C. Patil: Bull. Environ. Contam. Toxicol. 4, 262 (1969).

Matsumura, F., K. C. Patil, and G. M. Boush: Science 170, 1206 (1970).

Matsumura, F., K. C. Patil, and G. M. Boush: Nature 230, 325 (1971a).

Matsumura, F., V. G. Khanvilkar, K. C. Patil, and G. M. Boush: J. Agr. Food Chem. 19, 27 (1971b).

Matsumura, F., Y. Gotoh, and G. M. Boush: Science 173, 49 (1971c).

Menzel, D. W., J. Anderson, and A. Randtke: Science 167, 1724 (1970).

Miles, J. R., C. M. Tu, and C. R. Harris: J. Econ. Entomol. 62, 1334 (1969).

Miskus, R. P., D. P. Blair, and J. E. Casida: J. Agr. Food Chem. 13, 481 (1965).

Munakata, K., and M. Kuwahara: Residue Rev. 26, 13 (1969).
Pape, B. E., M. F. Para, and M. J. Zabik: J. Agr. Food Chem. 18, 490 (1970).
Patil, K. C., F. Matsumura, and G. M. Boush: Environ. Sci. Technol. In press (1972).
Plimmer, J. R., and P. C. Kearney: 3,4-Dichloroaniline transformations in soils and light. Abstr. 158th Nat. Meet. Amer. Chem. Soc., N. Y., Paper No. 29 (1969).
Reinert, R. E.: Limnos. 2, 3 (1969).
Robinson, J.: Chemical Fallout. M. W. Miller and G. G. Berg, Eds. Charles C. Thomas Publishers, Springfield, Ill., p. 113 (1969).
Robinson, J., A. Richardson, B. Bush, and K. E. Elgar: Bull. Environ. Contam. Toxicol. 1, 127 (1966).
Rosen, J. D., D. J. Sutherland, and G. R. Lipton: Bull. Environ. Contam. Toxicol. 1, 133 (1966).
Rosen, J. D., and D. J. Sutherland: Bull. Environ. Contam. Toxicol. 2, 1 (1967).
Rosen, J. D., D. J. Sutherland, and M. A. Q. Khan: J. Agr. Food Chem. 17, 404 (1969).
Rosen, J. D., M. Siewierski, and G. Winnett: J. Agr. Food Chem. 18, 494 (1970).
Schwemmer, B., W. P. Cochrane, and P. B. Polen: Science 169, 1087 (1970).
Sethunathan, N., E. M. Bautista, and T. Yoshida: Can. J. Microbiol. 15, 1349 (1969).
Stewart, N. E., R. E. Millemann, and W. P. Breese: Trans. Amer. Fish. Soc. 96, 25 (1967).
Tsai, Shan-Ching, G. M. Boush, and F. Matsumura: Unpublished data (1971).
Tsukano, Y., and A. Kobayashi: Private communication (1971).
Tucker, R. K., and D. G. Crabtree: Handbook of Toxicity of Pesticides to Wildlife. Fish Wildl. Serv., U.S. Dep. Interior, Resource Publ. No. 84 (1970).
Wedemyer, G.: J. Appl. Microbiol. 16, 661 (1968).

Weisgerber, I., W. Klein, and F. Korte: Tetrahedron 26, 779 (1970).
Wodicka, V. O.: A letter sent to W. M. Upholt of the Environ. Protection Agency, Aug. 23 (1971).
Yule, W. N., M. Chiba, and H. B. Morley: J. Agr. Food Chem. 15, 1000 (1967).
Zabik, M. J., R. D. Schuetz, W. L. Burton, and B. E. Pape: J. Agr. Food Chem. 19, 308 (1971).

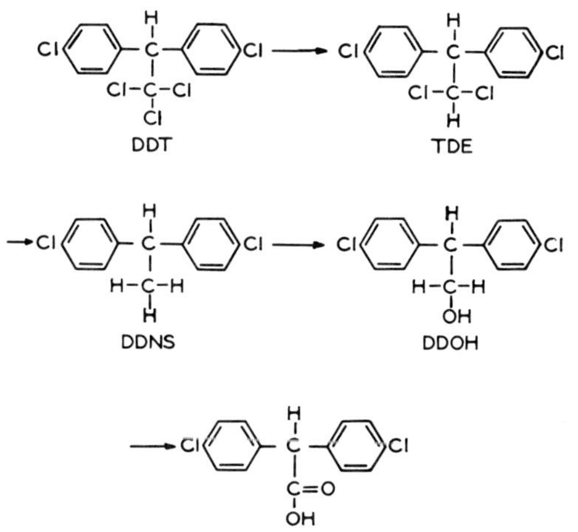

Fig. 1. Major metabolic routes of DDT by microorganisms.

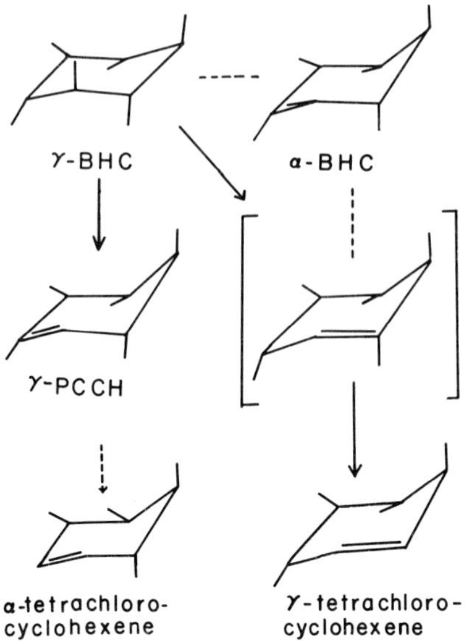

Fig. 2. Basic pattern of microbial metabolism of BHC.

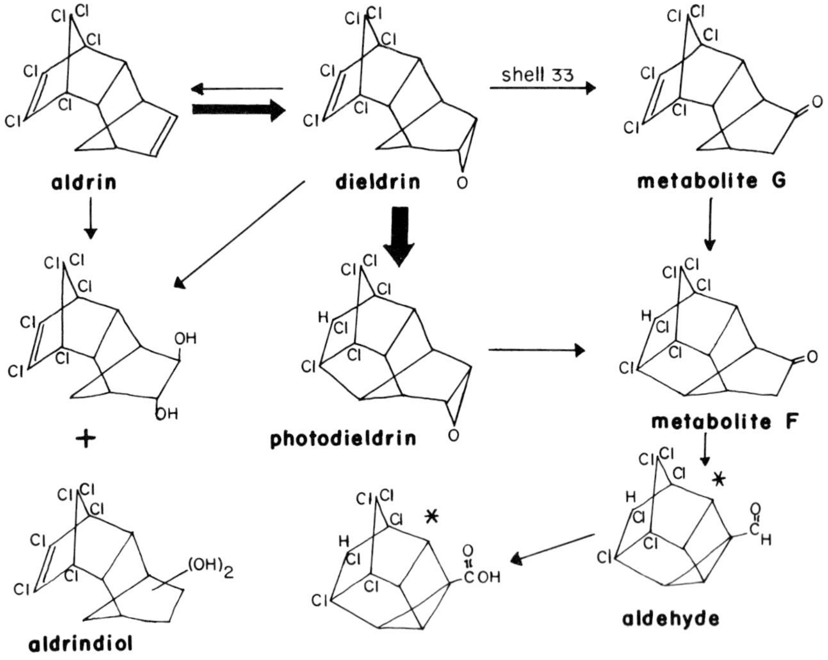

Fig. 3. Microbial degradation of aldrin and dieldrin. Unconfirmed metabolites are shown by asterisks.

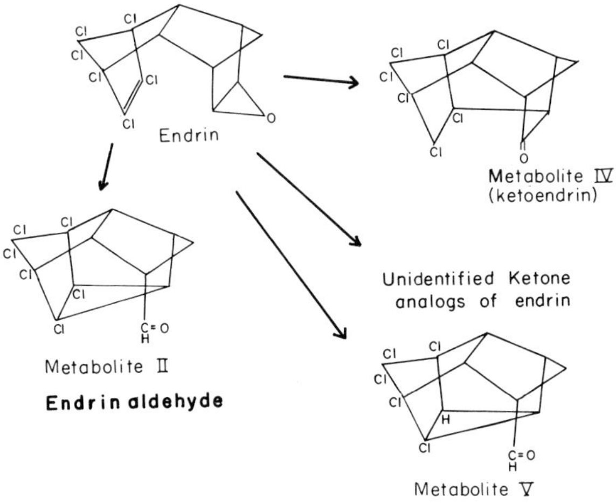

Fig. 4. Microbial degradation of endrin.

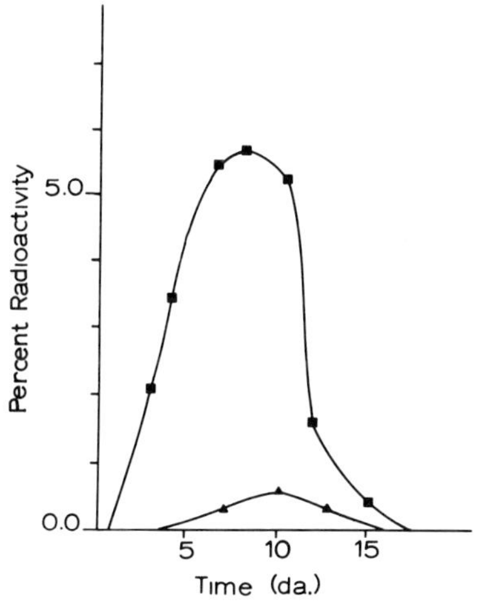

Fig. 5. Evolution of $^{14}CO_2$ from a culture of the fungus Trichoderma koningi that has been incubated with ^{14}C-UL-glucose (represented by square symbols) or ^{14}C-dieldrin (triangles).

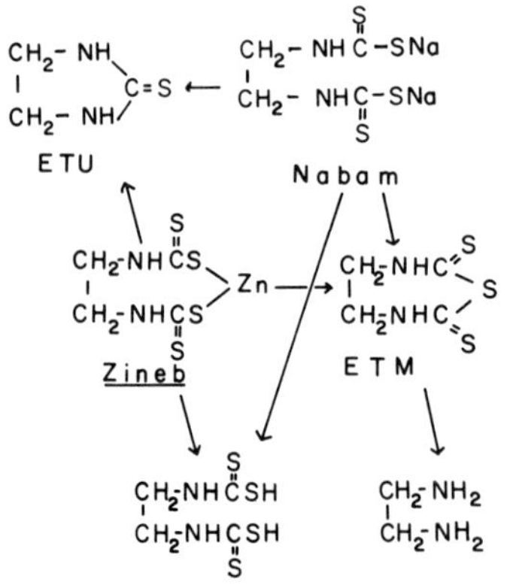

Fig. 6. Formation of ethylenethiourea (ETU) from two fungicidal chemicals.

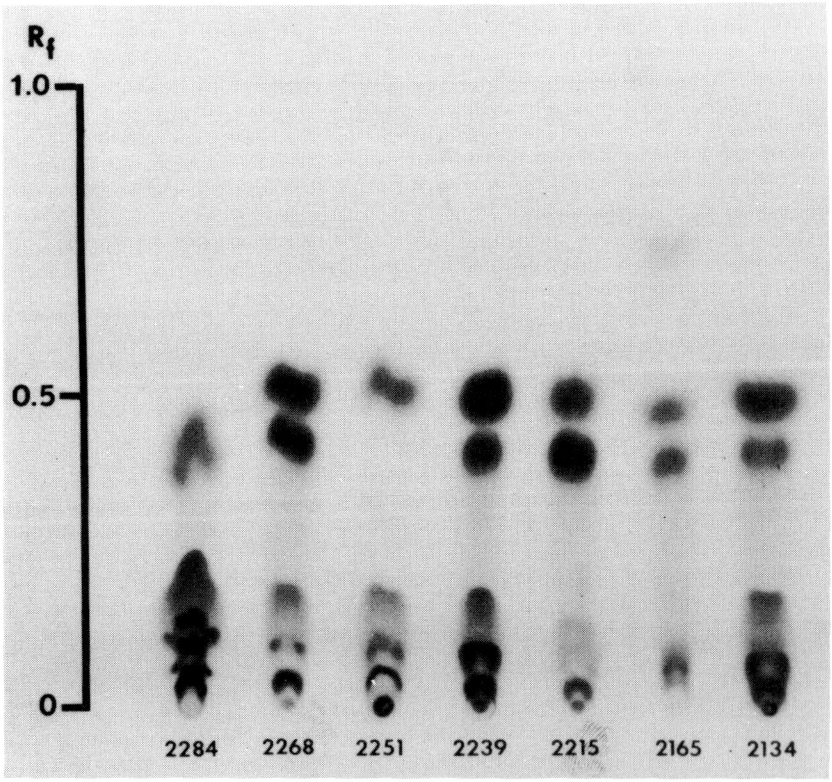

Fig. 7. Autoradiographic representation of the metabolic conversion of ^{203}Hg-labeled phenylmercuric acetate by microorganisms. The metabolites were separated by thin-layer chromatography on silica gel.

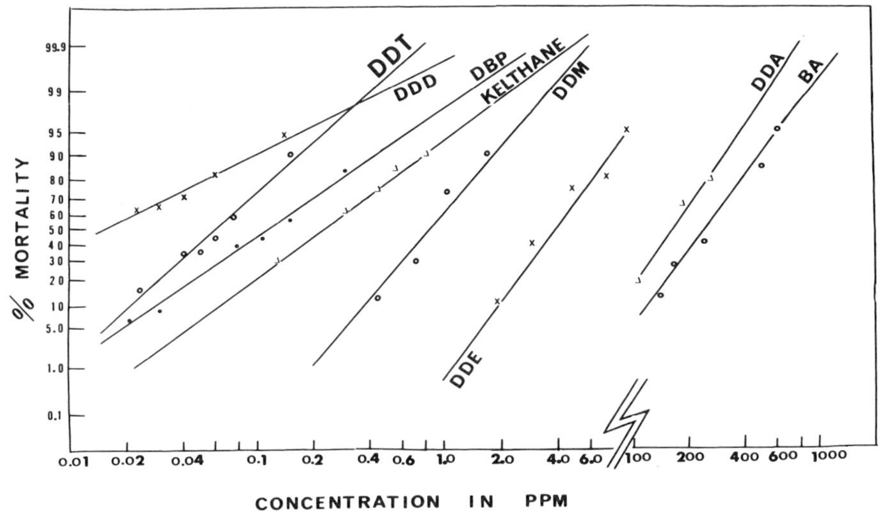

Fig. 8. Differential toxicities of DDT analogs and metabolites on brine shrimp, Artemia salina; 24 hours exposure at 24°C.

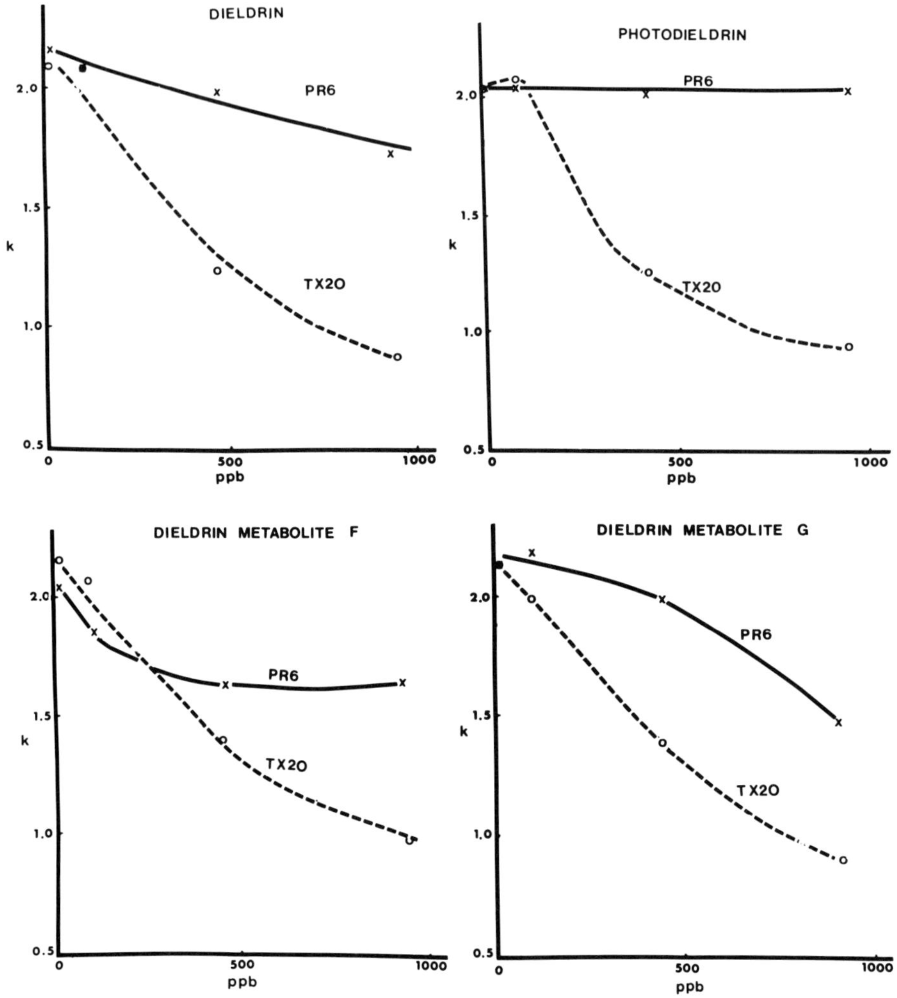

Fig. 9. Toxic effect of various conversion products of cyclodiene insecticides on two species of blue-green algae.

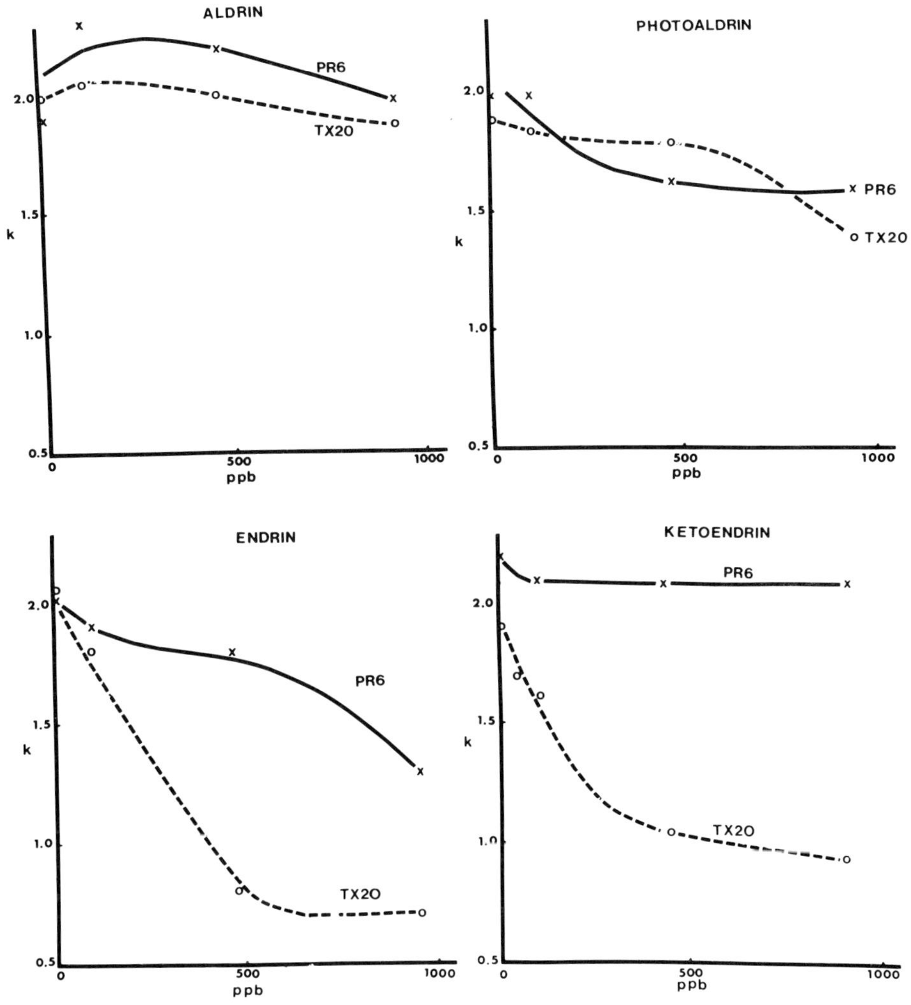
ALDRIN
PR6
TX20
k
2.0
1.5
1.0
0.5
0
500
1000
ppb
PHOTOALDRIN
PR6
TX20
ENDRIN
PR6
TX2O
KETOENDRIN
PR6
TX2O

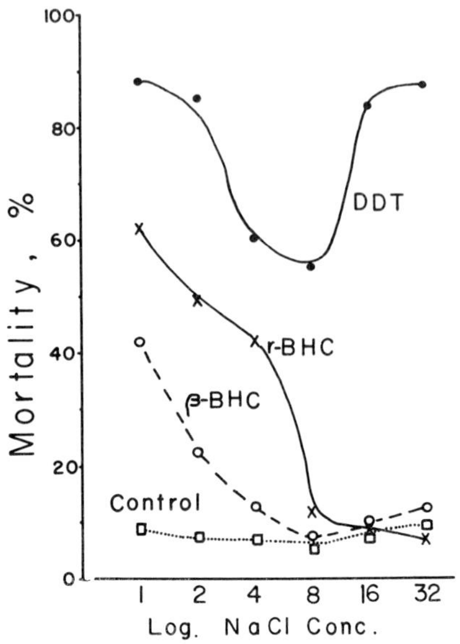

Fig. 10. Differential susceptibility of Artemia salina against three chlorinated hydrocarbon insecticides under various NaCl concentrations.

SPECTRAL CHANGES OF CHLOROPLASTS AS A MEANS TO EXAMINE PHYTOTOXIC EFFECTS OF PESTICIDES

Yorinao Inoue, Nakatsugu Yaginuma, Teruo Ogawa, Katsuko Konishi, and Kazuo Shibata

Laboratory of Plant Physiology,
Institute of Physical and Chemical Research
(Rikagaku Kenkyusho), Wakohshi, Saitama
and
Laboratory of Biological Chemistry,
Tokyo Institute of Technology, Meguroku, Tokyo, Japan

Introduction

Some herbicides inhibit photosynthesis strongly and specifically, and chloroplasts, the photosynthetic apparatus, undergo spectral changes when treated with those herbicides. A few of them have been applied extensively for photosynthesis studies because they inhibit a specific site(s) of the electron transport or photophosphorylation system in chloroplasts. The spectral changes of chloroplasts caused by treatment with herbicides and some other chemicals for plants were surveyed in the present study in order to classify them in terms of spectral change and to see if there exists any correlation between spectral change and herbicidal activity.

In parallel with the above study of spectral changes of chloroplasts and sub-chloroplast particles, an instrument with a duochromator and fiber optics was devised and constructed to observe the spectral changes directly on intact leaves. Difference spectrophotometry by dual wavelength measurements with this instrument, which we named "plant examiner," minimized the errors in absorbance readings due to scattering differences. This is a problem with common spectrophotometry of translucent leaves with a monochromator. In addition, one can follow, in this way, the spectral changes of the same spot on the same leaf, so that errors due to differences in sample leaves or due to loss or bleaching in extraction can be eliminated.

Experimental Approach

Whole chloroplasts were isolated from leaves of spinach (*spinacia oleracea*) according to the method reported by Itoh et al. (1963), purified by repeated centrifugation, and resuspended in 0.04 M phosphate buffer (pH 7.0). Grana and sonicated grana were prepared from this chloroplast preparation by the following procedure. Chloroplasts were blended with a Waring Blendor in the above phosphate buffer at $0^{o}C$ for 2 min. The suspension of broken chloroplasts was centrifuged at 5,000 x *g* for 10 min and the sediment discarded. The supernatant was centrifuged at 15,000 x *g* for 15 min, and the sediment was resuspended in the same phosphate buffer and used as the grana preparation. The supernatant was sonicated for 5 min at 10 Kc with a Toyo Riko sonicator model UF50-6 at $0^{o}C$. The sonicated suspension was centrifuged at 15,000 x *g* for 15 min and the pellet was discarded. The supernatant thus obtained was used as the sonicated grana preparation.

A solution (0.1 ml) of a chemical in ethanol was added to 5.0 ml of a chloroplast suspension, and the mixture was left standing for 30 min in the dark at room temperature before measurement. Addition of ethanol alone, without chemicals, showed no measureable effect on the spectrum and activities of chloroplasts. Difference spectrophotometry was applied in order to observe minute spectral changes. After incubation, the above mixture was placed in the sample compartment along with the reference mixture without the chemical, in the reference compartment of a spectrophotometer, and the difference spectrum between these mixtures recorded. The measurement was made using 1.0-cm cells, with a Shimadzu recording spectrophotometer model DF-40 designed particularly for difference spectrophotometry. The concentration of chloroplasts in the mixture was equivalent to 20 μg of chlorophyll per ml of suspension, and the concentration of chemicals used throughout the experiments was 500 μM.

The activity of "photosystem I" was estimated from the rate of methyl viologen-mediated O_2-linked photo-oxidation of $DPIPH_2$ (reduced form of 2,6-dichlorophenol indophenol) measured by the method reported by Izawa (1968). A chloroplast suspension (equivalent to 20 μg chlorophyll/ml) containing 20 μM methyl viologen, 1 mM ascorbate, 50 μM DPIP, and 10 μM DCMU was illuminated by red light (10^5 erg/cm^2 sec) from a projector lamp through a fan-cooled, heat-absorbing filter and a Toshiba VR-67 filter. The rate of oxygen uptake by illuminated chloroplasts was measured polarographically at 25°C with a Yanagimoto oxygen consumption recorder model PO-100. The activity of "photosystem II" was measured from the rate of DPIP photoreduction. A chloroplast suspension (equivalent to 20 μg chlorophyll/ml) containing 50 μM DPIP was illuminated by the same red light as above, and the rate of reduction was determined by photometry at 600 mμ. The activity of photo-bleaching was estimated from the decrease of absorbance at 678 mμ. A chloroplast suspension treated similarly with a chemical was illuminated with a 20 W white fluorescent lamp for 30 min at room temperature and the difference in absorbance at 678 mμ was measured between the illuminated and non-illuminated chloroplast suspensions. Both of these suspensions had been pre-treated with a chemical.

Distribution curves of chloroplast volume were determined with a Coulter counter model A. A sample suspension of chloroplasts was diluted 170 times with 0.04 M phosphate buffer (pH 7.0) and subjected to counting with an orifice at 100 μ in diameter. Coincidence count loss and non-linear response of the pulse height were corrected as described recently (Shibata, 1971).

Spectral Changes of Whole Chloroplasts

Spectral changes in the visible region caused by treatment of chloroplasts or sub-chloroplast particles with a number of chemicals examined were found to originate from two elemental phenomena, and the three types of spectral changes found for whole chloroplasts were in superposition of the changes resulted from these two elements. One of the elements is the conformational change of chloroplasts or sub-chloroplast particles, and the other is the change in the *in vivo* state of chlorophylls or carotenoids bound in a variety of states to structural proteins and lipids in thylakoid membranes. As described below, a conformational change induces intensification or flattening of absorption bands in the whole visible region without any shift of bands, while a change of a pigment from one form *in vivo* to the other induces transformation or shifts of a particular band(s).

Type I - A typical example of spectral changes, denoted as type I in this paper, is shown by the curves in Fig. 1. The absorption bands (curve A) of intact chloroplasts in suspension were intensified (curve B) over the whole visible region by treatment with pentachlorophenol (PCP), so that the difference between the spectrum (curve C) directly observed for the PCP-treated chloroplasts against non-treated chloroplasts is similar in shape to curve A of intact chloroplasts. An increase in the efficiency of light absorption by suspended particles is the cause for such a change over the whole spectral region, and swelling of the particles or disintegration of the particles into a larger number of particles is responsible for the efficiency increase as predicted theoretically (Duysens, 1952, 1956) and demonstrated experimentally (Itoh et al., 1963; Yamashita et al., 1968). In fact, the volume-distribution curve of PCP-treated chloroplasts measured with a Coulter counter as compared with the distribution curve of intact chloroplasts indicated swelling of chloroplasts by the treatment. The degree of swelling was approximately two times in terms of volume. The total number of chloroplasts in the suspension was not altered by the treatment so that disintegration was not involved in the intensification. The chemicals classified as type I in TABLE 1 induced intensification, although the degrees of intensification found with these chemicals were different, as discussed later in relation to their effects on activities.

Flattening of absorption bands due to shrinkage of chloroplasts, which may be classified also as a type I change, was found in a previous study (Itoh et al., 1963) when chloroplasts incubated in the dark were illuminated under phosphorylation conditions. Treatment of chloroplasts with cationic chain compounds was found in a separate experiment to cause flattening, which may be interpreted as due to either shrinkage or aggregation of chloroplasts.

Type II - The spectral changes designated as type II involve transformation of bands, as shown by curves B and C in Fig. 2, which are the difference spectra obtained with 3,4-dichloropropionanilide (Propanil) and 3,4-dichlorovaleranilide (DCVA), respectively. In the change shown by curve B, a chlorophyll a form having its red band at 684 mμ is transformed into a form having a band at 666 mμ, and the bands in the blue region are modified and enhanced. The ratio, $-\Delta \underline{A}_{684}/\underline{A}_{678}$, of the absorbance decrease at 684 mμ to the absorbance at 678 mμ of intact chloroplasts before the treatment was approximately 10%. Lutein, neoxanthin, and beta-carotene, when added to a suspension of intact chloroplasts, were found to cause the same transformation of the red band to the same extent (Konishi et al., in preparation).

Curve C obtained with DCVA is certainly different from curve B. This change seems, however, to include the same type of change as curve B, because the difference spectra obtained at earlier stages, e.g., 1 - 5 min after the addition of 500 μM DCVA, were similar in shape and relative band heights to curve B. This is shown in Fig. 3 by the difference spectra in the red region observed at different times of incubation with DCVA. After prolonged

TABLE 1

Types of Chemicals Classified according to their Abilities to Induce Changes in Spectral Pattern[a]

Type in spectral change	Chemical
Type I	Pentachlorophenol (PCP), atrazine, propanil, swep, 3-(3, 4-Dichlorophenyl)-1, 1-dimethylurea (DCMU), 2, 4-bis(i-Propylamino)-6-methylthio-1, 3, 5-triazine (Prometryne), Methyl-N-(3-chloro-4-ethoxyphenyl)-carbamate, Methyl-N-(2-methyl-4-chlorophenyl)-carbamate, 2, 4, 5-Trichloropropionanilide, 3, 4-Dichloroacetoanilide (DCAA), 3, 4-Dichloro-i-butyranilide (DCIBA), N-(3, 4-dichlorophenyl)-cyclopropanecarboxamide (Cypromid).
Type II	3, 4-Dichloro-chloroacetoanilode (DCCAA), Propanil, 3, 4-Dichlorobutyranilide (DCBA), 3, 4-Dichlorovaleranilide (DCVA), 3, 4-Dichlorobenzanilide (DCBeA), Ethyl-N-(3, 4-dichlorophenyl)-carbamate, i-Propyl-N-(3, 4-dichlorophenyl)-carbamate (DIPC), Fatty alcogols with 8 – 12 carbon atoms,[a] Carotenoids (Lutein, Neoxanthin, β-carotene).[a]

TABLE 1 - Continued

Type in spectral change	Chemical
Type III	Diphenylether (DE), 3-Methylphenyl-4'-nitrophenylether (HE-314), 3, 5-Dimethylphenyl-4'-nitrophenylether (HW-40187), 2-Nitro-4-chlorophenyl-4'-nitrophenylether (KK-60), 2, 4-Dichlorophenyl-4'-nitrophenylether (Nitrophen), 2-Fluoro-4, 6-dichlorophenyl-4'-nitrophenylether (MO-500), 2, 4-Dichloro-3'-methoxy-4'-nitrophenylether (X-52), 2, 4, 6-Trichlorophenyl-4'-nitrophenylether (CNP).
No spectral change	Propionanilide, 2, 4-D, Amitrol, α-NAA, 3-Chloropropionanilide, 4-Chloropropionanilide, 2, 3-Dichloropropionanilide, 2, 5-Dichloropropionanilide, Methyl-N-(3-methylphenyl)-carbamate, Methyl-N-(4-methoxyphenyl)-carbamate.

[a] From Konishi et al. (1971).

incubation, the spectrum changed further, to be the one such as shown by curve C in Fig. 2. The change from curve B to curve C will be called the second step to distinguish it from the change shown by curve B in the first step. The decrease of absorbance at 684 mμ on curve C relative to the absorbance at 678 mμ before the treatment was about 30%, so that the relative decrease in the second step is approximately 20% if one can assume that curve C is merely a superposition of two independent processes of the first and the second steps. In the first step, until 10% lowering of the red band, the negative red peak in the difference spectrum was present at 684 mμ and a clear isobestic point was observed at 674 mμ, but the curves obtained in the second step deviated and passed through a different isobestic point at 672 mμ and the negative peak shifted to a slightly shorter wavelength of 683 mμ. If we designate the chlorophyll a forms having their red band at 684, 683, and 666 mμ as C684, C683, and C666, respectively, according to the nomenclature by Duysens (1952), the above change in the first step is described as the transformation of C684 into C666, and the second step is the transformation of C683. The second step seems, however, to involve complex processes, because the bands enhanced in the first step were lowered in an intricate manner. In fact, the changes in the blue region in the second step showed some variations depending on the chemicals. For example (Shibata, 1959), fatty alcohols with 8 to 12 carbon atoms induce the same two steps of spectral change, but the lowering of the bands in the blue region in the second step was accompanied by a drop of absorbance near 505 mμ so that the difference spectrum showed an additional negative peak at this wavelength. TABLE 1 summarizes the chemicals which cause the changes of type II including those variations in the second step.

Type III - Another type of spectral change was obtained with diphenylether derivatives. The changes obtained for chloroplasts treated with diphenylether (DE) and 3-methylphenyl-4'-nitrophenylether (HE-314) are shown in Fig. 2 by curves D and E, respectively. These difference spectra indicate negative peaks at 682 and 700 mμ, although their relative heights are different. A maximum was found at 665 mμ, slightly above the zero level with these diphenylethers but, with 2,4-dichlorophenyl-4'-nitrophenylether (Nitrophen) as the reagent, the maximum was enhanced to a sharp positive peak at 670 mμ.

Plant hormones and regulators - Three other representative chemicals were examined to see their effects on the spectrum of chloroplasts. They are 2,4-D, a hormonal herbicide; alpha-NAA, a plant growth regulator; and Amitrol, an inhibitor for chloroplast development. These chemicals did not cause any change of the spectrum. The other seven chemicals listed in the last row of TABLE 1 are known to inhibit the Hill reaction only at very high concentrations, and to show low herbicidal activities. These chemicals did not change the spectrum at all.

Spectral Changes of Grana and Sonicated Grana

Spectral changes of sub-chloroplast particles treated with chemicals were more or less different from the changes found for whole chloroplasts and, in some cases, a qualitatively different change was superimposed.

Type I - Three typical variations are shown in Fig. 4. Changes qualitatively similar to that found for whole chloroplasts were obtained for grana and sonicated grana treated with 2-chloro-4-ethylamino-6-i-propylamino-1,3,5-triazine (Atrazine) as shown by the curves in the left column. The extent of intensification was, however, different, depending on the sample. The degrees of intensification determined at 680 mμ for whole chloroplasts, grana, and sonicated grana were approximately in a ratio of 4:2:1. The fact that the intensification is observable for grana and sonicated grana indicates that these sub-chloroplast particles undergo swelling or conformational changes leading to an increase in the yield of light absorption. Some reagents, which caused a change of type I for whole chloroplasts, induced additional changes for sonicated grana, and some others caused additional changes both for grana and for sonicated grana. An example of the former case is shown by the curves in the middle column of Fig. 4, which are the difference spectra obtained with N-(3, 4-dichlorophenyl)-cyclopropanecarboxamide (Cypromid), and an example of the latter is shown by the curves in the right column which are the spectra obtained with 3, 4-dichloro-i-butyranilide (DCIBA) or 3, 4-dichloroacetoanilide (DCAA). An additional negative peak was found at 684 mμ with a shoulder at 700 mμ and a positive peak appeared at 665 mμ. Such transformation of chlorophyll forms was thus found to take place in addition to swelling. The data in Fig. 4 indicate that the activity to induce the spectral changes is greater in the order of Atrazine < Cypromid < DCIBA, DCAA.

Type II - The change of the red band found for chloroplasts treated with DCVA proceeded further when the reagent was applied to grana or sonicated grana. The drop of absorbance at 684 mμ, which was 27% for whole chloroplasts (TABLE 2), was 32% for grana and 36% for sonicated grana (the curves in the right column of Fig. 5). A different change occurred when Propanil was applied to these sub-chloroplast particles. A weak negative shoulder appeared at 700 mμ on the difference spectrum obtained for grana, and the shoulder was greatly intensified in the case of sonicated grana. This is shown in the left column of Fig. 5.

Type III - In the case of sub-chloroplast particles, a positive peak appeared at 674 mμ with diphenylether as the reagent (the curves in the left column of Fig. 6) and at 670 mμ with Nitrophen (the curves in the right column). The high positive peak at 670 mμ obtained with Nitrophen masked the negative peak at 682 mμ which was clearly observed for whole chloroplasts.

Spectral Changes Related to Activities

TABLE 2 lists the activities of photosystems I and II of the chloroplasts treated with chemicals, the degrees of

TABLE 2

Spectral Changes Related to Photochemical and Photobleaching Activities[a]

Type	Chemical	$\Delta A/A$ in %	$-\Delta A_{678} \times 10^2$	A_I	A_{II}
Type I	PCP	8	13	200	3
	Swep	10	7	196	0
	DCMU	3	0.1	86	0
	Simazine	1	0	86	0
	Simetone	3	0	77	0
Type II	Propanil	9	14	153	0
	DCVA	27	12	164[b]	0
	DIPC	13	9	144[b]	0
Type III	Nitrophen	10	10	94	0
No spectral change	2,4-D	0	0	100	102
	α-NAA	0	0	97	99
	Amitrol	0	0	100	100

[a]The spectral changes, $\Delta A/A$, were estimated from the increase or decrease of absorbance at 680 mμ for type I, 684 mμ for type II, and 682 mμ for type III relative to the absorbance at 678 mμ of intact chloroplasts before chemical treatment, and photobleaching of chlorophylls was estimated from the absorbance decrease, ΔA_{678}, at 678 mμ after 30 min of illumination. The realtive activities of photosystems I and II are expressed by A_I and A_{II}, respectively.

[b]All experiments were performed at 500 μM of chemicals except the measurements of the photosystem I activities with DCVA and DIPC marked with "b" in the table. In this case, a concentration of 30 μM was used.

chlorophyll bleaching upon illumination of those treated chloroplasts, and the spectral changes measured as an increase or decrease of absorbance, $\Delta \underline{A}/\underline{A}$, relative to the absorbance at 678 mμ before treatment. The chemicals which caused the changes of type I seem to be further classified into two groups according to the relative absorbance increase. The increase of the order of 8-10% was obtained with 500 μM PCP and Swep as the reagent, whereas the increase obtained with DCMU, Simazine, and Simetone at the same concentration were only 1-3%. This is more distinctly reflected in the data of photobleaching of chlorophylls. Practically no bleaching was found with the latter three chemicals, while marked bleaching occurred in the chloroplasts treated with PCP or Swep. The activity of photosystem I was doubled by the treatment with PCP or Swep, which indicates uncoupling of photophosphorylation. By contrast, the activity was slightly suppressed by the treatment with the latter three chemicals. The activity of photosystem II was almost completely abolished by both sub-groups of chemicals. It is interesting to note that the symptom on plants by the effect of the latter group of herbicides appears rather slowly, while the symptom by the former group appears rapidly. Propanil, DCVA, and DIPC, which induced the spectral changes of type II, caused photobleaching of chlorophylls, inhibited the activity of photosystem II, and uncoupled photophosphorylation, promoting the activity of photosystem I. It may be noteworthy that these reagents caused photobleaching, and are rapid type herbicides. Diphenylethers, which induced changes of type III, also caused photobleaching to some extent and inhibited the photosystem II activity, but did not significantly affect photosystem I activity. The three other chemicals in the table, 2,4-D, α-NAA, and Amitrol, during 30 min of incubation, did not cause any spectral change of chloroplasts, were not effective in bleaching chlorophylls, and did not affect the two photochemical activities. All of the chemicals in the table having herbicidal activity, except these three hormone and regulator compounds, were effective in causing spectral change, photobleaching of chlorophylls, and inhibiting photosystem II activity.

Spectrophotometry of Intact Leaves with "Plant Examiner"

The instrument which we developed and named "plant examiner" illuminates a leaf alternatively with two light beams of different wavelengths, λ_1 and λ_2, so that the difference, $\Delta \underline{A}$, in absorbance between these wavelengths can be read directly on a meter or recorded on a recorder;

$$\Delta \underline{A} = \underline{A}(\lambda_2) - \underline{A}(\lambda_1)$$

The advantage of this dual-wavelength difference spectrophotometry is seen from comparison of the scanning profiles of a thin-layer chromatogram with a single beam and with dual beams. Curve A in Fig. 7, obtained with a single beam of 530 mμ, shows two bands of red pigments separated on the chromatogram, but the reading on the base line fluctuates because of scattering by the white material on the chromatogram. Such fluctuation masks low bands, if present, and introduces some errors to the reading of the band height. The dual-wavelength scanning method is based on the principle that the reading at the reference wavelength, λ_1, where the pigment does not absorb light, fluctuates in the same manner along the line of scanning as seen from curve B obtained at 622 mμ for the same chromatogram. Therefore, curve C, obtained by recording the absorbance difference between the two wavelengths, shows a straight base line without fluctuation, and the reading on the peak is not affected by scattering differences. With the "plant examiner," one can scan a leaf on a moving stage with the dual beams or read the absorbance difference of a spot on a leaf by dual-wavelength illumination and sampling through optical fibers (Fig. 8). TABLE 3 shows reproducibility of repeated measurements of $\Delta \underline{A}$ of a spot on a leaf by the procedure shown in Fig. 8B.

For scanning by the procedure of Fig. 8A, leaves were placed on an opal glass plate or filter paper

(Shibata, 1959; Kurozumi and Shibata, 1960). This eliminates the edge effect due to diffraction of the parallel illuminating light on the leaf margin. When a leaf is scanned without such an opalescent diffuser, the curve shows a negative peak on the leaf margin due to the artifact (Kurozumi and Shibata, 1960). The curves on the right side of Fig. 9 show the profiles of normal and Amitrol-treated what leaves on an opal glass plate scanned

TABLE 3

Reproducibility of the Measurement[a] of Dual-Wavelength Absorbance Difference, $\Delta A = A(\lambda_2) - A(\lambda_1)$, of an Intact Leaf with the "Plant Examiner"

$\lambda 2$	ΔA		
506	1.08	1.05	1.08
550	0.45	0.45	0.45
622	0.88	0.85	0.82
678	1.26	1.26	1.22
721	0.14	0.14	0.14

[a]The reference wavelength (λ_1) was fixed at 750 mμ, and λ_2 was varied.

[b]From Itoh et al. (1963).

with the plant examiner. The scanning was made toward the direction perpendicular to the leaf vein. These curves clearly show the effect of Amitrol during germination. The variations of $\Delta \underline{A}$ on these curves of A to F are due to true changes of absorbance when the light beam is passing across the leaf veins and not due to electrical noise. This is seen from curve E on the left side, which was recorded in an expanded scale for the same sample of curve E shown on the right side. The curves in Fig. 10 show the profiles of an Amitrol-treated leaf scanned along the leaf vein. One can thus observe the changes of pigment content along the leaf vein (Inoue et al., in preparation).

The dual wavelength spectrophotometry has so far been applied to observations of small spectral changes during biochemical reactions in intact tissue. It is based on the principle that the absorbance reading at the wavelength of the isobestic point of reaction fluctuates with time in the same manner as the reading at the wavelength where an absorbance change due to the reaction takes place. The same technique was found in the present study to effectively cancel out the scattering variations at different positions of intact plant tissues. Applications of the plant examiner to see the effects of chemicals directly on intact leaves or for observing various biological processes in leaves may be contemplated.

Acknowledgment

This work was supported by a grant given by the Ministry of Education, Japan.

References

Duysens, L. N. M.: Thesis, University of Utrecht (1952).

Duysens, L. N. M.: Biochim. Biophys. Acta 19, 1 (1956).

Itoh, M., S. Izawa, and K. Shibata: Biochim. Biophys. Acta 66, 319 (1963 a).

Itoh, M., S. Izawa, and K. Shibata: Biochim. Biophys. Acta 69, 130 (1963 b).

Inoue, Y., T. Ogawa, and K. Shibata: In preparation (1971).

Izawa, S.: Comparative Biochemistry and Biophysics of Photosynthesis. K. Shibata, A. Takamiya, A. T. Jagendorf, and R. C. Fuller, Eds. University Park Press, Pa., p. 140 (1968).

Konishi, K., N. Yaginuma, and K. Shibata: In preparation (1971).

Kurozumi, T., and K. Shibata: Cytologia 25, 229 (1960).

Shibata, K.: Methods of Biochemical Analysis. D. Glick, Ed. Academic Press, N.Y., Vol. VII, p. 77 (1959).

Shibata, K.: Methods in Enzymology. A. San Pietro, Ed. Academic Press, N.Y., Vol. XXIII, Part B (1971).
Yamashita, K., M. Itoh, and K. Shibata: Biochim. Biophys. Acta 162, 610 (1968).

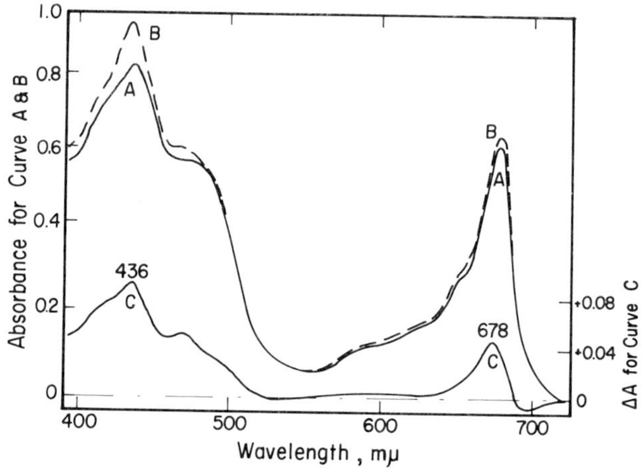

Fig. 1. Absorption spectra of whole chloroplasts before (curve A) and after (curve B) the treatment with 500 μM PCP for 30 min, and the difference spectrum (curve C) directly observed between PCP-treated and non-treated chloroplasts.

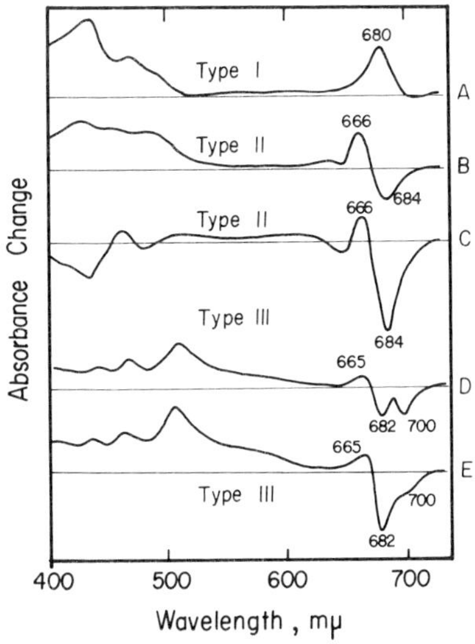

Fig. 2. Three types of difference spectra and their variations. The reagent applied to chloroplasts was PCP for curve A, Propanil for curve B, DCVA for curve C, DE for curve D, and HE-314 for curve E.

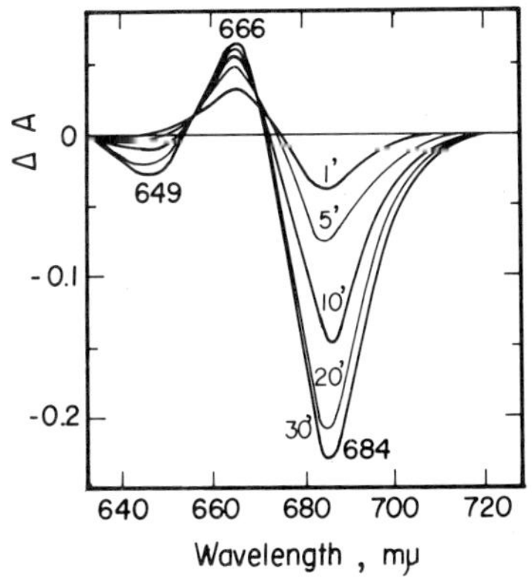

Fig. 3. Difference spectra of DCVA-treated minus non-treated chloroplasts in the red region observed at different incubation times. The figures on the curves indicate the observation time in minutes.

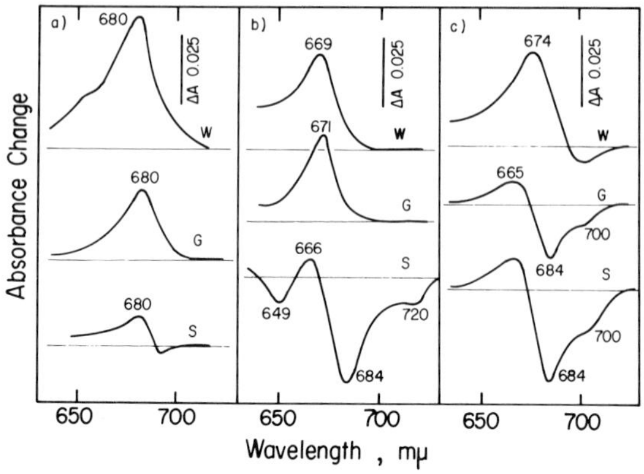

Fig. 4. Variations of type I spectral change dependent on the sample; whole chloroplasts (W), grana (G), and sonicated grana (S); the reagents were (a) Atrazine, (b) Cypromid, and (c) DCIBA and DCAA.

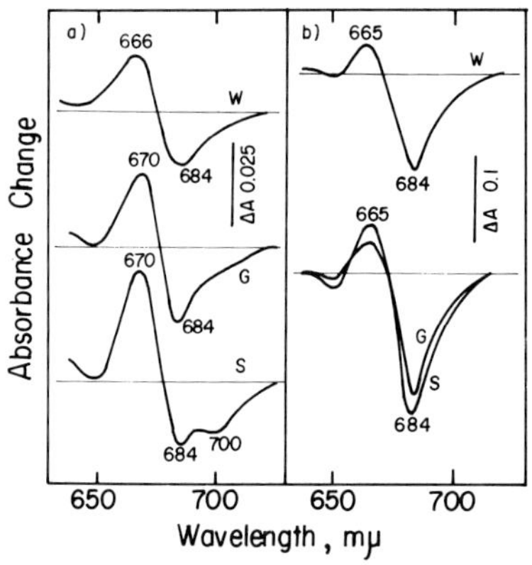

Fig. 5. Variations of type II spectral change dependent on the sample; the reagents were (a) Propanil, and (b) DCVA.

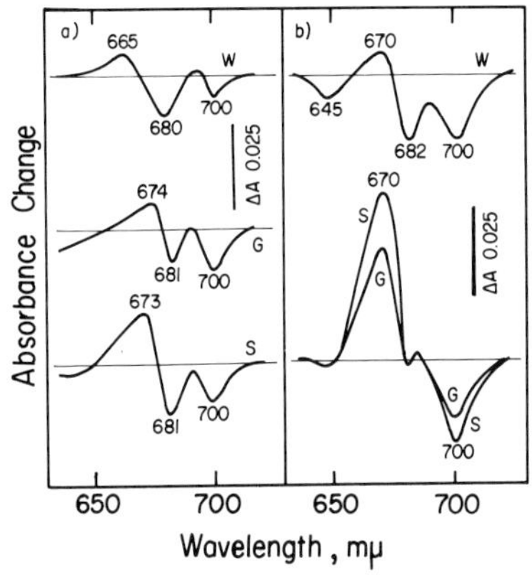

Fig. 6. Variations of type III spectral change dependent on the sample; the reagents were (a) DE, and (b) Nitrophen.

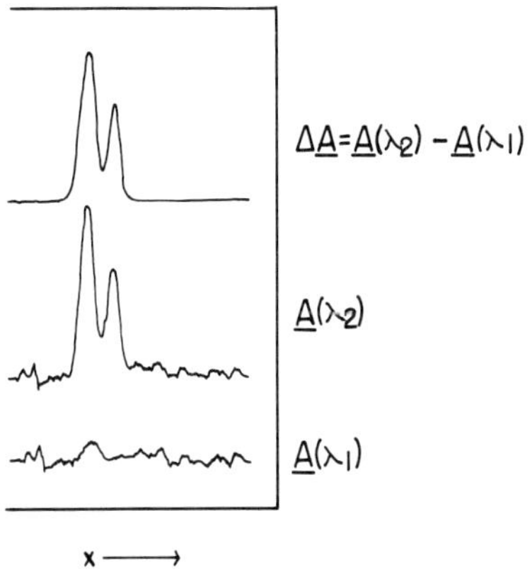

Fig. 7. The profiles obtained by scanning a thin-layer chromatogram; curves A and B obtained at a single wavelength of 530 and 622 mμ , respectively, and curve C obtained by the dual-wavelength scanning of the $\Delta \underline{A}$ value between these wavelengths.

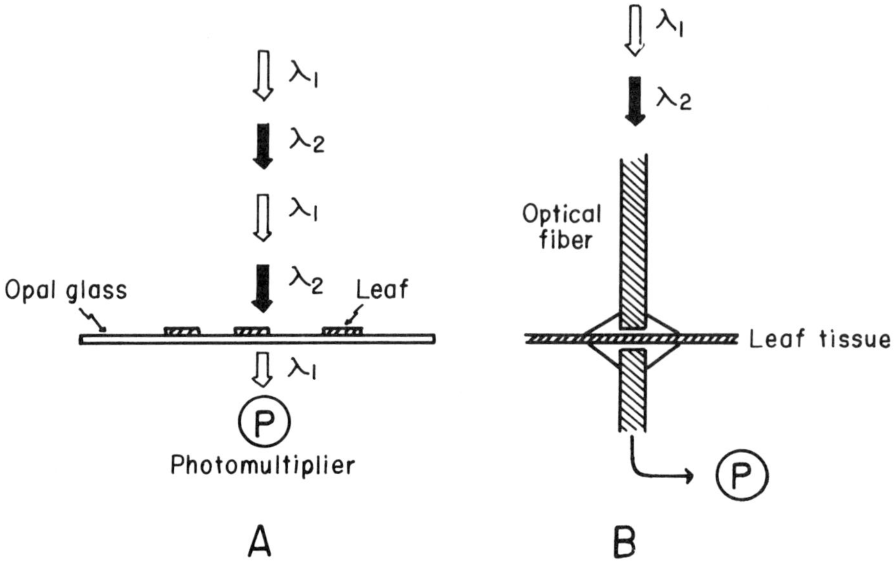

Fig. 8. Schematic diagrams showing (A) the way of scanning leaves by the dual-wavelength difference spectrophotometry, and (B) the measurement of the $\Delta \underline{A}$ value of a spot on a leaf with optical fibers.

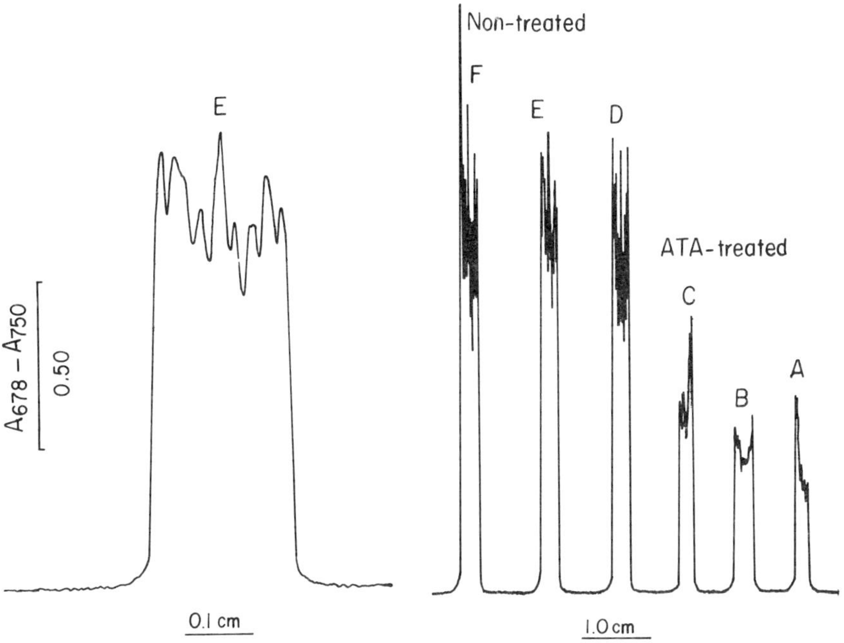

Fig. 9. Scanning profiles of normal leaves (curves D, E, and F) and Amitrol-treated leaves (curves A, B, and C). The scanning was made toward the direction perpendicular to the leaf veins. Curve E on the left side is the profile scanned in an expanded scale of the same leaf shown by curve E on the right side.

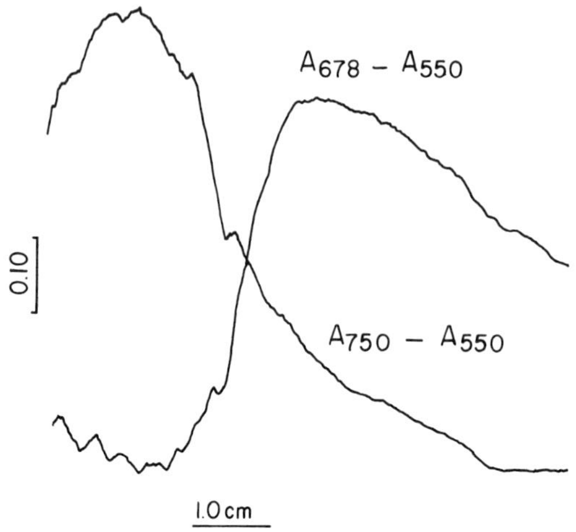

Fig. 10. Scanning profiles of an Amitrol-treated leaf along the leaf vein. The treatment with Amitrol transformed some chlorophyll forms into a form showing a maximum at 742 mμ[11]. This transformation which took place near the leaf base is clearly seen from the different profiles obtained by observing the two absorbance differences, $\Delta \underline{A} = \underline{A}_{678} - \underline{A}_{550}$ and $\Delta \underline{A} = \underline{A}_{750} - \underline{A}_{550}$. The left and right ends of these profiles are the leaf base and apex, respectively.

PART VIII

DESIGN OF NEW PESTICIDES

THE FEASIBILITY OF INSECT CONTROL UTILIZING A SEX PHEROMONE: STUDY ON FEMALE SEX PHEROMONES OF *CADRA*, *PLODIA*, AND *ANAGASTA*

Hiroshi Fukami

Pesticide Research Institute
College of Agriculture, Kyoto University, Kyoto, Japan

Introduction

The possibility of using sex pheromones in controlling insects has stimulated efforts to isolate and identify these substances. So far, more than 26 pheromones have been isolated from female insects. There are still a number of problems remaining in bringing these chemicals into practical use. A number of reviews have appeared in recent years discussing such endeavors.

Control of the cabbage looper, *Trichophisia ni* Hübner, by the use of male inhibition technique to reach their females has been successfully demonstrated in field trials (Shorey, 1970). Knipling and McGuire (1966) proposed model systems to calculate the expected theoretical effects of direct trapping of males by using a sex-attractant. Roelofs et al. (1970) have modified one of the above authors' population models (model II), and calculated the effects of various parameters on the degrees of theoretical control of red-banded leaf rollers. They compared the results of actual pheromone (synergized) trapping and the theoretical control values. Where the field populations were low, these workers were able to obtain good degrees of reduction in the actual levels of fruit damages, indicating that the effects of the synergized attractants were profound enough to reduce the population. There are a number of field attempts now in progress to examine the possibility of utilizing pheromones for insect control. The case for the Gypsy moth (Beroza and Bierl, 1971), for instance, is worthy of attention in view of its economic implications.

This article deals with the sex pheromone of three stored grain Lepidoptera: the almond moth, *Cadra cautella* (Walker); the Indian meal moth, *Plodia interpunctella* (Hübner); and the Mediterranean flour moth, *Anagasta kuehniella* (Zeller). All are widely distributed in the Temperate and Tropical zones of the world and are serious economic pests of stored food products.

Isolation and Identification of Sex Pheromones of *Cadra*, *Plodia*, and *Anagasta* Females

The presence of a sex pheromone in female moths of these three species has been reported by several investigators (see Jacobson, 1965), and recently it was reported by three independent research groups that each of these female moths emits the same compound (cis-9, trans-12-tetradecadienyl acetate) (Brady et al., 1971; Dahm et al., 1971; Kuwahara et al., 1971a and 1971b).

Our studies on the isolation of the sex pheromone associated with the *Cadra* female are summarized here. As mentioned in the preliminary report (Nakajima, 1970), each fraction from the isolation process was monitored easily by a simple bioassay which was conducted as follows: Approximately 30 three-day-old males were confined in a glass vial (11 cm diam x 7 cm deep) covered with a glass plate having a hole (3 cm diam) in the center. An end of a glass rod was immersed into a test solution to coat the last 1 cm of the tip with the testing material, and then inserted through the hole into the glass vial. A positive response was recorded if the male vibrated its wings vigorously and hovered about the glass rod within 30 seconds. Test solutions were prepared by successive 10-fold dilutions of 1 mg, or 1 female equivalent sample, in 1 ml of n-hexane, and were assayed beginning with the most dilute solution until a positive response was observed.

About 1,200,000 unmated females of *C. cautella* were employed using the FT-strain (Takahashi and Mutuura, 1964). Of interest, females emerge from rice bran at temperatures below 20°C, while both sexes emerge at temperatures of 28 to 30°C. The extract secured from these females was subjected to the isolation procedures shown in Fig. 1. The isolated pheromone elicited excitatory sexual response from the males at 10^{-9} mg/ml by the glass rod test, and showed a single peak having a relative retention time of 1.167 with myristyl acetate as an internal standard (46.9 min) in gas-liquid chromatography (GLC) on a capillary column (45 m x 0.25 mm inside diam) coated with Ucon Oil LB-550X (R-45) at 160°C. The high-resolution mass spectrum of the pheromone gave a molecular ion peak at m/e 252.2059, indicating the molecular formula $C_{16}H_{28}O_2$ (the calculated value 252.2089). On catalytic hydrogenation on paladium black the pheromone gave myristyl acetate, which was confirmed by co-GLC with the authentic sample. This result shows that the pheromone molecule is an acetate of a straight-chain C_{14} alcohol with two double bonds. The presence of the $CH_3CH{=}CH{-}CH_2{-}CH{=}CH{-}$ structural moiety in the molecule was concluded from the NMR spectral data; signals at τ 4.57 (4H, multiplet), 7.27 (2H, broad), and around 8.3 (3H, complicated) were assigned to $-CH{=}CH-$, $-C{=}C{-}CH_2{-}C{=}C-$, and $CH_3{-}C{=}C-$, respectively. These assignments were supported by the experimental result that microozonolysis of the pheromone gave ω-acetoxy-nonanal as a main product, which was identified by GLC coupled with mass spectrometry. Therefore, the structure of the pheromone was concluded to be 9,12-tetradecadienyl acetate. To assign the geometrical structure on the two double bonds, four possible isomers were prepared according to

TABLE 1

Separation of Geometric Isomers of the Pheromone by Gas-Liquid Chromatography

	Isomer of 9, 12-tetradecadienyl acetate				
	cis-cis	cis-trans	trans-cis	trans-trans	Standard C_{14}-OAc
Packed column 160° 10% PEG-20 M 150 cm	1.635	1.542			1.000 (12.8 min)
Goley column 160° BDS, 45 m x 0.5 mm	1.865	1.721	1.835	1.721	1.000 (35.2 min)
Goley column 160° R-45, 45 m x 0.25 mm	1.224	1.167	1.253	1.183	1.000 (46.9 min)

the synthetic route shown in Fig. 2. Of these isomers the cis-9, trans-12-isomer was identical to the natural pheromone with respect not only to physicochemical properties, such as NMR spectrum and GLC on the capillary column (TABLE 1), but also to the pheromonal activity (TABLE 2).

TABLE 2

Activity[a] of Isolated Pheromone by Glass-Rod Method

	Pheromone derived from			
	C. caut.	*P. interp.*	*A. kueh.*	Synthetic
C. cautella	10^{-10}	10^{-10}	10^{-10}	10^{-11}
P. interpunctella	10^{-10}	10^{-10}	10^{-10}	10^{-11}
A. kuehniella	10^{-9}	10^{-9}	10^{-9}	10^{-10}
E. elutella	10^{-9}	10^{-7}	10^{-9}	10^{-10}

The pheromone from *P. interpunctella* females was isolated utilizing a procedure similar to that employed on the *Cadra* pheromone. However, in this case we used 670,000 moths, including both sexes (the sex ratio about 1:1), and obtained a yield of 0.5 mg. No trouble was encountered in isolating the pheromone without separating females from males prior to the extraction. Also, using the same procedure 0.8 mg of the pheromone of *Anagasta kuehniella* females was isolated from about 570,000 male and female moths.

Both the pheromone isolated from *Plodia* and *Anagasta* females was identified as cis-9, trans-12-tetradecadienyl acetate, in comparison with the authentic sample on GLC with capillary column, as well as data secured from mass spectrogram, and microozonolysis. Thus, the isolation and identification of the sex pheromones of these three insects indicates that each of these females emits the same sex pheromone compound. The pheromonal activity of the isolated pheromones, as well as comparable synthetic compounds, is shown in TABLE 2. In TABLE 3 it is interesting to note that male moths of the three species responded to geometric isomers of the pheromone in different degrees.

TABLE 3

Pheromonal Activity[a] of Geometric Isomers by Glass-Rod Method

	Isomer of 9, 12-tetradecadienyl acetate			
	cis-cis	cis-trans	trans-cis	trans-trans
<u>C</u>. <u>cautella</u>	10^{-9}	10^{-11}	10^{-9}	10^{0}
<u>P</u>. <u>interpunctella</u>	10^{-4}	10^{-11}	10^{-9}	10^{-2}
<u>A</u>. <u>kuehniella</u>	10^{-3}	10^{-10}	10^{-4}	10^{-2}
<u>E</u>. <u>elutella</u>	10^{-5}	10^{-10}	10^{-4}	-

[a]Unit, mg/ml.

In our work, as shown in TABLE 4, only a few milligrams of the pheromone was isolated from more than a million nonmated female moths. Brady et al. (1971) reported that extracts of filter paper, in contact with nonmated females of <u>Plodia</u> for 3 days, contained many times more sex pheromone than extracts of abdominal tips of the same number of nonmated 3-day-old females. They isolated 0.2 mg of the pheromone from filter paper in contact with 19,000 females. These results are in agreement with the general findings that sex pheromones, if present, are in very small amounts in most species of insects. It requires great effort to isolate enough sex pheromone from a given insect species to identify its chemical structure, even with the development of modern techniques and instrumentation.

TABLE 4

Yield of Sex Pheromone from Three Species of Phycitid Moth

	Female moth used	Yield	Yield/ Female moth
C. cautella	1,200,000	6.1 mg	5.08 ng
P. interpunctella	335,000	0.5 mg	1.49 ng
A. kuehniella	283,000	0.8 mg	2.83 ng

Species Specificity

It has been stated that sex pheromones of closely related lepidopterous insects are not always specific in their biological activity (Shorey et al., 1968, 1970). The Cadra female itself, or its crude extracts, elicit excitatory sexual response from Plodia males and also from Anagasta males, and vice versa (Nakajima, 1970). Also, each sex pheromone associated with each of these pyralid moths has been shown to be identical. These facts are not too surprising, as there have been similar instances where two or more lepidopterous insects respond sexually to the same compound (TABLE 5). Roelofs and Comeau (1970), in systematic screening tests in the field with 26 monounsaturated long-chain acetates and alcohols, present another example. A portion of their results are summarized in TABLE 6, showing that a number of insects, belonging to various families, are attracted to the same compound. This presents the question as to how reproductive isolation is effected between insects releasing the same sex pheromone. Roelofs and his colleagues (1968 and 1970) also noted that the addition of dodecyl acetate to cis-11-tetradecenyl acetate resulted in the drastic increase of attractancy for males of the red-banded leaf roller and a decrease in attractancy for the oblique-banded leaf roller, although both of these males are attracted in the same degree to

TABLE 5

Examples of Nonspecies Specificity of Sex Pheromones

Sex pheromone	Insect
cis-9-tetradecenyl acetate	The fall armyworm *Spodoptera frugiperda* The southern armyworm *Prodenia eridania** The smaller tea tortrix *Adoxophyes fasciata** The summerfruit tortrix *Adoxophyes orana**
cis-11-tetradecenyl acetate	The red-banded leaf roller *Argyrotaenia velutinana* The oblique-banded leaf roller *Choristneura rosaceana* The European corn borer *Ostrinia nubilalis* The smaller tea tortrix *A. fasciata** The summerfruit tortrix *A. orana**
cis-9, trans-12-tetradecadienyl acetate	The southern armyworm *P. eridania** The almond moth *Cadra cautella* The Indian meal moth *Plodia interpunctella* The Mediterranean flour moth *Anagasta kuehniella*

TABLE 6

Lepidopterous Insects Attracted to Alkenyl Acetates in the Field[a]
(10 μl in long-lasting polyethylene wicks)

	C_{12}-OAc	C_{14}-OAc
cis-5	Gelechiidae (2) Tortricidae (1)	---
trans-5	---	---
cis-7	Gelechiidae (1) Pyralidae (1) Noctuidae (5) *Tricoplusia ni* Noctuidae[a] (4)	Noctuidae (1)
trans-7	Tortricidae[a] (1)	Noctuidae (1)
cis-8	Tortricidae (2) *Grapholita molesta* *G. prunivora*	
trans-8	Tortricidae (1) *G. packardii*	
cis-9	---	Gelechiidae (1) *Bryotopha similis* Noctuidae (2) *Spodoptera frugiperda*
trans-9		Gelechiidae (1) Pyralidae (1) Noctuidae (1)
cis-11	---	Tortricidae (2) *Argyrotaenia velutinana*
trans-11		Tortricidae (1) Phaloniidae (1)

[a]From Roelofs and Comeau (1970).

[b]Not attracted in this test but reported in the literature. Numbers in parenthesis show the number of species attracted.

their female sex pheromone, cis-11-tetradecenyl acetate, alone. It is likely that emission of different secondary compounds by females of each species, along with the sex pheromone, affects reproductive isolation in the field. Brady et al. (1971) observed that the pure sex pheromone shows less attractancy in a large chamber for the *Cadra* males than for the *Plodia* males (TABLE 7), and they suggested that a secondary compound is contained in the *Cadra* female and is inactive by itself but causes a definite increase in orientation of the *Cadra* male on combination with the sex pheromone. Several other examples have

TABLE 7

Attraction of *Plodia* and *Cadra* Males to Synthetic cis-9, trans-12-Tetradecadienyl Acetate and the Active Sex Pheromones of *Plodia* and *Cadra* Females

Male	Males trapped[a] per night (%)		
	Synthetic pheromone	Pheromone from *Plodia*	Pheromone from *Cadra*
Plodia	46	53	51
Cadra	27		25

[a]In an isolated and dark room (42.4 m^3). Each test 100 unmated males, with 10 ng of compound. Results are an average of 4 to 7 tests. From Brady et al. (1971).

been reported indicating that secondary compounds may affect the activity of a sex pheromone. Propylure, the sex pheromone isolated from the pink ballworm *Pectinophora gossipiella*, attracted the male moths in a laboratory test but did not in the field. When combined with N,N-diethyl-*m*-toluamide, which was isolated from *Pectinophora* females, propylure showed a definite attractancy to male moths in the field (Jones and Jacobson,

1968). Jacobson et al. (1970) isolated two sex pheromones from the female southern armyworm Prodenia eridania and identified the compounds as cis-9-tetradecenyl acetate and cis-9, trans-12-tetradecadienyl acetate. They stated that perhaps the latter acetate serves as a distance attractant for the male in nature, whereas the former acetate sexually excites the male when the sexes are in close proximity. Recently Tamaki et al. (1971a, 1971b) reported that both the female smaller tea tortrix Adoxophyes fasciata and the female summerfruit tortrix A. orana emit as their sex pheromone(s) a mixture of two compounds, cis-9- and cis-11-tetradecenyl acetate, each of which alone, however, elicits neither a mating dance nor copulatory attempts by the males. However, when the two compounds are mixed, maximum male response is observed in the ratio 4:1 (the cis-9- to the cis-11-isomer) for A. fasciata and in 9:1 for A. orana (Fig. 3). These facts emphasize that in a number of cases the activity for sexual attractancy of a pheromone is not always based on a single compound, but rather on a complicated combination of two or more compounds.

Other complicated features of sex pheromone activity are shown in our following experimental results. The characteristic behavioral response of the Cadra male is always to both the crude extract of the female and to the synthetic pheromone (cis-9, trans-12-tetradecadienyl acetate), although the degree of response varies. However, in a large chamber (1 x 1.5 m, 1.6 m in height) the activity of the synthetic pheromone for attraction of the male approaches maximum with exposure of around 10^{-6} mg of the compound, and decreases remarkably with increasing amount of the compound, finally down to near inactivity (Fig. 4). When the test is conducted with a crude extract of the female under the same condition, activity is still observed even with such a large amount as 10 female equivalents.

The information presented herein suggest that further research on mating behavior of insects, as well as associated pheromones is necessary in order to make use of sex pheromones in insect control.

References

Beroza, M., and B. A. Bierl: 2nd Int. Congr. Pesticide Chem. Astr. Tel-Aviv, Israel, p. 59 (1971).

Brady, U. E., J. H. Tumilson III, R. G. Brownlee, and R. M. Silverstein: Science 171, 802 (1971).

Dahm, K. H., I. Richter, D. Meyer, and H. Roller: Life Science 10, 531 (1971).

Jacobson, M.: Insect Sex Attractants. Wiley, New York (1965).

Jacobson, M., R. E. Redfern, W. A. Jones, and M. H. Aldridge: Science 170, 542 (1970).

Jones, W. A., and M. Jacobson: Science 159, 99 (1968).

Knipling, E. J., and J. U. McGuire: U.S. Dep. Agr. Inform. Bull. 308, 2 (1966).

Kuwahara, Y., C. Kitamura, S. Takahashi, H. Hara, S. Ishii, and H. Fukami: Science 171, 801 (1971a).

Kuwahara, Y., H. Hara, S. Ishii, and H. Fukami: Agr. Biol. Chem. 35, 447 (1971b).

Nakajima, M.: Control of Insect Behavior by Natural Products. D. L. Wood, R. M. Silverstein, and M. Nakajima, Eds. Academic Press, New York, p. 209 (1970).

Roelofs, W. L., and A. Comeau: Nature 220, 600 (1968).

Roelofs, W. L., and A. Comeau: J. Econ. Entomol. 63, 969 (1970).

Roelofs, W. L., and J. P. Tette: Nature 226, 1172 (1970).

Roelofs, W. L., E. H. Glass, J. P. Tette, and A. Comeau: J. Econ. Entomol. 63, 1162 (1970).

Shorey, H. H.: Control of Insect Behavior by Natural Products. D. L. Wood, R. M. Silverstein, and M. Nakajima, Eds. Academic Press, New York, p. 249 (1970).

Shorey, H. H., L. K. Gaston, and R. N. Jefferson: Advances in Pest Control Research. R. L. Metcalf, Ed. Interscience Publishers, New York, p. 8 (1968).

Takahashi, F., and A. Mutuura: *Japan. J. Appl. Entomol.* 8, 129 (1964).

Tamaki, Y., H. Noguchi, T. Yushima, and C. Hirano: *Appl. Entomol. Zool.* 6, 139 (1971a).

Tamaki, Y., H. Noguchi, T. Yushima, C. Hirano, K. Homma, and H. Sugawara: "*Kontyu*" (*Japan. J. Entomol.*) (In press, 1971b).

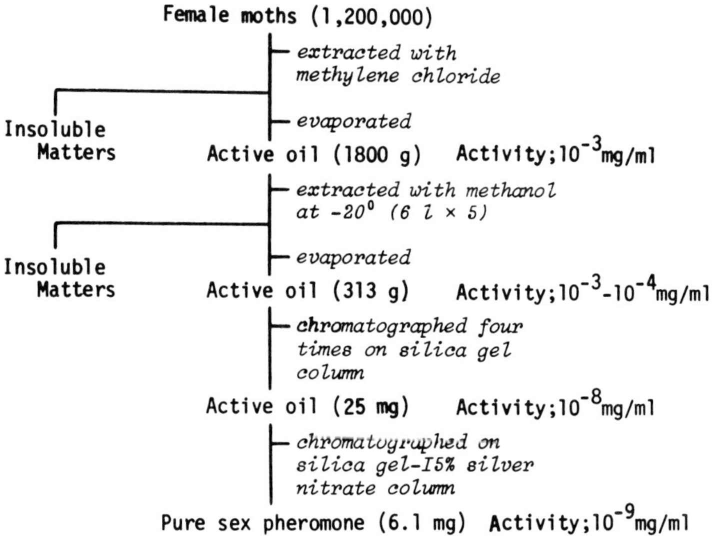

Fig. 1. Isolation of the sex pheromone of *Cadra cautella*.

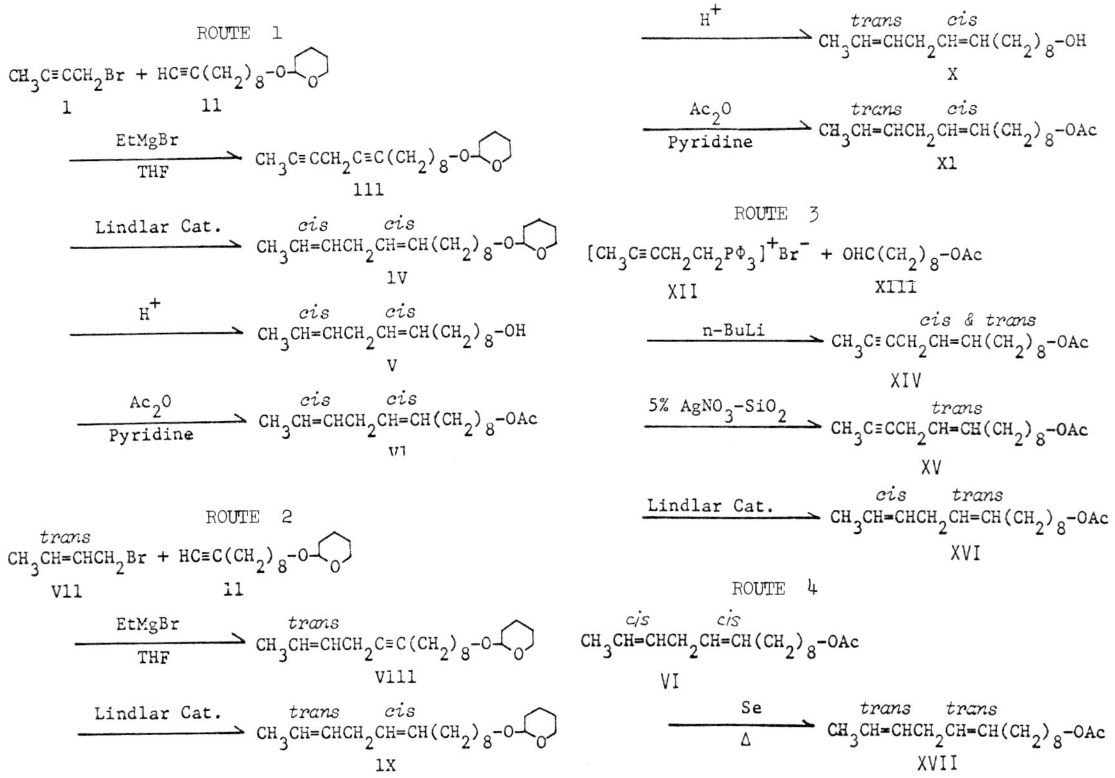

Fig. 2. Synthetic routes of the sex pheromone of
Cadra cautella and its geometric isomers.

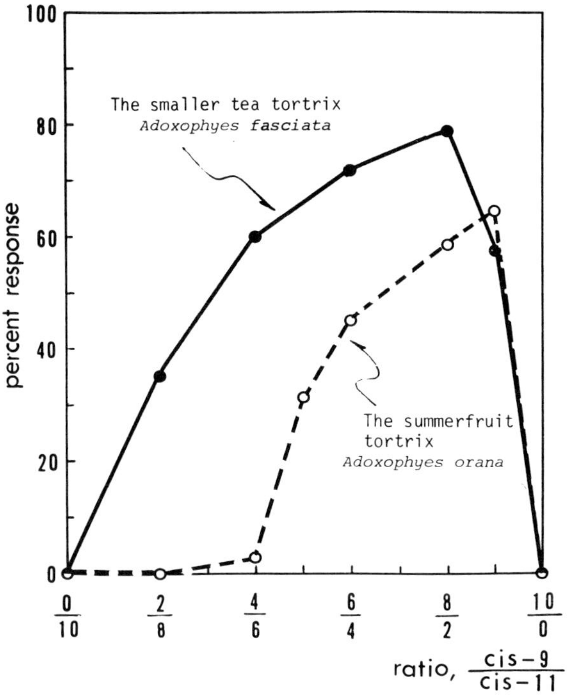

Fig. 3. Pheromone activity of cis-9-, and cis-11-tetradecenyl acetate in a laboratory bioassay. Applied amount; 1 x 10^{-7} mg. (From Y. Tamaki et al., 1971)

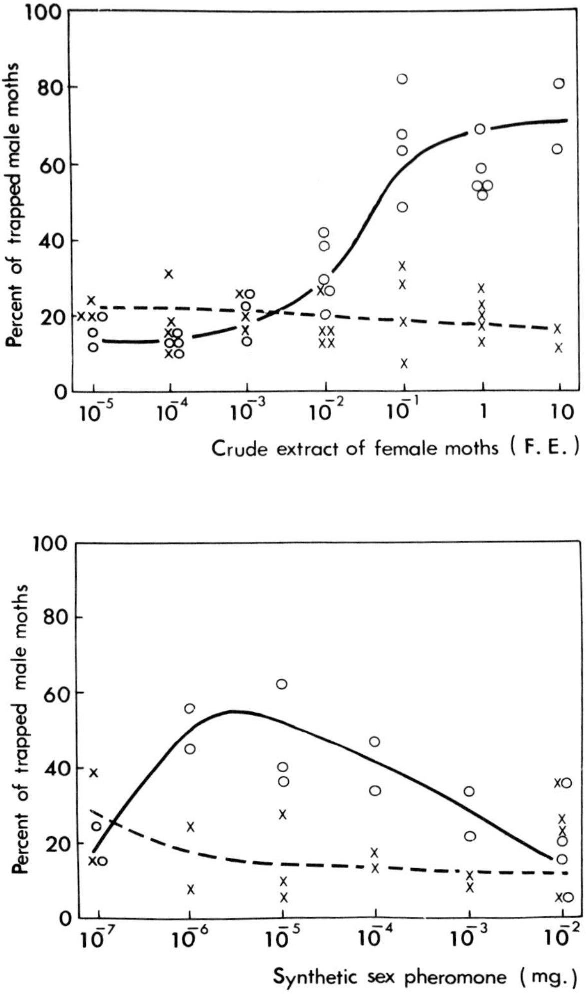

Fig. 4. Response of males of the almond moth in a chamber (1 x 1.5 x 1.6 m).

A NEW APPROACH IN THE DEVELOPMENT OF BIODEGRADABLE PESTICIDES

Tomomasa Misato, Isamu Yamaguchi, and Yasuo Homma

The Institute of Physical and Chemical Research
Wako-shi, Saitama, Japan

Introduction

Pollution-free pesticides should be stable chemically but sensitive for photolysis and degradation by microorganisms. DDT, BHC, and plastics have chemical structures which do not exist in the natural world and this is a reason why these substances are resistant to the degradation by microbes. It is supposed that naturally-occurring compounds, or their derivatives with similar structures, can be degraded by microbes. Thus, their practical use may not carry with it the possibility of environmental pollution. From these viewpoints, we have studied two groups of biodegradable chemicals and their utilization as fungicides: They are antibiotics as microbial products and complexes of amino acids and fatty acids, which are essential elements of all organisms (Takeuchi et al., 1958; Misato et al., 1959; Suzuki et al., 1965; Umezawa et al., 1965; Seto et al., 1966a; and Seto et al., 1966b).

Antibiotics had been used only on a very limited scale for the practical control of plant diseases, until the curative effect of blasticidin S on rice blast was discovered by the authors in 1958 under the guidance of Dr. Sumiki, Honorary Professor of Tokyo University (Takeuchi et al., 1958; Misato et al., 1959). The successful application of blasticidin S against rice blast has stimulated the development of agricultural antibiotics, and led to the discovery of several excellent agricultural antibiotics such as Kasugamycin (Umezawa et al., 1965), Polyoxin (Suzuki et al., 1965), etc. The amounts of formulated antibiotics used in 1970 in Japan for the control of plant diseases are given in TABLE 1.

TABLE 1

Agricultural Antibiotics used practically in Japan

Antibiotics	Disease	Amount used in 1970 (ton, K)
Blasticidin S (Dust)	Rice Blast	5.211
Kasugamycin (Dust)	Rice Blast	24.596
Polyoxin (Dust)	Rice Sheath Blight	2.094
Streptomycin (Solution)	Bacterial Disease of Fruits and Vegetables	215
Cycloheximide	Onion Downy Mildew Shoot Blight of Japanese Larch	2
Griseofulvin	Fusarium Wilt of Melon Apple Blossom Blight	4
Chloramphenicol	Rice Bacterial Leaf Blight	2

Two years ago we started studies on the application of common naturally-occurring substances such as amino acids, fatty acids, etc. as potential agricultural chemicals which are expected to be free from environmental contamination. As the result of our research we have found that N-Lauroyl-L-Valine, a complex of lauric acid and valine combined by N-acyllinkage, shows an excellent preventive effect against rice blast.

In this paper, we would like to report metabolism and degradation in environments of an agricultural antibiotic, blasticidin S, which has been used for rice blast control, and an amino acid derivative, N-Lauroyl-L-valine, which is promising as a pollution-free pesticide.

Blasticidin S

Blasticidin S is a fairly stable chemical, but in the field in practical use the antibiotic is subjected to degradation or inactivation by the rice plant, sunlight, or microorganisms. Concerning metabolism and degradation of antifungal chemicals or medical antibiotics, many papers have been reported, but studies on the agricultural antibiotics are few in number. In order to investigate the behavior and fate of blasticidin S in the paddy field, radioactive blasticidin S was prepared biosynthetically, and we studied its behavior by using a model system.

Preparation of labeled blasticidin S - Procedure of preparation of ^{14}C-labeled blasticidin S (Fig. 1) was described in a previous paper (Seto et al., 1966a). Cytosine-2-^{14}C (60 mCi/mM) and L-methionine (methyl-^{14}C) (62.5 mCi/mM) were purchased from the Radiochemical Centre, Amersham, England. The labeled blasticidin S prepared was recrystallized with distilled water, then further purified by thin-layer chromatography on AVICEL (chloroform:methanol:17% ammonia = 2:2:1), and the radioactivity was counted on a Beckman liquid scintillation spectrometer. The specific radioactivities of cytosine labeled-blasticidin S and N-methyl labeled blasticidin S are 5.63 mCi/mM and 8.40 mCi/mM, respectively.

Stability of blasticidin S sprayed on rice blast - As previously reported (Misato et al., 1959), the activity of blasticidin S sprayed on rice plants decreased gradually. TABLE 2 shows the stability of blasticidin S on leaves of rice plants in various conditions examined by the cup method using *Cladosporium fulvum* as the test organism. When plants were exposed to sunlight directly or on young plants, blasticidin S on leaves seemed to lose its activity

TABLE 2

Stability of Blasticidin S sprayed on Rice Plant

Days	In greenhouse			Outdoors	
	4th leaf stage	Milk-ripe stage		Milk-ripe stage	
	10 ppm	10 ppm	20 ppm	20 ppm	100 ppm
	μg/g (%)	μg/g (%)	μg/g (%)	μg/g (%)	μg/g (%)
0	3.2 --	3.1 --	5.7 --	5.9 --	28.7 --
1	2.4 (75.0)	2.7 (87.1)	5.1 (89.5)	4.4 (74.6)	23.3 (81.2)
3	2.1 (65.6)	2.4 (77.4)	4.6 (80.7)	2.7 (45.8)	19.5 (68.0)
5	1.5 (46.8)	2.1 (67.7)	4.2 (73.7)	1.8 (30.5)	17.0 (59.2)

more rapidly.

Translocation of labeled blasticidin S - Radioactive blasticidin S was applied to rice plants (designated as Jukkoku) of the fourth leaf stage or the milk-ripe stage at concentrations of 10 or 20 ppm with a sticker, neoesterine (0.1%). The treatment was carried out with an implement which was equipped with a small piece of sponge or rubber at the top of forceps. In the case of water cultures, the test plants of the fifth leaf stage were dipped in culture medium containing 100 mg $(NH_4)_2SO_4$, 50 mg Na_2HPO_4 $12H_2O$, 25 mg KCl, 15 mg $MgCl_2 \cdot 6H_2O$, 6 mg $FeCl_3$, 4 mg $CaCl_2$ and 0.13 mg $MnCl_2 \cdot 6H_2O$ in 1 liter distilled water. Labeled blasticidin S was dissolved in 50 ml of this medium at a final concentration of 5 μg/ml. After a given period of time, the plants were harvested and flattened by ironing, then radioautograms were taken using Fuji X-ray film.

Neither cytosine labeled-blasticidin S, BcS(*cyt.), nor N-methyl labeled-blasticidin S, BcS(*met.), were translocated over a period of 1 week. Figure 2 shows the translocation of blasticidin S on the leaves of healthy and infected young rice plants. It was found that only a small amount of ^{14}C-blasticidin S was translocated to the upper portions of the leaves of healthy plants, but in the case of infected plants, the radioactive antibiotic was incorporated and translocated more easily and seemed to be accumulated in the infected parts.

The translocation of ^{14}C-blasticidin S in water culture is shown in Fig. 3. This radioautogram indicates that blasticidin S was taken up by the roots and translocated to the entire plant in considerably high quantities.

Diffusion of blasticidin S into plants - Labeled blasticidin S (20 ppm) was sprayed or smeared on the rice plant in the milk-ripe stage. Three weeks later, ripened grains were harvested and separated into glumes and unpolished rice with forceps. The glumes were then washed with water to remove adsorbed radioactivity. The leaves were divided into halves in the midribs and one of the halves was washed thoroughly with distilled water. Then they were smashed or cut into pieces and homogenized with a definite volume of water, and the radioactivity was counted on an Aloka gas flow counter (model PDC801U). The results are presented in TABLE 3. Blasticidin S permeated into leaves at most only 15% after 10 days, and less into the stems of the plant. Also, the compound was located on the surface of rice grains and little diffused into unpolished rice (TABLE 4). When 20 ppm ^{14}C-blasticidin S was sprayed on the ears of plants, the unpolished rice contained less than 0.01 ppm radioactive compound as blasticidin S.

TABLE 3

Permeation of Blasticidin S into Rice Plant

Days	Not washed cpm/mg	Washed cpm/mg	Permeability (%)
0 (leaf)	725	41.1	5.65
1	720	69.7	9.70
3	565	65.0	11.50
5	572	72.0	12.60
10	518	81.5	15.73
5 (stem)	77.5	1.63	2.10

TABLE 4

Migration of Labeled Blasticidin S[a] into Rice

Glume	2,742 cpm/grain	97.9%
Unpolished rice	59.4	2.1

[a] Three weeks after treatment. Blasticidin S labeled at cytosine moiety.

Extraction of metabolites from rice plant - The procedure for extraction of metabolites from rice plants is shown in Fig. 4. Water cultured rice plants described above were divided into leaves and roots. Leaves were used mainly in the present experiment. In this procedure the radioactive compounds could be extracted with good recovery, but various other water-soluble compounds, such as proteins and sugars, were contaminated in this preparation and, therefore, Sephadex G-25 and IRC 50 (H^+) were used to eliminate the impurities. Most radioactivity (82%) adsorbed to the resin was eluted with 5% ammonia. This fraction was identified as blasticidin S. Small amount of cytomycin (4.8%) and trace of deaminohydroxyblasticidin S (1.2%) were identified by the combination of thin-layer chromatography and radioautography.

Degradation of blasticidin S on leaves - Radioactive-blasticidin S was applied on the leaves of rice plants and on glass plates and then exposed to sunlight, and covered with Kf film which transmitted ultraviolet as well as other solar rays. This film was kindly supplied by Kureha Chemical Co., Ltd. Degradation products were obtained by washing the leaf surface with distilled water. As shown in Fig. 5, blasticidin S located at the plant surface was decomposed efficiently by sunlight, and comparison with the products on the glass plate indicated that decomposition was not enzymic but physical. Cytosine was identified as the major product of decomposition. Cytomycin and other unknown substances were also observed on the thin-layer chromatogram. Isolation and characterization of the products are under investigation.

Adsorption of blasticidin S by soil - Several types of soil were washed thoroughly with water and poured onto the glass filter. To a soil column (2 x 6.5 cm), 1 ml of labeled blasticidin S (20 μg) was applied and developed with 80 ml distilled water. After development, the soil was cut into ten segments. To each segment, 10 mg cold blasticidin S was added as a carrier, and the labeled blasticidin S was extracted with alkaline 60% acetone. The results are presented in TABLE 5. It is concluded that

TABLE 5

Adsorption of Blasticidin S by Soil as Tested by Column Elution Method

Segment Number[a]	Soil species				
	KT 1-1	NO 1-1	M 9-1	US 8-1	K 36-1
	cpm (%)	cpm (%)	cpm (%)	cpm (%)	cpm (%)
1	3,847 (93.9)	3,530 (97.7)	4,612 (97.9)	4,510 (96.3)	3,789 (95.2)
2	59 (1.4)	22 (0.6)	49 (1.0)	61 (1.3)	67 (1.6)
3	64 (1.6)	5 (0.2)	17 (0.4)	30 (0.6)	57 (1.4)
4	36 (0.9)	10 (0.3)	11 (0.2)	25 (0.5)	16 (0.4)
5	33 (0.8)	0 (0)	9 (0.2)	7 (0.1)	7 (0.2)
6	12 (0.3)	7 (0.2)	7 (0.1)	5 (0.1)	10 (0.2)
7	6 (0.2)	2 (0.1)	2 (0)	3 (0.1)	5 (0.1)
8	17 (0.4)	7 (0.2)	0 (0)	13 (0.3)	7 (0.2)
9	6 (0.2)	10 (0.3)	1 (0)	1 (0)	6 (0.2)
10	0 (0)	5 (0.2)	1 (0)	14 (0.3)	11 (0.3)
Wash	17 (0.4)	5 (0.2)	3 (0.1)	15 (0.3)	3 (0.1)

[a] Top layer is represented by "1".

blasticidin S is adsorbed firmly on the surface of soil and not eluted with large volumes of water, as is normal.

Degradation of blasticidin S by microorganisms - Screening of microorganisms capable of altering the biological activity of blasticidin S was carried out according to the method reported previously (Seto et al., 1966b), except that the soils were obtained from paddy fields and the culture media were glucose-bouillon and yeast-extract (0.5%) - (2%). Microbes previously identified were also selected in the same way. The ability of newly isolated and previously identified microorganisms to degrade blasticidin S was examined by bioassay and thin-layer chromatography. *Bacillus cereus* was used as a test organism for bioassay. Thin-layer chromatograms were scanned on an Aloka thin-layer chromatogram scanner (model TRM IB), and radioautography was carried out.

TABLE 6 shows the alteration of biological activity of blasticidin S with cultures of various microorganisms. Several strains of fungi and bacteria were cultivated for 24 and 48 hr before blasticidin S was added. Following this, they were further incubated for 20 hr. Prior to addition of blasticidin S, pH of the medium was adjusted to 7.0 with N/10 HCl or NaOH. As indicated in TABLE 6, microbes usually found in paddy fields, such as *Pseudomonas marginalis*, *P. ovalis*, and *Fusarium oxysporum* depressed the activity of the compound under these conditions, and the major product of decomposition was cytomycin. However, washed cells or mycelia of these microbes did not show significant degradative ability. However, a fungal strain isolated from soil showed marked inactivation of blasticidin S by washed mycelia (Fig. 6). The major compounds produced were identified as deaminohydroxyblasticidin S and deaminocytomycin by TLC and UV spectra. Detail studies on identification of this fungal strain and biochemical characterization of the transformations are now in progress. Degradative pathways of blasticidin S are summarized in Fig. 7.

TABLE 6

Microbial Degradation of Blasticidin S

Number	Organisms	Activity remained
		$\mu g/ml$
0	Control	1,000
1	Pseudomonas marginalis	530
2	Ps. ovalis 1002	598
3	Ps. aeruginosa	698
4	Bacterium from soil (1)	690
5	Bacterium from soil (2)	700
6	Bacterium from soil (3)	720
7	Fusarium oxysporum	650
8	Phytophythora parasitica	690
9	Fungus from soil (1)	50
10	Fungus from soil (2)	630

L-Lauroyl-L-Valine

An amino acid pesticide, N-Lauroyl-L-valine, is a complex of lauric acid and valine combined by N-acyl linkage. It has been found by our research group that this compound shows excellent preventive action against rice blast disease. Since lauric acid and valine are common naturally-occurring compounds, it is considered that there are no residue problems. If only the N-acyl bond is split, it is assumed that this pesticide can be rapidly decomposed by microorganisms. In order to confirm this possibility, the carbon atom of the N-acyl linkage was labeled using lauric acid-1-^{14}C, and decomposition rates of the radioactive amino acid complex thus prepared by soil microorganisms was investigated.

Utilization of N-Lauroyl-L-Valine as carbon and nitrogen sources by bacteria - In order to select microorganisms capable of easily degrading N-Lauroyl-L-valine, culture media containing this compound as a main carbon and nitrogen source were used. The media was composed of 0.1% KH_2PO_4, 0.04% $MgSO_4 \cdot 7H_2O$, 3.0% $CaCO_3$, 0.05% Yeast extract, and 1.0% Sodium N-Lauroyl-L-valinate (pH 7.0). Microorganisms tested were nine strains of aerobic bacteria maintained at the Central Research Laboratory of Ajinomoto Co., Ltd., as listed in TABLE 7. Cells grown on a bouillon agar slant for 24 hr at 30°C were inoculated by a loop to a test tube containing the medium above mentioned, and incubated for 48 hr at

TABLE 7

Bacterial Growth in the Medium Containing N-Lauroyl-L-valine as Sources of Carbon and Nitrogen

Microorganisms	AJ No.	Other No.	Growth
Pseudomonas aeruginosa	2116	ATCC 10145	++++
Pseudomonas ovalis	2011	IAM 1002	+++
Achromobacter delmarvae	2390	IAM 2390	-
Escherichia coli	2617	ATCC 11775	++
Flavobacterium aquatile	2513	ATCC 11947	±
Bacillus subtilis	1315	ATCC 6051	-
Bacillus megaterium	1273	IAM 1073	-
Arthrobacter globiformis	1422	ATCC 8010	-
Micrococcus luteus	1003	ATCC 398	+

28°C. The degree of bacterial growth was measured by naked eye, and classified into six groups, indicated by the following signs (-, ±, +, ++, +++, ++++). As shown in TABLE 7, bacteria usually found in paddy fields such as Pseudomonas aeruginosa, P. ovalis, and Escherichia coli grew well in this medium.

Degradation of N-Lauroyl-L-valine by Pseudomonas aeruginosa - Since growth response of P. aeruginosa to the medium was the best among strains tested, this strain was used for studies on the degradation of N-lauroyl-L-valine. The colorimetric Pinacyanol chloride-$CHCl_3$ method modified by Kamimura was used for the determination of N-lauroyl-L-valine. This method was not interfered by lauric acid and valine. As shown in TABLE 8, after 96 hr more than 50% of this compound was decomposed. Additions of 0.03% $(NH_4)_2SO_4$ and 0.03% $(NH_4)_2CO_3$ to the medium stimulated the decomposition of the compound.

Production of $^{14}CO_2$ from N-Lauroyl-L-valine-^{14}C - N-Lauroyl-L-valine-^{14}C (Fig. 8) was prepared from lauric acid-1-^{14}C and L-valine.

Labeled N-lauroyl-L-valine and *P. aeruginosa* cells in phosphate buffer were mixed in vessels employed for Warburgs manometry, and incubated for 18 hr at 35°C (Fig. 9). Labeled CO_2 produced was trapped by ethanoalamine in the center well of the vessel, and carbon dioxide dissolved in the buffer was gasified by the addition of H_2SO_4 to the buffer. After further incubation, the radioactivity in the ethanolamine was counted on a liquid scintillation counter. The result is presented in TABLE 9. It was found that labeled CO_2 was released from the pesticide by the action of microorganisms such as *P. aeruginosa*. The generation of radioactive CO_2 would indicate cleavage of the N-acyl linkage. It was presumed that the freed fatty acid was oxidized through TCA cycle, and that radioactive CO_2 was generated in this process.

TABLE 8

Degradation of N-Lauroyl-L-valine by *Pseudomonas aeruginosa*

Experiment No.	Hr	Growth (O.D. at 562 mμ)	Residual N-Lauroyl-L-valine (mg/dl)	Ratio of consumption (%) of N-Lauroyl-L-valine
1[a]	0	0	2820	100.0
	96	1.900	680	75.9
2	0	0	2664	100.0
	96	1.500	1246	53.3

[a] 1 is added with $(NH_4)_2SO_4$ and $(NH_4)_2CO_3$

TABLE 9

Production of $^{14}CO_2$[a] from N-Lauroyl-L-valine (^{14}C) by *Pseudomonas aeruginosa*

N-Lauroyl-L-valine (^{14}C)	+	+	+	-
Cells	+	-	+	-
C.P.M.	33409 (as $^{14}CO_2$)	93135 (as substrate)	61623 (except $^{14}CO_2$)	40 (background)

[a]Ratio of produced $^{14}CO_2$ = $\frac{33409 - 40}{93135 - 40} \times 100 = 35.8\%$

Recovery = $\frac{61623 - 40 + 33409 - 40}{93135 - 40} \times 100 = 101.8\%$

References

Misato, T., I. Ishii, M. Asakawa, Y. Okimoto, and K. Fukunago: Ann. Phytopathol. Soc. Japan 24, 302 (1959).

Seto, H., I. Yamaguchi, N. Otake, and H. Yonehara: Tetrahedron Letters 32, 3793 (1966a).

Seto, H., N. Otake, and H. Yonehara: Agr. Biol. Chem. 30, 877 (1966b).

Suzuki, S., K. Isono, J. Nagatsu, T. Mizutani, Y. Kawashima, and T. Mizuno: J. Antibiotics (Tokyo), Ser. A 18, 131 (1965).

Takeuchi, S., K. Hirayama, K. Ueda, H. Sakai, and H. Yonehara: J. Antibiotics (Tokyo), Ser A 11, 1 (1958).

Umezawa, H., Y. Okami, T. Hashimoto, Y. Suhara, M. Hamada, and T. Takeuchi: J. Antibiotics (Tokyo), Ser. A 18, 101 (1965).

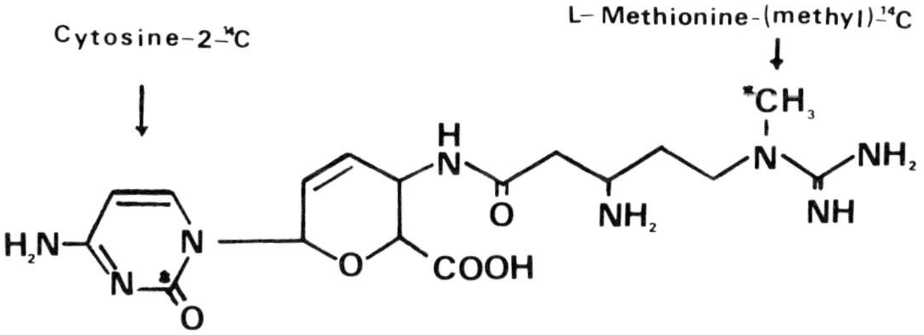

Fig. 1. Biosynthesis of blasticidin S.

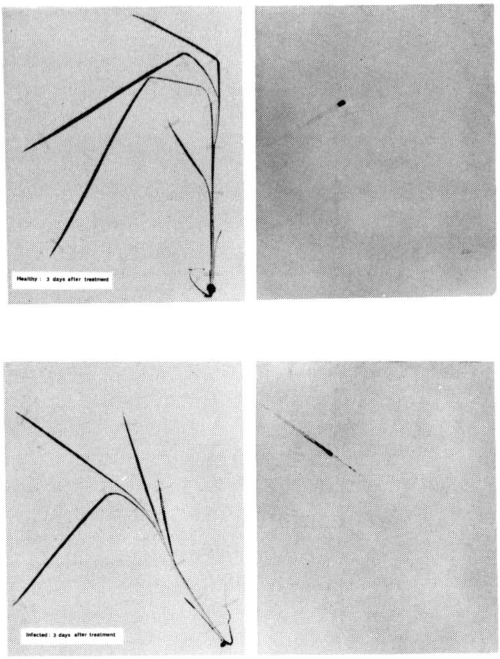

Fig. 2. Translocation of blasticidin S in the 4th leaf stage rice plant.

Upper: healthy plant, lower: infected plant

Blasticidin S (*cyt.) was applied on the infected plant 40 hr after infection.

Radioautograms were made 3 days after application of blasticidin S (*cyt.).

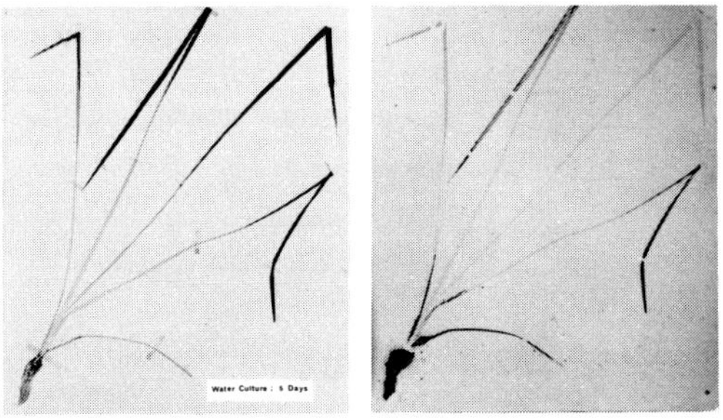

Fig. 3. Translocation of blasticidin S in the water culture.

Blasticidin S (*cyt.) was contained at concentration of 0.5 $\mu g/ml$ in the medium.

Radioautogram was made after 5 days cultivation.

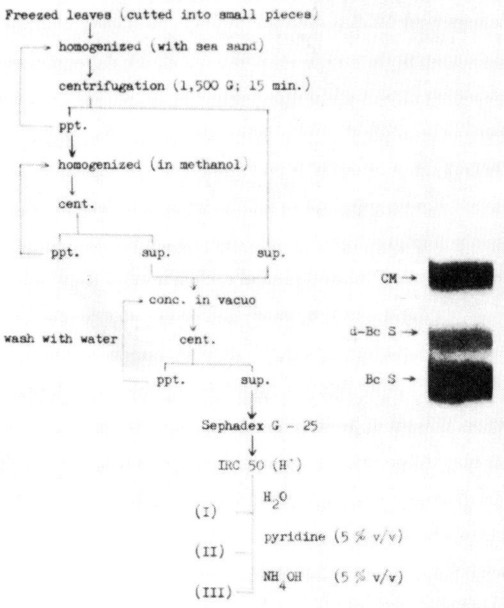

Fig. 4. Extraction of metabolites from rice plant.

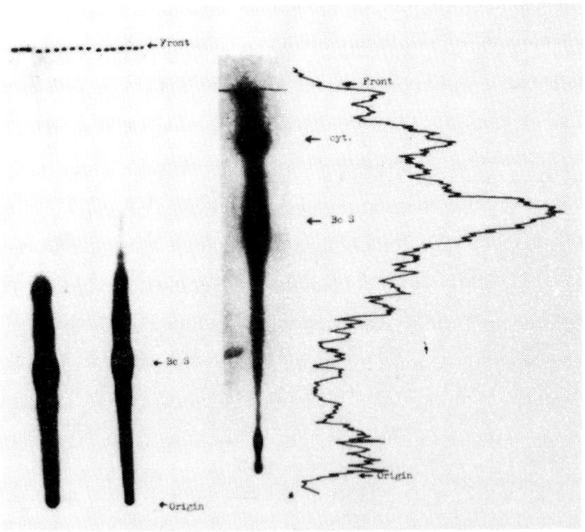

Fig. 5. Degradation of blasticidin S on leaves.

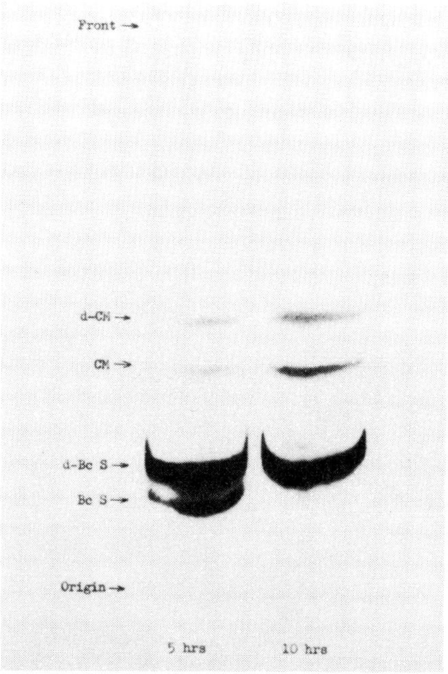

Fig. 6. Radioautogram of TLC of microbial metabolites.

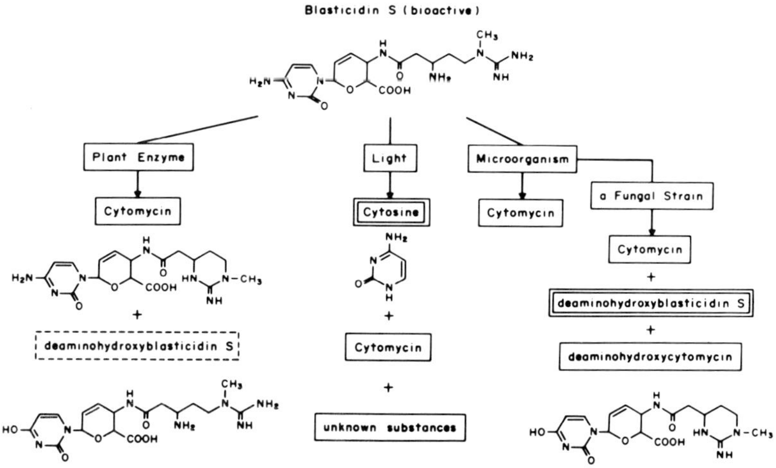

Fig. 7. Degradative pattern of blasticidin S.

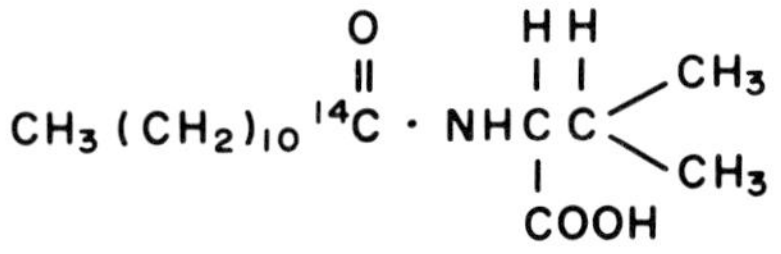

Fig. 8. Labeled N-Lauroyl-L-valine.

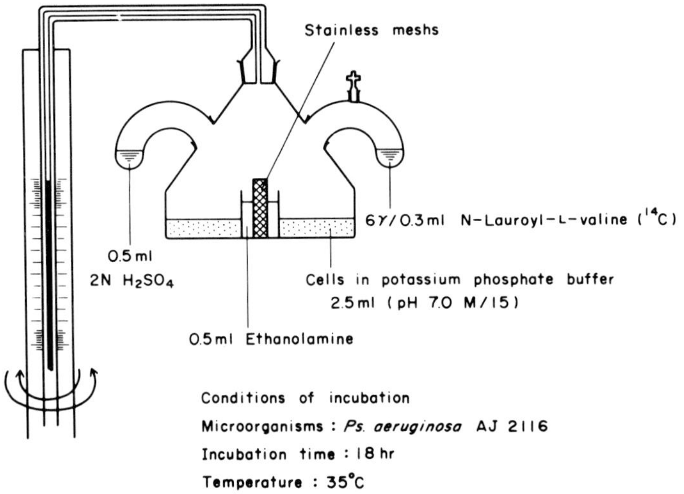

Fig. 9. Procedure of analysis of $^{14}CO_2$ produced from N-Lauroyl-L-valine (^{14}C) by Pseudomonas aeruginosa.

ENVIRONMENTAL EVALUATION OF SYSTEMIC FUNGICIDES

Teruhisa Noguchi

Nisso Institute for Life Science
Oiso, Kanagawa, Japan

Introduction

It is very important that we evaluate the impact of pesticides on the environment and that we are especially aware of ecological food chains and toxicant interrelationships. Therefore, in developing new pesticides, more knowledge is required as to the environmental toxicity of these compounds. Recently, we invented a novel type of systemic fungicide and are now evaluating this group from the viewpoint of environmental toxicology.

It is well known that sulfur plays an important role in an organism in maintaining life. Use of sulfur as a biocide, on the other hand, has been noted from ancient times. For instance, Homer tells how Odysseus, after slaying the suitors for his wife's hand, cleaned and purified his house by burning sulfur. Pliny, writing around A.D. 50, told of four kinds of sulfur having 14 medicinal virtues. In more recent times, some 35 years ago, dithiocarbamates were introduced as organosulfur fungicides by the patent of Tisdale and Williams (1934).

Turning to the problems of plant protection, more selective pesticides are desirable as man becomes more reluctant to add synthetic substances of broad and lasting effectiveness to the environment. Thus the interesting selective toxicity of the arylthionocarbamates (Noguchi et al., 1962) encouraged us to extend a similar type of investigation to the closely related arylthiocarbamides (Noguchi and Ohkuma, 1971).

Arylbisthiocarbamide was chosen as a prototype and it and related synthetic analogues were developed. They were evaluated against tobacco mosaic virus (TMV) since compounds of this type might affect the metabolic regulatory system of *de novo* synthesis of polynucleotides. Monoalkylthioallophanate derivatives were found more active than bis-derivatives against TMV. Bis-alkylthioallophanate, however, manifested potential activity *in vivo* against phytopathogenic fungi, while meta- and para-bis-substituted analogues were inactive.

As a result of these investigations a novel type of systemic fungicides was found in 1966 (Noguchi et al., 1969a, 1969b). This group was designated as the thiophanates: dialkyl 4,4'-o-phenylenebis (3-thioallophanate). Among the active compounds were diethyl 4,4'-o-phenylene bis (3-thioallophanate) (thiophanate, NF-35), and its dimethyl analogue (thiophanate-methyl, NF-44). These compounds contribute greatly in the effective control of several major important phytopathogenic fungi including powdery mildew of apple, apple scab, postharvest rot, and mold of citrus, pear scab, rice blast, rice sheath blight, and *Cercospora* leaf spot of sugar beet.

Furthermore, extensive field trials in Japan and other countries confirmed that thiophanates were highly effective in the control of many kinds of fungus diseases (Ishii, 1970). They were also found to be effective against some postharvest diseases of citrus, peach, and banana. In addition, no sign of phytotoxicity was observed with spray foliar application of thiophanates (2000 ppm) on 48 species of plants in 13 families. In addition to low toxicity to host plants, thiophanates were found to increase the chlorophyll content of the leaf. They were also of high ovicidal activity against the eggs of the spider mite (*Tetranychus urticae* Koch).

Toxicological Evaluation

TABLE 1 summarizes items to be studied for the full toxicological and pharmacological evaluation of a pesticide. Almost all of these experiments have been completed with thiophanate-methyl with a few still currently under investigation.

TABLE 2 shows the low toxic properties of thiophanate-methyl in tests for acute toxicity to several species of animals by different routes of administration. Neither did the compound exhibit any significant activity in various kinds of pharmacological tests (TABLE 3).

TABLE 4 summarizes the results of the subchronic toxicity studies of thiophanate-methyl in mice following 6 months of administration. According to the pathological observations (TABLE 5) it was concluded that the "no effect" level of this compound was 1600 ppm, corresponding to 250 mg/kg per day in mice.

Similar results were obtained with rats. From the histopathological point of view, the "no effect" level of thiophanate-methyl in rats (6 months administration) was considered to be 1600 ppm, which corresponded to 120 mg/kg per day.

The reproductive performance of mice receiving thiophanate-methyl during pregnancy was studied (TABLE 6). There were no significant effects between the control and treated groups on body weight, fetal mortality, mean body weight of fetuses, and average litter size of live fetuses. Also, microscopic observations of cleared skeletons showed no significant difference in incidence of malformations.

Three types of mutagenic studies were performed: the dominant lethal mutation test, host mediated assay, and *in vitro* or *in vivo* cytogenetic studies. No serious effects of thiophanate-methyl upon these mutagenicities was noted.

Dynamic transformation patterns of thiophanates were studied with ^{14}C-labeled compounds and with ^{35}S-labeled compounds (Fig. 1).

TABLE 1

Toxicological Evaluation on the Safety of Pesticide

1) Acute Toxicity
   (mouse, rat, rabbit, cat, dog, monkey, etc.
   (oral, dermal, s.c., i.m., i.p., i.v., etc.
2) Subacute Toxicity
   (mouse, rat, dog, etc.
   (various routes of administration
3) Chronic Toxicity
   (rat, dog, etc.
   (dietary feeding, etc.
4) Carcinogenicity
   (mouse (sensitive & non-sensitive)
   (dietary feeding, etc.
5) Teratogenicity
   (mouse, rat, rabbit, etc.
   (various routes
6) Three-generation Reproduction
   (mouse, rat, etc.
   (dietary feeding, etc.
7) Mutagenicity
   (a) dominant lethal assay: mouse, rat, hamster
   (b) host-mediated assay: mouse, rat
   (c) cytogenetics: rat, rat kangaroo, hamster
8) Synergistic Toxicity
   a) acute, b) subacute, c) chronic
9) General Pharmacology: various species of animal
10) Metabolism
   a) man & animals (mouse → cow)
   b) plants, c) soil & water
11) Toxicity on the Metabolites
   a) acute, b) subacute, c) chronic, etc.
12) Toxicity to Wildlife
   a) fish: carp, trout, bluegills, catfish, etc.
   b) birds: quail, pheasant, ducks, etc.
   c) honeybee, etc.
13) Toxicity to Domestic Animals: cow, pig, etc.
14) Residues and Methods of Analysis

TABLE 2

Acute Toxicity of Thiophanate-Methyl[a]

Species	Sex	Oral	Dermal	Intraperitoneal	Subcutaneous
Mouse (ddy)	M	3.51	>10.0	0.79	>10.0
	F	3.40	>10.0	1.11	>10.0
Rat (Wistar)	M	7.50	>10.0	1.64	>10.0
	F	6.64	>10.0	1.14	>10.0
Guinea-pig	M	3.64	10.0	-	-
	F	6.70	10.0	-	-
Rabbit	M	2.27	>10.0	-	-
	F	2.50	>10.0	-	-
Dog (Beagle)		4.00 (MLD)	>10.0	-	-
Japanese quail		>5.00	-	-	-
Carp (*Cyprionus carpio*)			TLM	75.0 ppm	
Rainbow trout (*Salmo iridens*)			TLM	8.8 ppm	

[a] LD_{50}; g/kg)

TABLE 3

Pharmacologic Evaluation of Thiophanate-Methyl

Central actions -	Body temperature (rat)
	Analgesic action (mouse)
	Sedative and hyponotic action (mouse)
Peripheral actions -	Mydriatic action
	Atropine- and Papaverine-like actions (mouse)
	Antihistamine action (guinea pig)
	Antiepinephrine action (guinea pig)
Action on the Cardiovascular and Respiratory System -	
	Respiration (rabbit)
	Blood pressure (with and without bleeding, rat)
	Heart rate (rabbit)
	ECG (rabbit)
Effect on Cholinesterase Activities - (rat)	
Hematologic studies -	Hemolytic action (rabbit)
	Anticoagulant action (rabbit)
	Red and white cells (rabbit)

TABLE 4

Subchronic Toxicity in Mice

No. of animals -	12 M and 12 F per group
Feeding levels -	0.0, 12.8, 64, 320, 1,600 and 8,000 ppm
Duration -	6 months
Mortality -	IF (320 ppm), IF (0.0 ppm); Pneumonia
Body weight -	Slight retardation (8,000 ppm, M & F)
Organ weight -	No significant difference from control of any dosage levels except 8,000 ppm-liver
Hematologic observation -	No significant changes
Histopathology -	Enlarged liver cells & acidophilic body of 8,000 ppm-liver
No effect level -	1,600 ppm

TABLE 5

Histological Examination of Mice Fed Diet Containing Thiophanate-methyl for 6 Months[a]

Group (ppm in diet):		8,000		1,600		320		64		12.8		0	
No. of examination:		12 M	12 F	12 M	12 F	12 M	12 F	12 M	12 F	12 M	12 F	12 M	12 F
Liver	cloudy swelling	3	2	1	4	3	1	2	2	2	1	3	2
	variation of size	5	3	0	0	0	0	0	0	0	0	0	0
	acidophilic body	0	1	0	0	0	0	0	0	0	0	0	0
Spleen	hemosiderosis	1	2	1	1	1	2	1	1	3	1	1	2
Lung	alveolar degeneration	0	1	1	1	1	1	1	1	1	1	2	1
	congestion/hemorrhage	1	1	1	1	1	1	1	1	1	0	1	0
	edema	1	0	0	0	0	0	0	0	0	0	0	0
Kidney	cell. prolif. of glomeruli	0	0	0	1	0	0	0	0	0	0	0	0
	focal cell. infiltration	2	2	0	1	1	0	0	1	1	0	1	0
	hyaline drop. in tubular epithel.	0	0	0	2	0	0	0	0	0	0	0	0

[a]Heart, gastrointestine, pancreas, testis, ovary, bone marrow, thymus, thyroid, adrenal, central nervous system showed no significant changes.

TABLE 6

Reproductive Performance of Mice Receiving Thiophanate-Methyl during Pregnancy

Dose mg/kg	No. of litters	Total implants	Litter size (mean±S.E.)	Living fetuses: Live (%)	Living fetuses: Mean body weight (g)	Living fetuses: Sex ratio M : F	Living fetuses: Immatured (%)	Living fetuses: Malformed (%)	Dead fetuses: Immatured (%)	Dead fetuses: Resorbed (%)
0	20	244	12.20±0.62	10.90 ±0.51	1.13 ±0.03	100: 95	0	2 (cp)[b] (0.77)	17 (6.97)	9 (3.69)
40	20	225	11.25±0.59	10.20 ±0.51	1.12 ±0.02	100:101	0	1 (cp) (0.44)	9 (4.00)	12 (5.33)
200	20	221	11.05±0.46	10.10 ±0.45	1.18 ±0.04	100:100	0	2 (cp) (0.90)	10 (4.53)	9 (4.07)
500	20	233	11.65±0.37	10.60 ±0.35	1.14 ±0.03	100:116	0	1 (cp) (0.43)	13 (6.45)	6 (2.57)
1000	20	225	11.25±0.51	9.70[a] ±0.49	1.10 ±0.04	100: 94	0	2 (cp) (0.89)	12 (5.33)	18 (8.00)

[a] P = 0.05 [b] cleff palate

Figures 2 and 3 show the results of a balance study on the radioactivities which were administered to mice in the form of labeled thiophanate-methyl. Slight differences among the four types of labeled compounds were found in cumulative dose curves of excreted radioactivity. The excretion of thiophanate-methyl was very fast and most of the radioactivity eliminated from the animal body about 24 hr after the administration of labeled thiophanate-methyl. The excretion pattern of methyl-^{14}C labeled thiophanate-methyl (Compound III) was found to be somewhat different from the others studied. Its excretion was faster than the other types of labeled compounds. It was assumed that the part of the molecule containing the methyl group was converted into another molecule which was rapidly absorbed through the intestine, and partially metabolized to expired gas (CO_2) by the animal.

Time-course distribution of radioactivity in the organs and tissues of mice was investigated by using all four types of labeled compounds of thiophanate-methyl. An example of such experiments is shown in Fig. 4. In summary, the following results were obtained: (a) The rate of incorporation of radioactivity in organs or tissue was different, and decreased in the following order with each labeled compound.

Thiourea-^{14}C ... Compound I

After 3 hr; Stomach > Liver = Kidney > Blood = Lung = Heart.

After 96 hr; Liver > Blood ≫ Kidney = Stomach.

Thiourea-^{35}S ... Compound II

After 3 hr; Liver > Stomach = Blood = Kidney > Lung = Heart.

After 96 hr; Bone = Stomach > Liver = Lung = Heart.

Methyl-^{14}C ... Compound III

After 3 hr; Liver = Stomach > Kidney > Blood = Heart = Lung.

After 96 hr; Liver > Thyroid = Kidney = Stomach = Lung = Heart.

Phenyl-^{14}C . . . Compound IV
After 3 hr; Liver > Stomach > Kidney > Heart = Blood = Lung
After 96 hr; Liver > Kidney = Stomach > Heart = Lung = Thyroid.

(b) The incorporation pattern between ^{14}C- and ^{35}S-labeled compounds showed quite significant differences. That is, only 3 hours after administration, the radioactivity of ^{35}S in the intestine and in blood became markedly higher than those of ^{14}C-labeled compounds, indicating that the C=S bond in thiophanate-methyl molecules had already been cleaved at the time of absorption. With the ^{35}S-labeled compound, a small amount of ^{35}S remained in some cartilage, indicating that organic sulphur was metabolized to inorganic sulphate and incorporated into cartilage as chondroitin sulphate.

According to the results (as shown in these figures), almost all the radioactivity disappeared, within 96 hours, from all organs and tissues with the phenyl-^{14}C-labeled compound (Compound IV), and its decay curve showed the quicker disappearance of radioactivity in organs and tissues than that of any other types of labeled compounds. It was concluded that thiophanate-methyl and its metabolites, having the major skeleton of the molecule, such as the phenyl group, disappeared rapidly from organs or tissues.

In order to identify the metabolites of the labeled thiophanate-methyl in urine, feces, organs, and tissues, thin-layer co-chromatographic methods were used, thus allowing comparison of the metabolites with those of authentic samples which had been synthesized via an independent route.

TLC of the Urine Not Treated With Any Hydrolytic Reagent

In the untreated urine of mice administered thiophanate-methyl, eight spots, representing metabolites, were obtained on the TLC radioautogram of phenyl-^{14}C

labeled thiophanate-methyl (Compound IV). The major metabolite (Metabolite D), and a minor metabolite (Metabolite A), and thiophanate-methyl itself were identified by thin-layer co-chromatography with authentic synthesized samples. Four other spots were also found. All fractions were separately collected and treated with N-HCl or β-glucuronidase.

TLC of the Urine Treated With N-HCl or β-glucuronidase

After treating the urine with β-glucuronidase or N-HCl, six radioactive spots were located on the TLC as indicated in Fig. 7. From these spots, the major metabolites were identified as Met-D and its hydroxy derivative (Met-E), which were conjugated with glucuronic acid, by co-chromatography with authentic standards. A minor metabolite, Met-A, was also detected on the TLC plate and identified utilizing an authentic standard. The other three spots, including Met-B, have not been identified. The results of the comparative studies on the metabolism of thiophanate-methyl in urine excreted from mice are shown in Fig. 5. It was clear that the major metabolite, Met-D or its 5-hydroxy derivative (Met-E), did not contain a sulfur atom in its molecule, but contained carbon atoms at three different parts of the parent molecule. These major metabolites were identified by co-chromatography with authentic standards, in different solvent systems. An intermediate metabolite, which contained a sulfur atom in its molecule, was detected on TLC plates of four types of labeled thiophanate-methyl. It is well known that two types of N-glucuronide are formed in the body, one being the N-glucuronide of aromatic amine, which is unstable in acidic condition, while another is that of carbamate or sulfonamide which is relatively stable in acid. These facts suggest that three types of glucuronide might be formed from the Met-D of thiophanate-methyl, while Met-E might be conjugated with glucuronate only at the 5-hydroxy group of Met-E. The main metabolic pathway of thiophanate-methyl in mice is shown in Fig. 14.

To identify the metabolites of the labeled thiophanate-methyl in rat liver microsomes, TLC autoradiograms of the incubated mixture were prepared and compared with authentic compounds of anticipated metabolites. As shown in Fig. 6, the chromatogram of Compound IV (phenyl-^{14}C) revealed four spots attributable to metabolites of thiophanate-methyl. Metabolite-A, -D, and -E were identified by means of the co-chromatography method, again utilizing authentic samples synthesized via an independent route. The results with four types of labeled compounds of thiophanate-methyl indicate that the main metabolite (Met-D) did not contain a sulfur atom, but did contain a carbon atom derived from thiocarbamide and an ester, suggesting that degradation took place at the C=S bond. A small portion of Met-D, 2-methoxycarbonylaminobenzimidazole, was further biotransformed into Met-E, 5-hydroxy-2-methoxycarbonylaminobenzimidazole. Met-A was identified as dimethyl 4,4'-o-phenylenebis (allophanate).

Evaluation of Environmental Effects

The fate of a pesticide in or on plants and crops must be surveyed in order to understand the residual activity to pests as well as to other organisms. The actual residue data is obtained by chemical analysis of the materials. The following study was undertaken in an effort to understand the fate of thiophanate-methyl when applied to leaf and glass surfaces under semicontrolled conditions. The persistence of thiophanate-methyl, dimethyl 4,4'-o-phenylenebis (3-thioallophanate), and its metabolites on plant leaves and glass plates was examined using radiolabeled compounds. After counting the radioactivity of surface washings and homogenate extracts, three solutions each were concentrated under reduced pressure to a small volume suitable for TLC analysis. Tentative identification of two metabolites, compounds D and A, was done by TLC-co-chromatography, with authentic standards, after extraction from TLC plates from which crude extracts were roughly separated (Fig. 7).

The percentage of applied radioactivity recovered decreased gradually on the plant leaves, and more rapidly on glass plates. These results are indicated graphically in Fig. 8 and 9. The percentages of radioactivity in leaf homogenates and in leaf residues seemed to be nearly constant during the experimental period. From the macro-autoradiogram of the apple leaves which had received ^{14}C-thiophanate-methyl, it was concluded that ^{14}C materials hardly migrated to the rest of the plant part. From these data it is presumed that ^{14}C materials could not penetrate into the leaf tissue rapidly, and the loss of ^{14}C materials, indicated by the total recovery, could be attributed to evaporation.

It was shown from TLC-autoradiograms (Fig. 7) of ^{14}C materials in each extract fraction that the major radioactive components were thiophanate-methyl, compound D (methyl 2-benzimidazolecarbamate), and compound A (dimethyl 4, 4'-phenylenebis (allophanate)). These compounds were identified by TLC co-chromatography. The half-life of thiophanate-methyl in apple and grape leaves was found to be around 15 and 12 days, respectively, and around 3 days on glass plates. Abundances of these two metabolites, methyl 2-benzimidazolecarbamate and dimethyl 4, 4'-o-phenylenebis (allophanate), versus applied thiophanate-methyl are shown in TABLES 7 and 8.

The persistence of thiophanate-methyl on apple leaves was nearly the same as on grape leaves, and the half-life of the compound was 15.1 days on the former substrate and 12.2 days on the latter. As shown in TABLE 7, the ratio of the corrected half-life was about one-twentieth.

The abundance of compounds D and A occurring by the decomposition of thiophanate-methyl, in or on apple leaves, accounted for about 12% and 6% of the applied thiophanate-methyl, respectively (Fig. 9). The maximum abundance in the experiment was around 1 month after treatment. Similarly, the maximum amount of compounds D and A in or on grape leaves accounted for about 12% and 3.5% of the

TABLE 7

Half-life of ^{14}C Materials when Thiophanate-Methyl-Phenyl-^{14}C Applied

	Treated area	^{14}C-recovery		Thiophanate-methyl	
		Obtained[a]	Corrected ratio of[b]	Obtained[a]	Corrected ratio of[b]
Apple leaf	ca. 19 cm^2	38.5	1.00	15.1	1.20
Grape leaf	ca. 28 cm^2	24.4	0.94	12.2	1.20

[a]Graphically obtained.

[b]Half-life obtained graphically were corrected based on the area treated.

TABLE 8

Metabolic Pattern of Thiophanate-Methyl in Plants

Compounds	% Abundance after 14 days on		
	Apple leaf	Grape leaf	Glass plate
Thiophanate-methyl	52.6	49.5	5.5
Methyl 2-benzimidazole-carbamate (D)	10.1	8.9	24.1
Dimethyl 4,4'-o-phenylene-bis (allophanate) (A)	6.4	2.5	2.4

thiophanate-methyl.

As serious influences of pesticide residues on the environment are recognized, it becomes essential to study the detailed behavior of pesticide in soil.

The results of such a study with thiophanate-methyl,

i.e., dimethyl 4,4'-o-phenylenebis (3-thioallophanate), follow. Three kinds of ^{14}C- or ^{35}S-labeled thiophanate-methyl were introduced into unsterilized soils, sandy or silty loam, in test tubes at 23 ± 3°C for 2 months, and the persistences of the mother compound, along with its metabolite, MCAB (2-methoxycarbonyl-aminobenzimidazole), were determined. Two kinds of soil, differing in organic matter content, were used.

The total recoveries, e.g., organic solvent extracts plus residual soil, of radioactivity for Compounds I, II, and III are shown in Fig. 10. They had been incubated for 60 days in sandy loam soil. In the case of the silty loam, data was not obtained because of the difficulty in determining the unextractable radioactivity in the soil. The typical TLC-autoradiographic patterns showing the metabolites of thiophanate-methyl are given in Fig. 16 for the extracts by ethyl acetate. The patterns of abundance of thiophanate-methyl, and its main metabolite in the case of Compound I, are shown in Fig. 12 for sandy loam soil, and in Fig. 18 for the silty loam soil. TABLE 9 shows the persistence of thiophanate-methyl in soils.

TABLE 9

Persistence of Thiophanate-Methyl in Soils[a]

	Sandy loam		Silty loam	
Days	23°C	33°C	23°C	33°C
0	64.7	-	58.0	-
7	1.2	1.1	1.3	1.1
14	2.0	1.2	3.6	1.7
28	1.3	0.8	1.0	0.1
60	0.7	0.6	0.6	0.1

[a]% for treated Thiophanate-methyl.

In the case of the sandy loam, similar decay curves showing the total recovery of radioactivity were obtained at 23 and 33°C. The persistence of ^{14}C in Compounds I and II were similar, but the decrement of the former was smaller than the latter, as shown in Fig. 10. On the contrary, the persistence of ^{35}S in Compound II did not show the apparent decrement and nearly 100% remained in the soil. Thus, it is quite clear that each of the three moieties in thiophanate-methyl have different characteristics. The thiocarbonyl moiety is unstable and thus a carbon atom might escape to the atmosphere from the soil with time. In fact, Compound I and Compound III produced a main metabolite, having a R_f value of 0.57 (Fig. 11) which was not obtained from Compound II. Thiophanate-methyl was very susceptible to degradation in soil and transformation to the main metabolite (metabolite D) which was identified as MBC, 2-methyl benzimidazolecarbamate, by means of thin-layer co-chromatography using three developing solvent systems.

Thiophanate-methyl degraded almost entirely within 7 days in both soil types (Fig. 12 and 13). The degradation appeared to be too rapid for the effect of the difference in temperature to be recognizable. On the other hand, the abundance of MBC kept increasing until the 7th day at 33°C, and the 28th day at 23°C. Both decreased after that time. The degradation of MBC was more rapid at 33°C than 23°C, and slightly faster in the silty loam than in the sandy loam. These data suggest that the activity of microorganisms might be an important factor in the degradation of thiophanate-methyl. The abundances of MBC 60 days after treatment were about 20% at 23°C and 2-15% at 33°C, based on treated thiophanate-methyl.

Thus it is concluded from these experiments that the persistence of thiophanate-methyl is very short in soil, and thiophanate-methyl is transformed to the metabolite MBC, which is also degraded at a moderate rate, and gradually disappears from the soil (TABLE 9).

The metabolic pathway in animals, plants, and soil is summarized in Fig. 14. Patterns of the pathway in these biosystems are basically similar, but in animals, the major metabolite (methyl 2-benzimidazolecarbamate) is hydroxylated at the 5-position, then eliminated as various forms of the conjugate.

The effect of thiophanates on soil microflora was examined using a percolation technique at 30°C. Even the application of thiophanate-methyl at 1000 ppm did not exhibit any serious effects on the population of bacteria, fungi, and actinomycetes (Fig. 15, 16, and 17).

References

Ishii, K.: VIIth Int. Congr. Plant Protection, Paris, Résumés, p. 200, September 21-25 (1970).

Noguchi, T., and K. Ohkuma: Relation of Structure and Activity in Thiophanates. Presented at the 2nd Int. Congr. Pesticide Chem., Tel-Aviv, Israel, February 22-29 (1971).

Noguchi, T., A. Kaji, Y. Igarashi, A. Shigematsu, and K. Taniguchi: Antimicrobial Agents Chemotherapy 259 (1962).

Noguchi, T., K. Khomoto, Y. Yasuda, S. Hashimoto, K. Kato, K. Miyazaki, and S. Kano: (Nippon Soda Co., Ltd.) Belg. Patent 722,080 (1969a).

Noguchi, T., Y. Yasuda, M. Asada, S. Hashimoto, K. Kato, K. Miyazaki, and S. Kano: (Nippon Soda Co., Ltd.) Belg. Patent 734,742 (1969b).

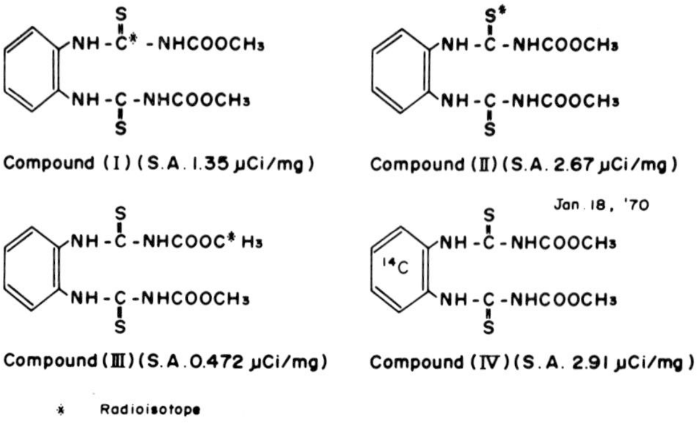

Fig. 1. Labeled compounds used in this study.

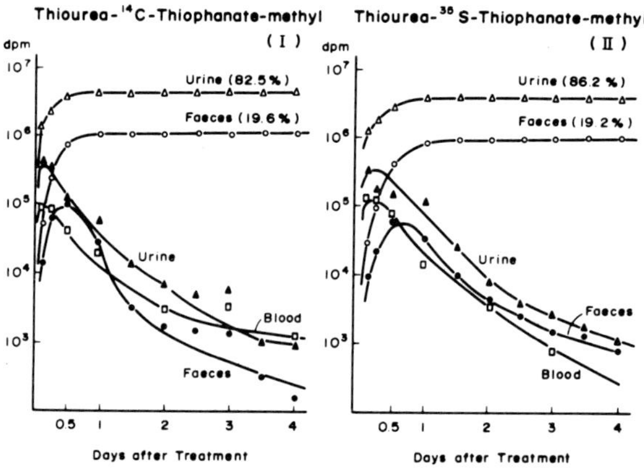

Fig. 2. Elimination of radioactivity in the urine, feces, expired gases, and blood of mice treated orally with 2.70 μci/head of compound (I) and 2.67 μci/head of compound (II).

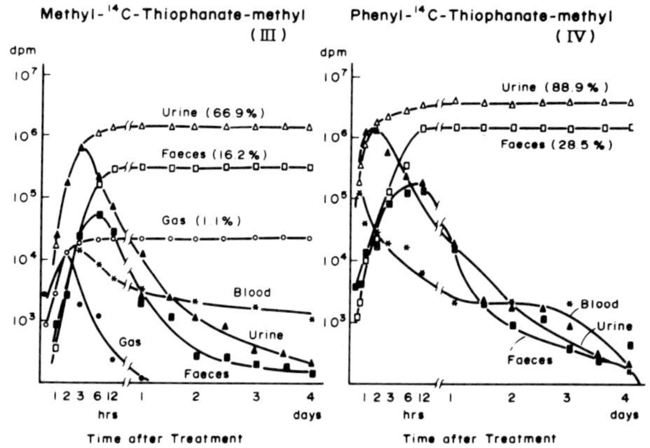

Fig. 3. Elimination of radioactivity in the urine, feces, expired gases, and blood of mice treated orally with 1.416 μci/head of compound (III) and 3.20 μci/head of compound (IV).

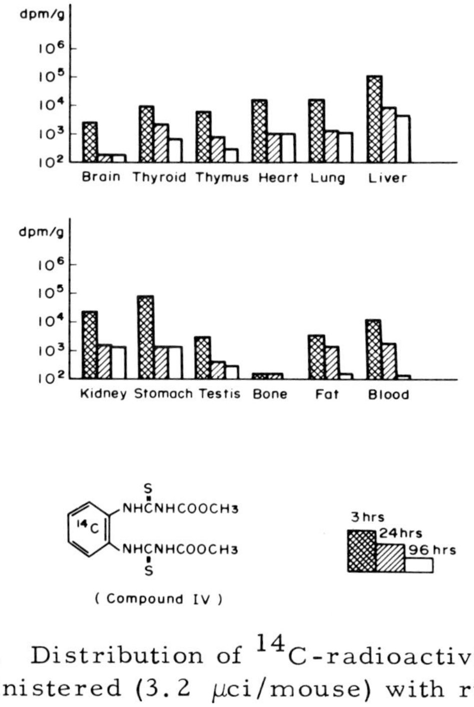

Fig. 4. Distribution of ^{14}C-radioactivity in mice orally administered (3.2 μci/mouse) with ring-labeled thiophanate-methyl.

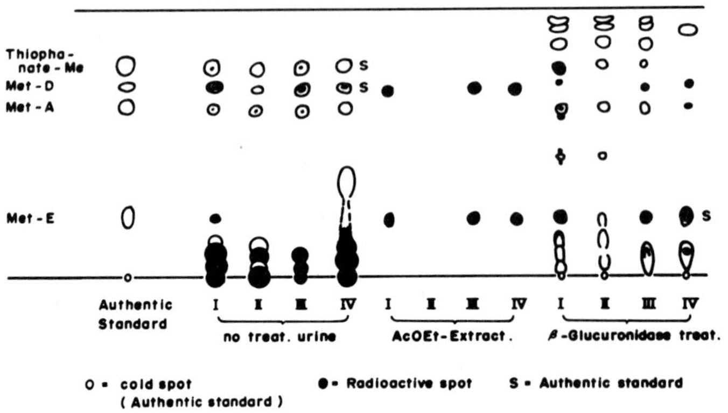

Fig. 5. TLC autoradiogram of metabolites in the urine of mouse administered thiourea-^{14}C (compound I), thiourea-^{35}S (compound II), methyl-^{14}C (compound III), and phenyl-^{14}C (compound IV) thiophanate-methyl.

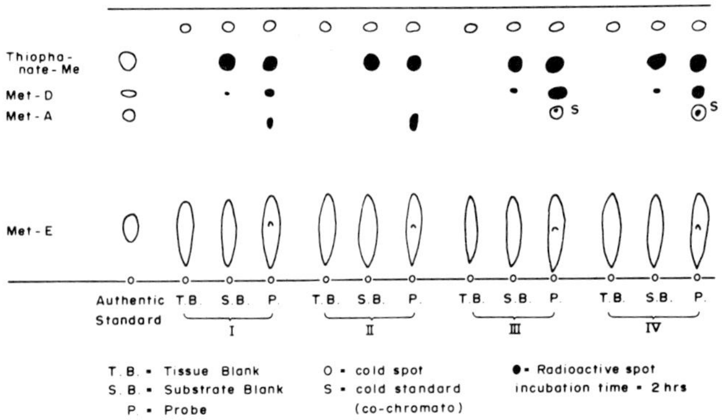

Fig. 6. TLC autoradiogram of metabolites of thiourea-^{14}C (compound I), thiourea-^{35}S (compound II) methyl-^{14}C (compound III), and phenyl-^{14}C (compound IV) thiophanate-methyl in rat liver microsome.

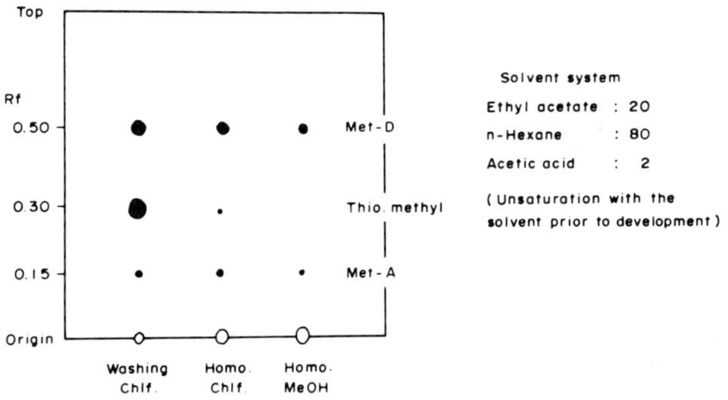

Fig. 7. Thin-layer chromatographic patterns of ^{14}C-compounds in each solvent extract when thiophanate-methyl-phenyl-^{14}C applied topically on the apple leaves.

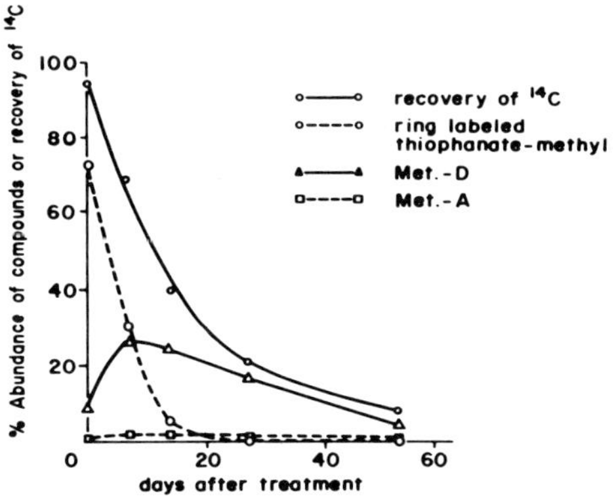

Fig. 8. Transformation patterns of thiophanate-methyl on glass plates.

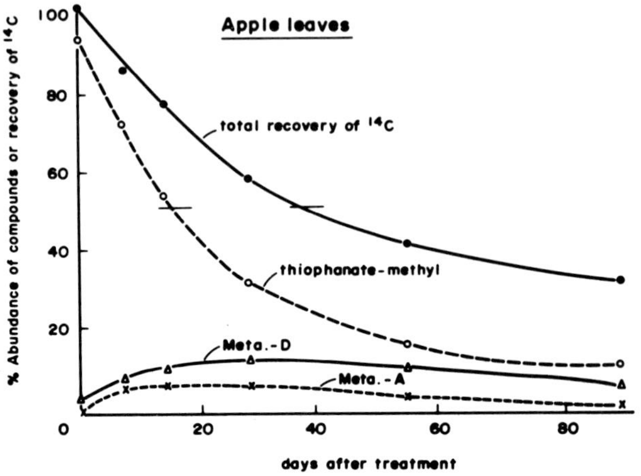

Fig. 9. Transformation patterns of thiophanate-methyl on plant leaves (apple leaves). Similar results were also obtained with grape leaves.

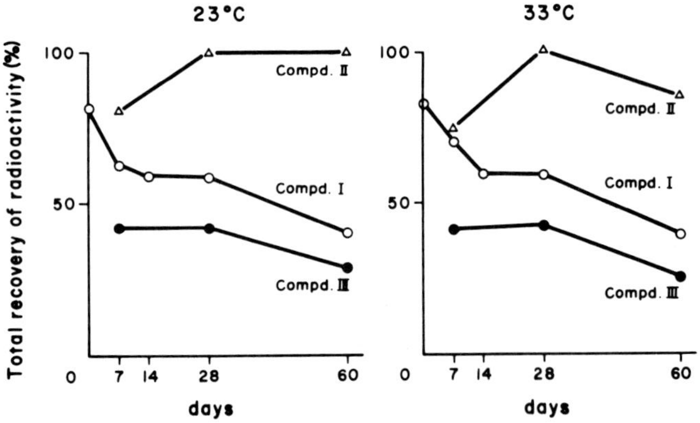

Fig. 10. Time course of total recovery of radioactivity in sandy loam treated with labeled thiophanate-methyl.

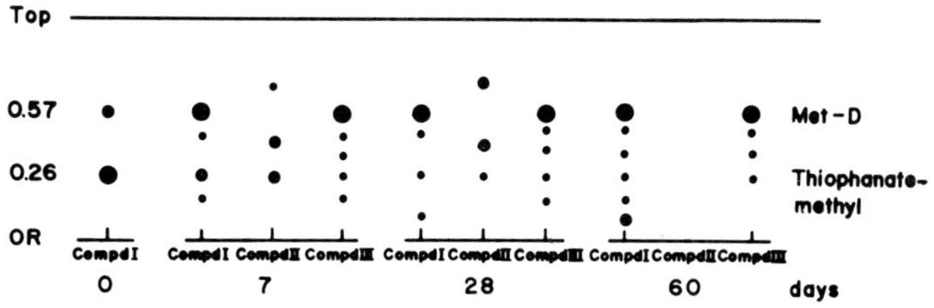

Fig. 11. TLC-autoradiograms of solvent extracts from sandy or silty loam. Ethyl-acetate:n-hexane: AcOH-20:80:2-(unsatd.)

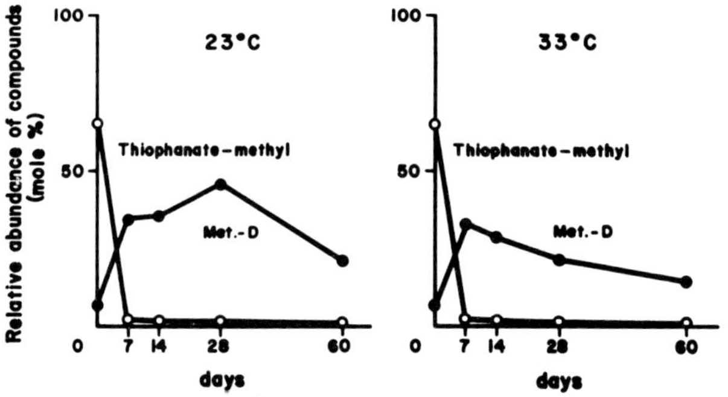

Fig. 12. Time course of relative abundance of compounds in sandy loam treated with thiophanate-methyl.

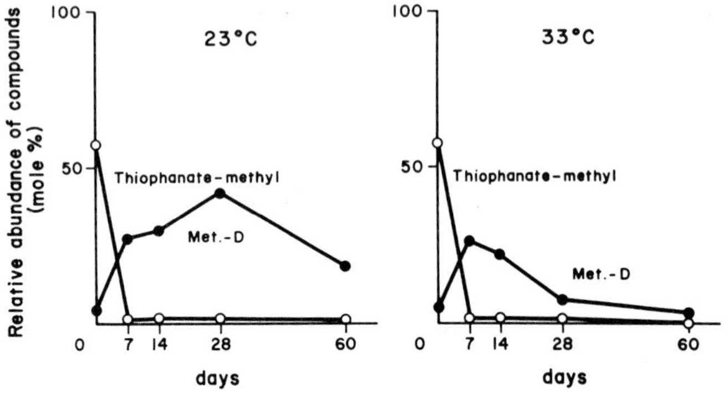

Fig. 13. Time course of relative abundance of compounds in silty loam treated with thiophanate-methyl.

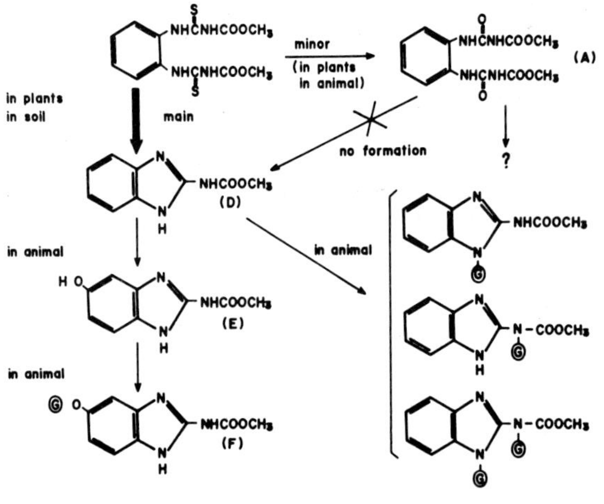

Fig. 14. Metabolic pathway of thiophanate-methyl.

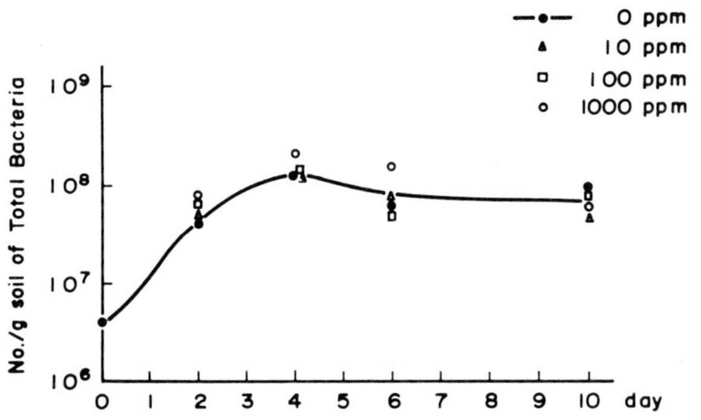

Fig. 15. Effect of thiophanate-methyl on soil microflora (total bacteria).

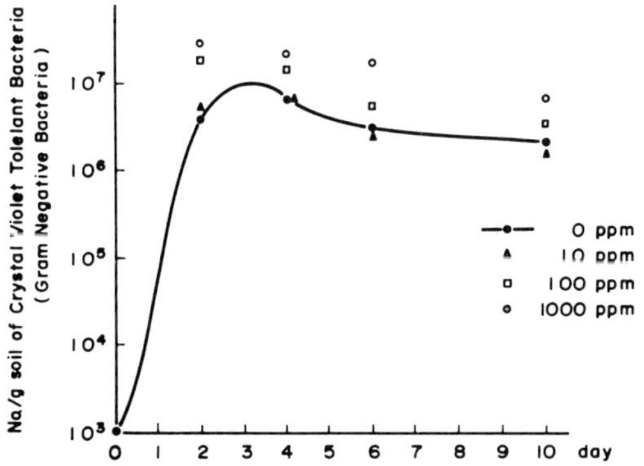

Fig. 16. Effect of thiophanate-methyl on soil microflora (gram-negative bacteria).

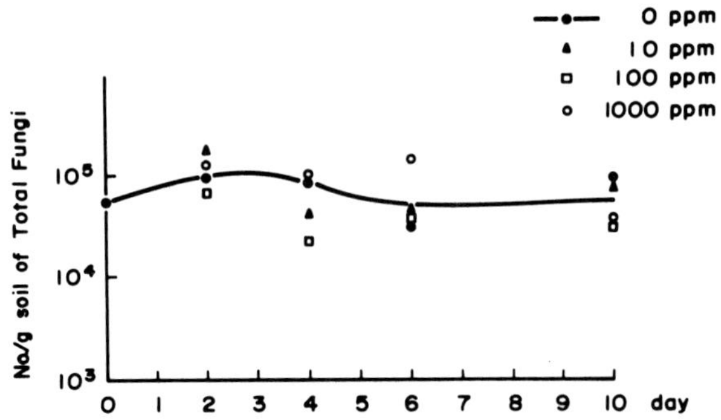

Fig. 17. Effect of thiophanate-methyl on soil microflora (fungi).

SUBJECT INDEX

R

S

T

U

W